Historical & Cultural Astronomy

The Historical & Cultural Astronomy series includes high-level monographs and edited volumes covering a broad range of subjects in the history of astronomy, including interdisciplinary contributions from historians, sociologists, horologists, archaeologists, and other humanities fields. The authors are distinguished specialists in their fields of expertise. Each title is carefully supervised and aims to provide an in-depth understanding by offering detailed research. Rather than focusing on the scientific findings alone, these volumes explain the context of astronomical and space science progress from the pre-modern world to the future. The interdisciplinary Historical & Cultural Astronomy series offers a home for books addressing astronomical progress from a humanities perspective, encompassing the influence of religion, politics, social movements, and more on the growth of astronomical knowledge over the centuries.

The Historical & Cultural Astronomy Series Editors are: Wayne Orchiston, Marc Rothenberg, and Cliff Cunningham.

Donald Campbell

The Arecibo Observatory: A History of Innovation and Discovery

Springer

Donald Campbell
Department of Astronomy
Cornell University
Ithaca, New York, USA

ISSN 2509-310X ISSN 2509-3118 (electronic)
Historical & Cultural Astronomy
ISBN 978-3-031-75224-7 ISBN 978-3-031-75225-4 (eBook)
https://doi.org/10.1007/978-3-031-75225-4

Physics
Astronomy, Astrophysics and Cosmology
History and Philosophical Foundations of Physics
Astronomy, Observations and Techniques
Earth Sciences, general
Civil Engineering
Electrical Engineering

Acknowledgement to David Parker: Science Photo Library for the cover photo.

Editorial Contact: Barbara Amorese

This Springer imprint is published by the registered company Springer Nature Switzerland AG
The registered company address is: Gewerbestrasse 11, 6330 Cham, Switzerland

This book is dedicated to William E. Gordon, who conceived of the Arecibo telescope, raised the funding to build it, oversaw its construction, and was its first director. It is also dedicated to all the staff at the Observatory and at Cornell University whose innovative ideas and hard work kept the Arecibo Observatory as a highly productive scientific enterprise for 57 years.

Preface

On 1 December 2020, the 900 ton suspended structure of the 1000 ft. (305 meter) Arecibo radio telescope collapsed, the end of an incredibly productive 57 years. Arecibo was unique among large telescopes for the broad range of its research capabilities, which stretched from the Earth's upper atmosphere to the large and small solid bodies in our solar system, our galaxy, and the far reaches of the universe. This wide-ranging research program was envisaged by William E. Gordon, Professor of Electrical Engineering at Cornell University in Ithaca, New York, when he proposed the construction of the telescope during a seminar he gave in May of 1958. In his talk, which centered on the specifications of a radar system that would be used to study the properties of the Earth's ionosphere, he pointed out that the proposed antenna and high powered transmitter would be capable of radar studies of the Sun and two of the terrestrial planets, Venus and Mars, and of "receiving radio waves from the cosmos." Bill Gordon's following two years were consumed by finding the funding, settling on a site for the telescope, a sinkhole south of the city of Arecibo, Puerto Rico, and refining the telescope design. Gordon and four other Cornell related families moved to Arecibo in the summer of 1960 to oversee the contracting and construction and to build the needed receiver and data acquisition systems. Construction began that September.

Once it was completed in late 1963, the telescope more than lived up to Gordon's predictions especially after successive major upgrades of the telescope in the 1970s and 1990s. The early 1970s upgrade, which replaced the original wire mesh reflector surface with aluminum panels, allowed observations up to 5 GHz, including the very important 21 cm wavelength spectral line of atomic hydrogen, the dominant gas constituent of our and other galaxies. This was accompanied by the installation of a high powered transmitter at 2.4 GHz for radar observations of solar system bodies, especially the imaging of the surface of Venus under its cloud cover. The 1990s Gregorian upgrade significantly increased the telescope's sensitivity for both radio astronomical observations and radar studies of solar system bodies. At the same time, a large visitor center was constructed that highlighted the observatory's research areas for over 100,000 visitors per year including tens of thousands of school children.

In setting out to write this book, I had intended to cover the period from Bill Gordon's May 1958 seminar until the mid-1970s. This period encompasses the telescope design effort, securing the funding to build the telescope, the construction period from September 1960 until October 1963, very early science, and the first major upgrading of the telescope. While this is still the emphasis of the book, there were multiple suggestions that the latter years should be included. I have done this although in lesser detail than the early years and the history now extends to the end of Cornell University's management of the Arecibo Observatory on 30 September 2011. The last chapter briefly covers the Arecibo Observatory's remaining years before the telescope's collapse and, of course, the collapse itself and its aftermath.

While several chapters cover the results of the observatory's research work, the emphasis of the book is on the original construction and the efforts to maintain the capabilities of the telescope and its instrumentation to do leading edge research in all three of its research fields. A secondary emphasis is on the interactions with the funding agencies, the Advanced Research Project Agency (ARPA) of the U.-S. Department of Defense, which funded the construction and operation of the observatory and, after 1969, the U.S. National Science Foundation and NASA. Keeping the observatory adequately funded was a major task for the observatory's management as it is for any research institution.

I first saw the Arecibo Observatory in Puerto Rico when I joined its research staff in early 1965 having recently graduated with a Master of Physics degree from the University of Sydney in Australia. At Sydney University, I had been working on some design features of a new radio telescope, the Molongo Cross, that was being built near Canberra, the nation's capital. At that time Australia was one of the leading countries involved with the relatively new field of radio astronomy. As you will read, a joint astronomy center was formed between Cornell and Sydney universities and this led to a letter in December 1964 from Professor Thomas Gold of Cornell offering me a position at the then recently completed Arecibo Observatory in Puerto Rico; the beginning of a lifetime association with the Observatory and Cornell University. After a year in Puerto Rico, I enrolled in Cornell's graduate program to complete my PhD is astronomy under Professor Frank Drake with the thesis based on observations with the Arecibo telescope. Upon completion, I returned to Puerto Rico remaining there except for a brief period until 1987. For the last six years I served as the observatory's director. I moved to a faculty position at Cornell University in Ithaca, New York, but maintained a very close association with the Arecibo Observatory until 30 September 2011, the last date of Cornell University's management of the observatory. During this period, I was deeply involved with a second major upgrading of the telescope, the Gregorian upgrade, served as associate director of the National Astronomy and Ionosphere Center (NAIC), the management organization for the observatory at Cornell University and, from 2008 to 2011, served as the director of NAIC. For much of my time at Cornell I made regular trips to the observatory keeping in touch with what was happening there and also pursuing my own research interests in the field of radar studies of the terrestrial planets, satellites, and small bodies in our solar system. The discussion of scientific results in the history perhaps overemphasizes Arecibo radar studies of solar system bodies and I apologize to all those astronomers and ionospheric physicists for not giving their work more of the emphasis that it deserves.

Building the world's largest radio/radar telescope and keeping it capable of cutting-edge research requires the involvement of talented and dedicated people. Many are mentioned in the history including William E. Gordon, William McGuire, Merle LaLonde, Miguel Feyjoo, Rolf Dyce, Allan Love, Hal Craft, Joseph Vellozzi, José Maldonado, Frank Drake, Tor Hagfors, Jon Hagen, Paul Goldsmith, and Lynn Baker. Of equal importance were the local staff, most from Puerto Rico, who maintained all the telescope's electronic observing equipment, much of which they built, the telescope control and analysis software, and the telescope's structure and mechanical systems. Again, many are mentioned in the history but to name a few, Domingo Albino, Rafael Amador, Hiram Cabassa, Elliot Mayo, Ron Tower, Edgar Castro, Santiago Rosado, Tom Dickenson, Antonio Crespo, Felipe Soberal, José Jimenez, Phillip Perillat, and Arun Venkataraman. NAIC's Ithaca instrumentation laboratory played a crucial supporting role primarily building cooled receiver systems and line feeds. Key long-term people were George Peter, Pete Mosher, William Genter, and Kurt Kabalac. General support staff are crucial to keeping any organization working smoothly. In Arecibo key people were Rey Medina, Rey Velez, Angel Vasquez, Maria Delgado, Carmen Segarra, John Pappas, Edith Alvarez, Aida Carrasquillo, Benjamin Santiago, Raul Garcia, Mercedes Vives, Carmen Torres, Lucy Lopez, Maria Rodriguez, Alice Hine, and Eva Robles. In Ithaca, Gene Bartell, Roz Arsenault, Jill Tarbell, David Howe and James Blair. Many of the scientific staff are mentioned in the history and their numbers over the almost fifty years of Cornell's management of the observatory are too large to single out any subset for mention. Finally, I would like to apologize to all those staff members that I have not mentioned, thank you for all your efforts in keeping the telescope as an exceptionally productive scientific instrument while maintaining a very positive work environment.

Writing this history would not have been possible without the assistance and support of many people. Of critical importance were a trove of relevant documents and a series of audio interviews that were assembled by Professor Joanna Rankin in the late 1970s in preparation for writing a history of the Arecibo Observatory. She generously provided them for my attempt to write the history. Dr. Herb Carlson provided very valuable background information about the observatory's ionospheric programs as did Dr. Shikha Raisada about studies of the Earth's neutral upper atmosphere using passive and active optical instruments. Dr. Miller Goss provided documents from the Australian CSIRO's archives related to Joseph Pawsey's visit to Arecibo in 1961. The family of Philipp Blacksmith provided photos that he had taken in his role overseeing the telescope's initial construction for the Air Force. The book title is based on a suggestion by Martha Haynes. Many thanks also to my wife, Anita Watkins, for her support during this long endeavor, and for reviewing the manuscript, correcting my grammatical mistakes, assembling the table of acronyms, and selecting index terms. Paul Goldsmith reviewed the final draft of the manuscript making many useful suggestions. Finally, a thank you to my editor at Springer, Ramon Khanna, for his help and patience while I worked on the history.

Ithaca, New York, USA
July 2024

Donald Campbell

Contents

Chapter 1
The Genesis of the Idea

After working on techniques for improving the distance that radio waves could propagate for communications purposes, in the mid-1950s William E. (Bill) Gordon, Professor of Electrical Engineering at Cornell University, turned his attention to the possibility of measuring the electron density and temperature of the Earth's upper atmosphere, the ionosphere, as a function of height using the backscattering of radio waves by the individual electrons present there. He laid out the specifications for such a radar system in a seminar in May 1958, pointing out that the radar system could also be used for studies of the Sun and nearby planets, and the very large radio antenna that would be needed could also be used for radio astronomical investigations. This is put into context by looking at the then state of our knowledge of the Earth's ionosphere and the terrestrial planets as well as the then current situation with the new field of radio astronomy.

1.1 Laying Out the Specifications

On 1 November 1963, a celebration took place on the edge of a sinkhole in the limestone karst terrain ten miles south of the city of Arecibo in Puerto Rico. Over the previous 3 years, the sinkhole had been transformed into the world's largest radio telescope, a stunning example of innovative civil, mechanical and electrical engineering. There was much to celebrate at the telescope's dedication. Attendees included the governor of Puerto Rico, Luis Muñoz Marin; the U.S. Assistant Secretary of Defense, Eugene Fubini; U.S. Congressman George Miller; Antonio Fernos Isern, the Resident Commissioner for Puerto Rico in the U.S. Congress; Brigadier General Holtzman, Director of the U.S. Air Force Cambridge Research Laboratories; Dale Corson, Provost of Cornell University; Thomas Gold, Chair of the Astronomy Department and Director of the Center for Radiophysics and Space Research at Cornell; and William (Bill) Gordon, Professor of Electrical Engineering at Cornell, who had conceived of the idea of the telescope and guided its construction. Bill

D. Campbell, *The Arecibo Observatory: A History of Innovation and Discovery*,
Historical & Cultural Astronomy, https://doi.org/10.1007/978-3-031-75225-4_1

Fig. 1.1 William E. Gordon in the 1950s. Credit: Cornell University

Gordon's talk on the capability of the telescope for investigating the Earth's upper atmosphere, the ionosphere, was short and did not refer to his role in the project. It was not until Dale Corson's talk that an uninformed member of the audience would have appreciated the very central role that Gordon played in bringing the project to fruition. This event would not have been taking place except for the single-minded drive of Bill Gordon. He was literally the right person in the right place, Cornell University, at the right time, post sputnik America.

The project to build what became known as the Arecibo telescope had its formal origin in a seminar given by Bill Gordon (Fig. 1.1) on 29 May 1958 in the School of Electrical Engineering at Cornell University. During the seminar he discussed his ideas for building a very large radar system to investigate the properties of the Earth's upper atmosphere, the ionosphere, and how they vary with height, time of day, the seasons and the level of solar activity. At that time little was known about the ionosphere, the region of the Earth's atmosphere above about 65 km (40 miles), where solar ultraviolet (UV) radiation breaks the atomic bonds of some of the constituent atoms and molecules forming positively charged ions and negatively charged electrons. The use of "incoherent scattering" in the announcement of the talk (Fig. 1.2) was the term for the type of measurements that Gordon proposed, reflections of radar pulses from the negatively charged free electrons in the ionosphere.

Based on estimates of the number of electrons per cubic centimeter (cc) at different heights in the ionosphere and the miniscule amount of power that each electron would reflect back to the transmitting telescope, Gordon calculated that he needed to build a radio telescope with a 305 m (1,000 ft) diameter reflector equipped with a 1 MW transmitter and an approximately 500 ft central tower to hold the

E E SEMINAR

DATE: Thursday, May 29, 1958

TIME: 4:45 PM (Following tea at 4:15 pm)

PLACE: Phillips 101

SPEAKER: W. E. Gordon, Cornell University

Free electrons in an ionized medium scatter radio waves incoherently so weakly that the power scattered has previously not been seriously considered. Calculations show that this incoherent scattering, while weak, is detectable with a powerful radar. A radar with components each representing the best of the present state of the art is capable of

1. measuring electron density and electron temperature as a function of height and time at all levels in the earth's ionosphere and to heights of one or more earth's radii

2. measuring auroral ionization

3. detecting transient streams of charged particles coming from outer space

4. exploring the existence of a ring current

The capabilities listed above depend on the incoherent scattering or radio waves by free electrons. In addition the instrument is capable of

1. obtaining radar echoes from the sun, Venus, and Mars and possible from Jupiter and Mercury, and

2. receiving from certain parts of remote space hitherto undetected sources of radiation at meter wavelengths.

Fig. 1.2 The announcement for William E. Gordon's 29 May 1958 seminar at Cornell University where he outlined the requirements for a radar system that could investigate the properties of the ionosphere, detect radar echoes from planets and study sources of radio waves in the cosmos. From the papers of William McGuire, Cornell University archives

transmitting and receiving feed at its focus (see Chap. 2). However, Gordon was very aware that such a large radio telescope and radar system could also be used for studies of the Sun and nearby planets and for radio astronomical studies of our galaxy and the universe. In the announcement for Gordon's May seminar (see Fig. 1.2), he lists the objectives of the ionospheric measurements, primarily the measurements of the density of electrons and their temperature as they change with height and time, but the last two items state that the telescope would also be capable of "obtaining radar echoes from the Sun, Venus, and Mars and, possibly, from Jupiter and Mercury" and, rather quaintly even then, "receiving from certain parts of remote space hitherto undetected sources of radiation at meter wavelengths". The desire to study the Sun and planets put two significant constraints on the future telescope. Some part of the ionosphere is always overhead no matter where you are on Earth so the study of the ionosphere only required a telescope with a parabolic reflector that could look straight up to the zenith. However, the Sun and planets are only overhead in the tropics and then only periodically at any given location. A telescope of the size that Gordon was contemplating would have to be fixed to the ground; it was too big to be steerable, so it would have to either be located in the tropics or be constructed with a tilt so that it could observe celestial bodies that are overhead in the tropics, a difficult and expensive option. It was this consideration that drove the decision to build the telescope in Puerto Rico. A second complication was that it takes 16 min for a radar signal to travel to the Sun and back to Earth at the speed of light and shorter times for the terrestrial planets, Mercury, Venus and Mars, when they are closest to the Earth. To observe the Sun and terrestrial planets, the telescope would need to be able to track at least for a bit over 16 min to allow time to receive the radar echo from the Sun. The Earth rotates one degree every 4 min, which requires being able to move the telescope up to two degrees from the zenith to compensate for the Earth's rotation, a distinct complication for the design of the telescope. Despite the complications, the tracking requirement was maintained and led, eventually, to a spherical rather than a parabolic reflector[1] and a vastly more complex and expensive telescope. It also led to a much more versatile telescope that made major contributions to the fields of ionospheric research, planetary studies and radio astronomy.

[1] A reflector with a parabolic shape focuses incoming radio waves to a point where they are collected by a device such as a horn antenna. However, a parabola has only one axis of symmetry, the principal axis, so it has to be pointed with this axis oriented in the direction of the incoming radio waves. A satellite dish is a common example of a parabolic antenna. For a spherical reflector every radial line from the center of the sphere is a line of symmetry. However, incoming radio waves are only partially focused and an additional device is needed to complete the focusing. If such a device can be built, then the telescope can focus radio waves coming from different directions by moving the focusing device rather than the entire telescope.

1.2 What We Knew in 1958

Gordon's proposal in the spring of 1958 to build a giant radar to study the Earth's upper atmosphere, the ionosphere, to investigate the Sun and planets and to study the sources of radio emission in the universe coincided with the dawn of the space age. The Soviet Union launched the first Earth satellite, Sputnik, on 4 October 1957 and the United States followed with Explorer I on 30 January 1958. It is often forgotten that the ostensible impetus for the development of satellite launch capabilities by both countries was the 1957/1958 International Geophysical Year (IGY), a world-wide collaboration to investigate geophysical phenomena from the interior of the Earth to the ionosphere and the Earth's magnetic field. Given the tremendous increases over the following decades in our knowledge of the ionosphere, our solar system and the Universe, it is worthwhile to look at what we knew in these fields in early 1958 to put Gordon's dream into context.

The Ionosphere: Guglielmo Marconi's successful transmission in 1902 of a radio signal at a frequency of 272 kHz from England to Nova Scotia, Canada, a distance of 1800 miles (2900 km), provided the first clear evidence of an ionized layer in the Earth's upper atmosphere. Radio waves, like visible light ones, travel in straight lines, so they should not be able to propagate around the curved surface of the Earth. Refraction in the atmosphere will bend the rays by a small amount but not enough for Marconi's transmission to be picked up in Nova Scotia. That it could be detected was indicative of an electrically conducting ionized layer at a height of about 50 miles (80 km) that acted as a reflector for Marconi's transmission. The ionosphere, the region of the Earth's atmosphere above about 65 km (40 miles), where solar radiation causes electrically neutral atoms and molecules to lose one or more electrons creating free positively charged ions and negatively charged electrons, had been discovered. Marconi's successful transmission demonstrated that radio waves could be used to study the ionosphere, and several groups in England and the United States pursued this. Using multiple frequencies, in 1927 Edward Appleton in England showed that there were three regions of charged particles in the lower ionosphere, which he labeled D (the lowest), E and F (see Fig. 1.3). Appleton received the Nobel Prize for this work.

In the United States, groups at the Naval Research Laboratory and the Carnegie Institution, both in Washington D.C., developed the first versions of ionosondes, commonly called ionospheric sounders. These devices transmit radio waves vertically at multiple frequencies one by one and for each frequency the signal undergoes a mirror-like reflection from the level in the ionosphere where the density of electrons is such that its plasma frequency[2] matches the transmitted frequency. By measuring the strength of the reflected signal and the time delay between its transmission and reception, it is possible to determine the height of the reflection and the density of electrons at that height. Using multiple frequencies the electron densities over a range of heights can be determined. Ionosondes continue to be used

[2]The plasma frequency in Hz is given by 9 $\sqrt{N}$ where N is the number of electrons per cubic meter.

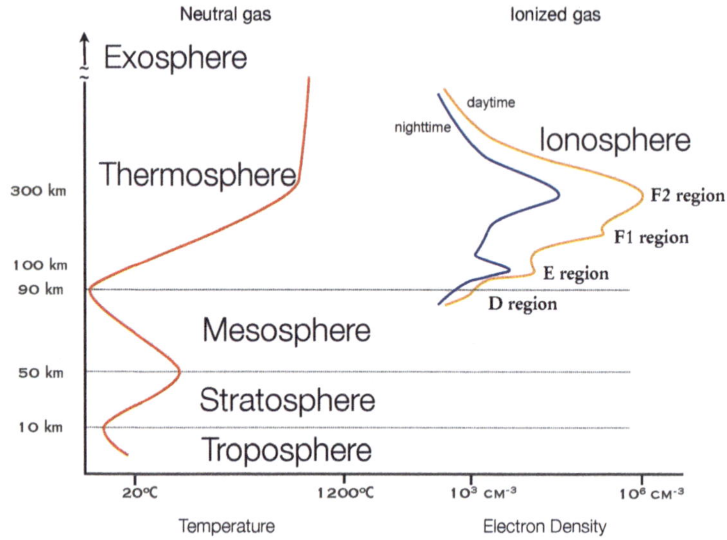

Fig. 1.3 The left side of the figure shows the layers of the Earth's neutral atmosphere defined by what is happening with the temperature. In the troposphere, the bottom-most layer in which we live, the temperature decreases with height until about 10 km (30,000 ft). Above this height is the stratosphere where the temperature increases due to absorption of ultraviolet (UV) radiation from the Sun until about 50 km (30 miles) when it starts to drop again with height in the mesosphere due to decreasing absorption of solar radiation. Above about 90 km (~60 miles) the temperature rises again in the thermosphere due to absorption of very short wavelength UV and X-ray radiation from the Sun. This short wavelength radiation also ionizes some of the atoms and molecules creating layers of ionization, ions and electrons (see the right hand plot showing the electron density with height) starting with the D-region at about 65 km (40 miles), the E-region at about 100 km (60 miles), the F1-region at about 200 km (~120 miles), and the F2 region, the peak electron density, at about 300 km (170 miles). At night the electron density at all heights drops by a large factor due to the absence of the ionizing radiation from the Sun. Credit: Figure courtesy of Eliana Nossa

to study the lower ionosphere but then and now can only measure the electron density up to the height of the maximum electron density, the F2-region peak density (see Fig. 1.3), where the plasma frequency is a maximum. Radio transmissions above this frequency pass right through the ionosphere and give us no information about its properties above the height of the F2-region peak.

In these pre-World War II years, many questions remained; what part of the spectrum of solar radiation was responsible for the ionization and was it different for the three regions, what was the composition and temperature of the ionosphere at different heights and what were the conditions, electron density, temperature and composition, above the height of the F2 region. Below about 100 km the constituents of the atmosphere, primarily oxygen (O_2) and nitrogen (N_2) molecules, are well mixed by turbulence. Above this, in the absence of turbulence, the constituents can separate out under the influence of Earth's gravity with the lighter elements such as

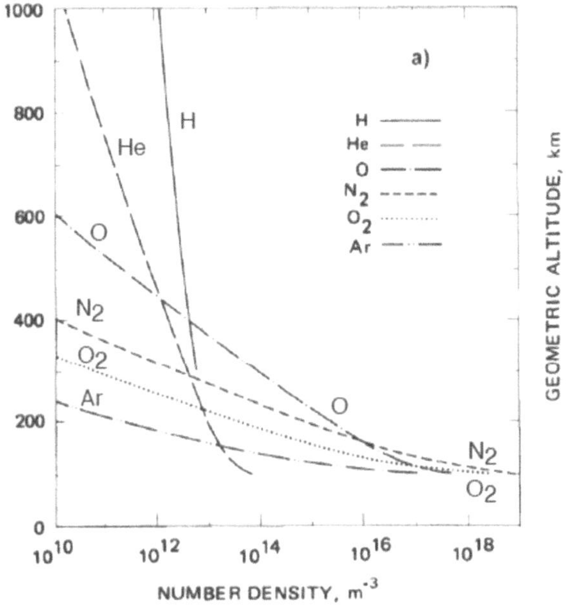

Fig. 1.4 The 1976 NASA standard composition of the atmosphere for periods of minimum solar activity as a function altitude above the surface of the Earth. Below 100 km (60 miles) turbulence keeps the atmospheric constituents, primarily nitrogen (N_2) and oxygen (O_2), well mixed so there is no variation with altitude. Above 100 km, in the absence of turbulence, the different atomic and molecular constituents (Argon, Ar), atomic oxygen (O), Helium (He), and atomic hydrogen (H)) have an altitude distribution that is proportional to temperature and inversely proportional to their atomic or molecular weight. An atom of oxygen weighs 16 times that of an atom of hydrogen so hydrogen distributes itself over 16 times the altitude of oxygen. Ionized oxygen (O^+) constitutes the major component of the ionosphere between about 200 km and 400 km with ionized hydrogen (H^+, a proton) dominating above ~500 to 700 km depending on the season (Figure 29c from NASA-TM-X-74335 US. Standard Atmosphere 1976, National Aeronautics and Space Administration)

helium (He) and hydrogen (H), rising higher than the heavier ones such as oxygen in a process called diffusive equilibrium (see Fig. 1.4).

In the post-World War II period, direct measurements of the properties of at least the lower ionosphere became possible using rockets, initially surplus German V2 ones and, in the 1950s, ones developed for this purpose. The tremendous advantage of rockets is that they could measure not only electron densities but also temperatures and the composition of the atmosphere or ionosphere up to the maximum altitude they could reach. However, their major contribution was the measurement of the energy emitted by the Sun at all wavelengths including x-ray and ultraviolet (UV), which have sufficient energy to strip electrons from the atoms in the atmosphere. The energy of radiation at different wavelengths is inversely proportional to the wavelength, and x-rays have very short wavelengths compared to those of visible light. This led to a better understanding of how the regions in the lower ionosphere are formed. Short wavelength solar UV radiation is responsible for the F-region ionization and x-rays are responsible for the E-region ionization. The D-region ionization remained a mystery until it was realized that short wavelength UV radiation from hydrogen in the Sun's chromosphere, the region just above the Sun's visible surface, could ionize nitric oxide, a very small constituent of this region of the atmosphere, but there is enough of it to account for the D-region ionization. At the maximum of solar activity in the Sun's eleven year cycle, very short wavelength x-rays can penetrate deep enough to enhance D-region ionization.[3]

Solar System: While we did know something about the ionosphere prior to Bill Gordon's seminar in May 1958, it is hard to remember how little we knew about the planets before the space age and the concurrent tremendous improvement in Earth based optical and radio telescopes. Even the scale size of the solar system as represented by the average distance between the Earth and the Sun, the Astronomical Unit, was not well enough known to accurately navigate spacecraft to other planets. Mercury, the closest planet to the Sun, was a blur of dark and light albedo features with a rotation period that was thought, incorrectly, to be the same as its orbital period about the Sun, 88 days. The beautiful white clouds covering Venus, our sister planet in size and orbiting within the habitable zone about our Sun, were assumed to be water clouds with science fiction novels depicting rain forests and a great place to visit in some future time. In fact, nothing was known about the planet's surface, its atmosphere below the clouds or the composition of the clouds. It was only in 1956 with the first detection of radio waves from Venus by Cornell Mayer using the then new 50 ft (15 m) radio telescope of the U.S. Naval Research Laboratory that reality started to intrude. If the radio waves came from the planet's surface, they indicated a temperature of about 330 °C (~630 °F), definitely not a habitable planet if true. Despite attempts at explanations that would allow Venus to be inhabitable, by the mid-1960s it was generally agreed that its surface temperature was about 460 °C (840 °F), the atmosphere was primarily carbon dioxide with a surface pressure about 90 times that of Earth's, and the white clouds were sulfuric acid droplets. After this there were virtually no science fiction novels about Venus as a habitable planet, but there was strong scientific interest in why the atmosphere of Venus evolved so very differently from that of Earth. The future Arecibo telescope would provide the first large scale look at the surface of the planet with good enough resolution to identify surface features.

Given the amazing level of our current knowledge of Mars from orbiting spacecraft and surface rovers, it is hard to realize that in 1958 what we knew about the planet had not progressed much beyond Percival Lowell's early 1900's claim that there were canals on Mars carrying water from the polar regions to the dry equatorial ones. Indeed, a Giant Colorprint Map of the Solar System published by the American Map Company, Inc. in 1959 and "Approved by Leading Boards of Education" still showed an image with Lowell's canals on Mars. We knew that there were bright and dark regions that changed with time (Fig. 1.5), which were still thought by many

[3]A comprehensive discussion of the early history of studies of the ionosphere including the 1957/1958 International Geophysical Year can be found in Friedman (1994).

Fig. 1.5 Mount Wilson 60 inch telescope color image of Mars made on 24 August 1956 by Robert Leighton of the California Institute of Technology showing albedo features. The reddish color is due to iron oxide in the soil. White polar caps were thought to be either water ice or frozen carbon dioxide. Credit: NASA/VRS/SCIENCE PHOTO LIBRARY

respected scientists to be evidence of "plant" life on Mars that changed with the seasons. It was not until imagery was obtained by the Mariner 9 orbiter in 1972 that it was clearly established that the changing albedo features were due to the movement of dust by surface winds. There were polar caps but whether of water ice or frozen carbon dioxide was not clear, and it was known that the atmosphere consisted primarily of nitrogen with some carbon dioxide at a maximum surface pressure of about 0.8% that of Earth's. Gordon's dream would result in the most powerful radar system in the world that would, in tandem with spacecraft missions,[4] contribute significantly to our current knowledge of the solar system.

Radio astronomy: Radio waves from our galaxy, the Milky Way, were first detected in the early 1930s by Karl Jansky and Jansky's discovery was followed up by Grote Reber in the later 1930s. However, the study of radio emission from the Sun, planets, the Milky Way and the universe, what is now called Radio Astronomy, was not seriously pursued until after World War II and then primarily in Australia, Great Britain and the Netherlands with the United States not getting serious about this new field of astronomy until the early 1950s. Much of the early work, including some in the U.S. at Cornell University and the Naval Research Laboratory, concentrated on studies of radio emission from the Sun and the Milky Way. However, in the late 1940s very strong radio emission was detected from several small areas of the sky, radio stars, sparking the development of larger telescopes and interferometer systems (arrays of telescopes working in tandem) to improve the resolution. The completion of the 250 ft (76 m) Jodrell Bank radio telescope near Manchester,

[4]The first spacecraft flyby of a planet, Venus, by the U.S.'s Mariner 2 was in 1962.

England in 1957 was a major advance in the sensitivity and resolution of large single dish radio telescopes. Clearly, a 1000 ft telescope as envisaged by Gordon even with limited steerability would enable forefront studies in this new field of astronomy.

1.3 A Preview of Arecibo's Scientific Discoveries

Did the Arecibo telescope live up to Gordon's hopes at the time of the seminar in 1958 and the optimism at the telescope's dedication in November 1963? Arecibo was, and remained until its collapse in December 2020, as unusual for the breadth of its research capabilities ranging from the Earth's upper atmosphere, the ionosphere, to the solar system to our Galaxy and the far reaches of the universe. Almost all radio telescopes are passive devices in that they only receive the radio wave emissions from hot gases and solid surfaces in the universe including our solar system, known as thermal emission, and from very energetic phenomena where high-energy electrons accelerated in magnetic fields radiate over a wide range of wavelengths, non-thermal emission. Arecibo was almost unique in that it had active systems, high-powered radar transmitters, that allowed it to investigate the Earth's ionosphere by bouncing radar signals from the free electrons and to obtain information about the solid bodies in our solar system by studying the properties of radar signals reflected from the surfaces of planets, their satellites and smaller bodies such as asteroids and comets.

Over the 57 years between the inauguration of the telescope and its collapse in 2020, it was used to make major advances in our understanding of the Earth's upper atmosphere, the ionosphere, to significantly increase our knowledge of our solar system and to expand our knowledge of our place in the Universe. Notable among very early work was the discovery that the nighttime ionosphere in winter was partially maintained by high energy electrons originating at Arecibo's magnetic conjugate point in Argentina, and the measurement of the rotational period of Mercury showing, unexpectedly, that the period was two thirds of its 88 day orbital period about the Sun rather than being the same as had long been believed. With the discovery of pulsars by Jocelyn Bell at Cambridge University in 1967, Arecibo came into its own as a radio astronomical telescope. Pulsars, neutron stars about 20 km in diameter with masses about 1.4 times that of the Sun, are composed almost entirely of neutrons and typically rotate about once per second. They are the remains of supernova explosions of massive stars. Like the Earth, the magnetic poles are offset from the rotational poles and beamed radiation from the magnetic polar regions sweeps across the sky and, if it is positioned correctly, over the Earth as the neutron stars rotate. This is similar to the periodic loom of a lighthouse, hence pulsars. The most dramatic early pulsar related discovery with the Arecibo telescope was in 1974 of the first binary neutron star system, two neutron stars, one a pulsar, orbiting about each other. Russell Hulse and Joseph Taylor received the 1993 Nobel Prize for this discovery, which led to the confirmation of the existence of the gravitational waves predicted by Einstein via measurements of the decay with time of the pulsar's orbit about the other neutron star. This helped to justify the construction of the Laser Interferometer Gravitational Observatory (LIGO) that in 2015 for the first time directly detected gravitational waves from two coalescing black holes.

A resurfacing of the reflector of the Arecibo telescope in the early 1970s allowed it to operate at frequencies much higher than the 600 MHz set by the initial wire mesh surface, and the installation of a second high power radar transmitter specifically for planetary studies resulted in the first high resolution radar images of a large area of the surface of Venus allowing insights into its volcanic/tectonic and impact cratering history. In the early 1980s, a new class of pulsars rotating with periods of milliseconds was discovered with the Arecibo telescope and the first three-dimensional view of part of the large-scale filamentary structure of our visible universe was obtained. In 1990 Arecibo made the first discovery of planets about another star, in this case a neutron star. A radical change to the optics of the telescope and the installation of an even higher powered transmitter in the mid-1990s significantly increased the telescope's sensitivity and that of the planetary radar system. Combined with the addition in 2004 of a multi-pixel receiver system at the wavelength emitted by atomic hydrogen, 21 cm, the telescope was used for further studies of the large scale structure of the universe, detailed mapping of the distribution and dynamics of atomic hydrogen, the main gas constituent of the Milky Way, studies of the magnetic structure of the Milky Way, and searches for new pulsars and the enigmatic Fast Radio Bursts, very short bursts of radio waves from seemingly random directions in the sky. Precise measurements of the arrival times of pulses from millisecond pulsars were aimed at detecting gravitational waves and Arecibo's key role in long baseline radio interferometry revealed the details of very high energy phenomenon in the universe. Radar studies of the planet Mercury gave stunning images of ice deposits in its polar regions, and radar studies of near-Earth objects have provided precise orbital information and images of these objects that may pose a hazard to Earth. The 1990s upgrade enhanced the telescope's ability to study ionospheric phenomenon by providing a two beam capability—radar echoes could be obtained from the ionosphere from two directions simultaneously—leading to improved measurements of winds in the lower ionosphere.

1.4 The Evolution of an Idea

Bill Gordon came to Cornell University in Ithaca, a small city in the Finger Lakes region about 250 miles (400 km) northwest of New York City, in July, 1948 as, simultaneously, a Research Associate in the School of Electrical Engineering and a graduate student pursuing a doctoral degree in electrical engineering. His initial job as a research associate was to supervise the newly formed Microwave (Radio) Astronomy Project that was establishing the first university based radio astronomical observatory in the United States (Campbell 2019). It was located a few miles from the Cornell campus close to the current airport. Gordon had been Associate Director of the Engineering Research Laboratory at the University of Texas at Austin

Fig. 1.6 William E. (Bill) Gordon (left) and Charles Seeger in Cornell University's Radio Astronomy Project's laboratory in September 1948. Cornell University archives

working with its Director, Professor Edward Hamlin, on studies of the propagation of radio waves in the atmosphere for communications purposes. Hamlin was recruited by Charles Burroughs, the Director of Cornell's School of Electrical Engineering, in 1947 as Professor of Electrical Engineering and Director of the newly formed Radio Astronomy Project. He persuaded Bill Gordon to follow him to Cornell. Unfortunately, Hamlin died suddenly in April 1948 before Gordon's planned arrival. Burroughs's renewed the invitation and Gordon arrived in Ithaca with his family in July. He was immediately appointed to succeed Hamlin as Director of the Radio Astronomy Project (Fig. 1.6). Most of the radio astronomical research was then being conducted by Charles Seeger, Assistant Professor of Electrical Engineering (and the elder brother of the folk music icon, Pete Seeger), and Martha Stahr, Assistant Professor in Cornell's Department of Astronomy and the first female faculty member in Cornell's College of Arts and Sciences. One of Bill Gordon's early tasks at Cornell was to help finish the construction of a 17 ft diameter radio telescope that was being built at the site close to the current airport.

However, Gordon's real interest was in the propagation of radio waves in the atmosphere to allow communications over long distances, the subject of his doctoral thesis, working initially with Charles Burroughs and, later, with Henry Booker, a prominent researcher in the area of radio wave propagation in the atmosphere from Cambridge University in England, who joined the Electrical Engineering faculty as a professor in December, 1948. Gordon continued his involvement with the radio astronomy project as it provided him with a salary. Having served in the military in

World War II, he was 31 when he came to Cornell and had a family to support. As he recounted in a self interview in 1979, "While I earned a living doing some observing and data reduction for that project (the Radio Astronomy project), my interest really was in the propagation of radio waves in the atmosphere and in communication by radio."[5] It was his interest in radio communication that, over the next ten years, led Gordon to the adventure of his life, the construction of the Arecibo telescope.

Prior to the satellite era, the only way to communicate over long distances was via overland or undersea copper wire cables or by radio waves, the propagation of radio signals through the atmosphere. In the absence of cables, wireless was the only option. However, as discussed earlier, like light waves radio waves travel in straight lines and don't bend around the curved shape of Earth, limiting the distance over which they can be detected to about 30 miles (50 km) depending on the height of the transmitting station. This distance is somewhat extended by refraction in the atmosphere but not enough to communicate over distances much larger than 30 miles (50 km). However, radio waves had been detected at large distances from the transmitter, so there was some mechanism that was allowing them to propagate around the curved surface of Earth. We are talking about frequencies near 3000 MegaHz (MHz), not the 272 kiloHz (kHz) of Marconi's first transmission across the Atlantic Ocean, which had a wavelength of about 1 km (0.6 miles) and underwent a mirror like reflection from the ionosphere's E-layer (Fig. 1.2). At kHz frequencies the size of the transmitting antennas is so large as to be impractical while at 3000 MHz small dish antennas similar to satellite dishes can be used. It turned out that small density irregularities in the atmosphere caused by turbulence scatter the transmitted radio wave into a wide range of angles. Gordon's doctoral thesis research involved reflection of radio waves in the forward direction, what is known as forward scatter, from density irregularities in the troposphere (Fig. 1.2), a technique that was of considerable utility for radio communications over distances of up to several hundred kilometers. He completed his PhD in June 1953 and was appointed to the Cornell faculty as an Associate Professor in the School of Electrical Engineering. While Gordon was 35 at this time, it is clear that his immediate appointment as an Associate Professor was an indication of his high standing within the School.

Having developed the theory for tropospheric scatter for communications, Gordon and Booker looked into the problem of communicating over even longer distances. This meant reflections at greater heights. Gordon concentrated on forward scatter from density irregularities in the stratosphere (i.e. heights above 10 km, Fig. 1.2) while Booker looked into forward scatter from electron density irregularities in the E-region of the ionosphere at heights of about 100 km (Fig. 1.2). Both of these techniques were used for communications over distances up to about 1000 km, but they had limited use due to the large antennas and high transmitter powers required.

What next? In Gordon's own words "Naturally, the question arose, was there anything that would happen at higher heights?" Reflections at heights of several

[5] A transcript of Gordon's self interview is available in the Cornell University archives.

hundred or more km in the ionosphere would allow communications over even larger distances. As discussed earlier, at these heights there is a mix of free negatively charged electrons, positively charged ionized atoms and molecules (i.e. missing one or more electrons, Fig. 1.3) and neutral atoms and molecules with the dominant ionized atoms being those of oxygen with one electron missing, singly ionized oxygen, O^+. Bill Gordon's first task was to find out how effective a single electron is in reflecting radio waves, what is called its cross section. Combining this with an estimate of the expected number of electrons per cubic centimeter (cc), the electron density, as a function of height, would allow him to compute the range of transmitting antenna sizes, transmitter powers and receiving antenna sizes required for a communications link. All of this was a slow evolution and it was not until early 1958 that Gordon did the calculation of the ability of an electron to reflect radio frequency power even though it was already reasonably well known having been done by J. J. Thompson, the discoverer of the electron. When Gordon calculated the range of sizes for the transmitting and receiving antennas and the transmitter power needed for a communications system based on forward scatter from the electrons in the ionosphere, it was clear that such a system was not realistic, the antenna sizes and transmitter powers needed were probably feasible but far too large for any practicable use.

One evening in early 1958 while, as he recounts in his self interview, Gordon was at home in Dryden, a small town about 13 miles from Ithaca, he turned the problem around and asked himself whether it was possible to use a radar system to study the properties of the ionosphere, especially those above the height of the F2-region at ~300 km. Gordon set about calculating the antenna size and transmitter power needed to study the ionosphere at altitudes up to about 1000 km using reflections from the individual electrons. Since the transmitter frequency would be very much higher than that used for the ionospheric sounders, the vast majority of the power would pass right through the ionosphere into space but a miniscule amount would be reflected from the electrons back to the transmitting antenna. If the antenna was big enough and the transmitter power large enough, then this reflected power could be detected. It takes 2 ms for a radio (or light) wave to go to 300 km and back again. By measuring the amplitude of the received power as a function of the time since a pulse of radio waves was transmitted, the electron densities at different heights could be measured in much the same way that airport radars can measure the distances to airplanes. Since the electrons move about randomly at an average speed determined by their temperature, the received echo has a spread of Doppler shifts,[6] small changes in frequency due to the differing velocities of the electrons away from or towards the radar. The range of velocities is determined by the temperature of the electrons and, so, a measure of the Doppler spread allows the temperature to be

[6]The Doppler shift is the small change in frequency of the reflected radar signal due, in this case, to the velocity of the electron towards or away from the radar. A typical example is the sudden change in the tone of a train whistle from higher to lower as it passes through a station. The electrons will have a random distribution of velocities so the signal echo will be spread out over a range of frequencies determined by the temperature, the Doppler spread.

Fig. 1.7 Ken Bowles in 1957 or 58. Courtesy of Ann Bowles

obtained. Two parameters of the ionosphere, the electron density and the temperature, can be measured over heights from the D region to 1000 km or more.

In his initial calculation Gordon assumed that the electrons are randomly distributed in space and that the ions play no role except to ensure the average electrical neutrality of the plasma. Since the miniscule amount of power reflected by each of the electrons are added together to calculate the total received power, the process is called "incoherent scatter" as opposed to the mirror like reflections of the low frequency ionospheric sounder signals. We will see later that the assumption by Gordon that the ions play no role in the reflection of the radar wave was overly simplistic; ions do play a role, a good thing as it allows considerably more information than just the electron density and temperature to be obtained.

In June 1958, Kenneth Bowles of the National Bureau of Standards visited Cornell. Bowles (Fig. 1.7) was a former graduate student of Henry Booker at Cornell obtaining his PhD in 1955. Both Gordon and Bowles recalled a discussion of Gordon's ideas for building a large telescope and radar system to do incoherent scatter measurements of the ionosphere. As recounted by Bowles in his December, 1958 paper (Bowles 1958) he recalled that he calculated that a new pulsed transmitter at 40.92 MHz that the U.S. National Bureau of Standards was acquiring for its Long Branch site in Illinois, when combined with a rather simple antenna array, should be capable of detecting incoherent scatter from electrons in the lower ionosphere. Bowles set about building the array near Long Branch (sometimes referred to as Havana).

Gordon was scheduled to give a talk at the yearly URSI-IRE (International Union of Radio Scientists - Institute of Radio Engineers) meeting at Pennsylvania State University on 22 October 1958. The talk was based on his May seminar at Cornell and a paper that was to be published in the Proceedings of the Institute of Radio Engineers (IRE now IEEE) in November 1958 (Gordon 1958a, b). Gordon gave his talk but he had an arrangement with Bowles that if Bowles succeeded in detecting an incoherent scatter signal, he would call Gordon and let him announce it at the meeting. In Gordon's own words from his self-interview in 1979, "Well, the paper I know, it's rather clear in my mind, was scheduled for an afternoon session, and I had arranged with Ken, who had at that stage not yet made any successful measurements, but he was still putting the system together out in Havana—I remember arranging to telephone Ken at the lunch hour, on the day I was to give the paper, and I did that and started the paper by saying to the audience something like the following, "My purpose is to tell you about incoherent backscatter from the ionosphere, and the possibility of building a tool to make use of it in studying the ionosphere. And then I want to tell you about a telephone call that I just had. And I went through the paper, that had been prepared, and then added the conversation that I'd just had on the phone with Ken Bowles, who had received the first incoherent backscatter echoes from the ionosphere, and it was a dramatic touch that was added to that meeting." Gordon went on to say, "I remember, it was in some ways one of the high points, I think, in my career."

It was not surprising that Ken Bowles detected an incoherent scatter echo from the F-region of the ionosphere. Bill Gordon had calculated that a 30 m antenna was all that was required; he was interested in studying much higher altitudes where the electron densities are much lower than in the F-region. However, while the radar echo strength measured by Bowles was close to that predicted by Gordon's theory, the Doppler spread was much smaller than predicted. In other words, it did not reflect the electron velocities corresponding to the expected temperature. In his first short report on the measurements, Bowles (1958) pointed out that the Doppler spread depended on the average spacing of the electrons relative to the wavelength of the radar. For a spacing of roughly more than a wavelength, the Doppler spread would, as Gordon predicted, reflect the temperature of the electrons. If less, then the Doppler spread would be significantly reduced. Bowles continued his measurements with the NBS radar system confirming his results that the Doppler spread was less than 10% of that predicted on the basis of Gordon's theory of a random distribution of electrons (Bowles 1959). The correct theory was detailed by a number of investigators in papers published in 1960. Rather than being randomly distributed, the electrons are actually transitorily bunched in the vicinity of the ions creating density fluctuations, which are responsible for the radar echo. In this theory, it is the velocities of the ions determined by their temperature that control the Doppler spread. Since the mass of an oxygen ion (o^+), the dominant species at heights less than about 400 km, is roughly 30,000 that of an electron and the electrons and ions have the same average kinetic energy (that is, they have the same temperature), then the velocity of the ions is only about 0.006 that of the electrons and the Doppler spread is reduced by this factor compared with that predicted by the random distribution of electrons theory that Gordon used. Because sensitivity of the measurement decreases as the square root of the Doppler spread, it is about 13 times higher than predicted by Gordon's simplified theory. A very readable discussion of this issue can be found in Cohen (2009).

Our interest here is in what impact Bowles results had on Gordon's plans for a radar system to study the ionosphere. It clearly could be made much smaller (and cheaper) and still achieve its objectives. The surprising answer is none. Remember this is now late 1958. As we will see, Gordon was already in talks with the newly formed Advanced Research Projects Agency (ARPA) of the Department of Defense about possible funding for the radar and preparing to submit a funding proposal to build the facility. It seems probable that by late 1958 the idea of building the world's largest radio telescope and radar system had taken hold at both Cornell and ARPA and neither institution was interested in building something smaller. ARPA may have also thought that the higher sensitivity would aid in meeting its research objectives.

Was Gordon the first to think of applying incoherent scatter at radio wavelengths to studies of the ionosphere? When he discussed the idea with Henry Booker, Booker told him that he had also calculated the cross section of an electron while a student at Cambridge University but had not pursued the idea. Ken Bowles in his 1958 paper states that the idea that "incoherent scattering from free electrons in the ionosphere, vibrating independently, might be observed by radar techniques has apparently been considered by many workers although seldom seriously, because of the enormous sensitivity required." Who the "many workers" were is not made clear. In a 10 August 1959 chronology of his ideas and experimental work related to the NBS proposal to build an incoherent scatter radar in Peru,[7] Bowles (1959) states that during his June, 1958 conversation with Bill Gordon on Gordon's ideas for a radar system to study the ionosphere, "I knew what he was talking about, having considered the same technique myself while still at Cornell over four years ago." Cohen (2009) recounts that Bowles told him that "he (Bowles) independently thought of the IS (incoherent scatter) idea." However, the coincidence of timing suggests that it was the June 1958 conversation with Bill Gordon that motivated Bowles to build the 41 MHz array at Long Branch and carry out the experiment in October 1958. There is no doubt that Gordon was the first to publicize the idea given his 29 May 1958 talk at Cornell, the paper submitted in June 1958 for publication in the Proceedings of the IRE, and his URSI-IRE talk in October, 1958. Bowles certainly deserves all the credit for being the first to detect an incoherent radar backscatter signal from the ionosphere and for the construction of the Jicamarca incoherent scatter radar in Peru at the same time that the Arecibo incoherent scatter radar was being built.

References

Bowles, K.L. 1958. Observations of Vertical-Incidence Scatter from the Ionosphere at 41 Mc/sec. *Physical Review Letters* 1 (12): 454–455.

———. 1959. *Incoherent Scattering by Free Electrons as a Technique for Studying the Ionosphere and Exosphere: Observations and Theoretical Considerations*. National Bureau of Standards Report 6070.

Campbell, D.B. 2019. Radio Astronomy at Cornell University: The Early Years 1946 to 1962. *Journal of Astronomy History and Heritage* 22 (3): 503–520.

[7] Bowles, K.L. 20 August 1959, Memorandum to R. D. Huntoon, Deputy Director of the National Bureau of Standards, Chronology of the idea for an Incoherent Scatter Radar to be built in Peru. Cornell University archives under Arecibo Ionospheric Observatory.

Cohen, M.H. 2009. Genesis of the 1,000-Foot Arecibo Dish. *Journal of Astronomical History and Heritage* 12 (2): 141–152.

Friedman, H. 1994. From Ionosonde to Rocket Sonde. *Journal of Geophysical Research* 99: 19143–19153.

Gordon, W.E. 1958a. Incoherent Scattering of Radio Waves by Free Electrons with Applications to Space Exploration by Radar. *Proceedings of the IRE* 46 (11): 1824–1829.

———. 22 October 1958b. Union Radio-Scientifique Internationale – Institute of Radio Engineers (URSI-IRE), Fall Meeting. Pennsylvania State University.

Chapter 2
The Evolution of the Telescope Design

The initial design of the telescope as laid out by William Gordon during his May 1958 seminar was a 1000 ft (305 m) fixed parabolic antenna with a central tower and equipped with a 1 MW transmitter and a receiving system. While the emphasis was on using the radar system to study the Earth's ionosphere, the strong desire to also use it for studies of the Sun and nearby planets dictated that it had to be sited in the tropics, resulting in the selection of two possible sites in Puerto Rico. A proposal to build the telescope and radar system was submitted to the Advanced Research Projects Agency (ARPA) of the US Department of Defense in November 1958. Driven by the desire to improve the telescope's capabilities for radar studies of the Sun and nearby planets, the decision was made to change the telescope's reflector from a parabolic shape to a spherical one despite the difficulties of devising a device that would complete the focusing of incoming radio waves. This change allowed the telescope to point up to 20° from the zenith. A new site for the telescope south of the city of Arecibo in Puerto Rico was selected. A revised proposal was submitted to ARPA in June 1959, which was given approval in September. As a result of a design competition, a contract for the design of a cable suspended structure and a cable supported reflector was awarded to the Von Seb Joint Venture, a consortium of companies, in April 1960.

2.1 Setting the Telescope's Detailed Specifications

During his 29 May 1958, talk Gordon discussed the concept of incoherent scattering for studying the ionosphere and the specifications of the antenna and transmitter that would be required. It is fortunate that Ralph Bolgiano, then Professor of Electrical Engineering at Cornell, took notes during the talk that have been preserved. In the second page of the notes (Fig. 2.1), Gordon gives the formulas that he used to

D. Campbell, *The Arecibo Observatory: A History of Innovation and Discovery*, Historical & Cultural Astronomy, https://doi.org/10.1007/978-3-031-75225-4_2

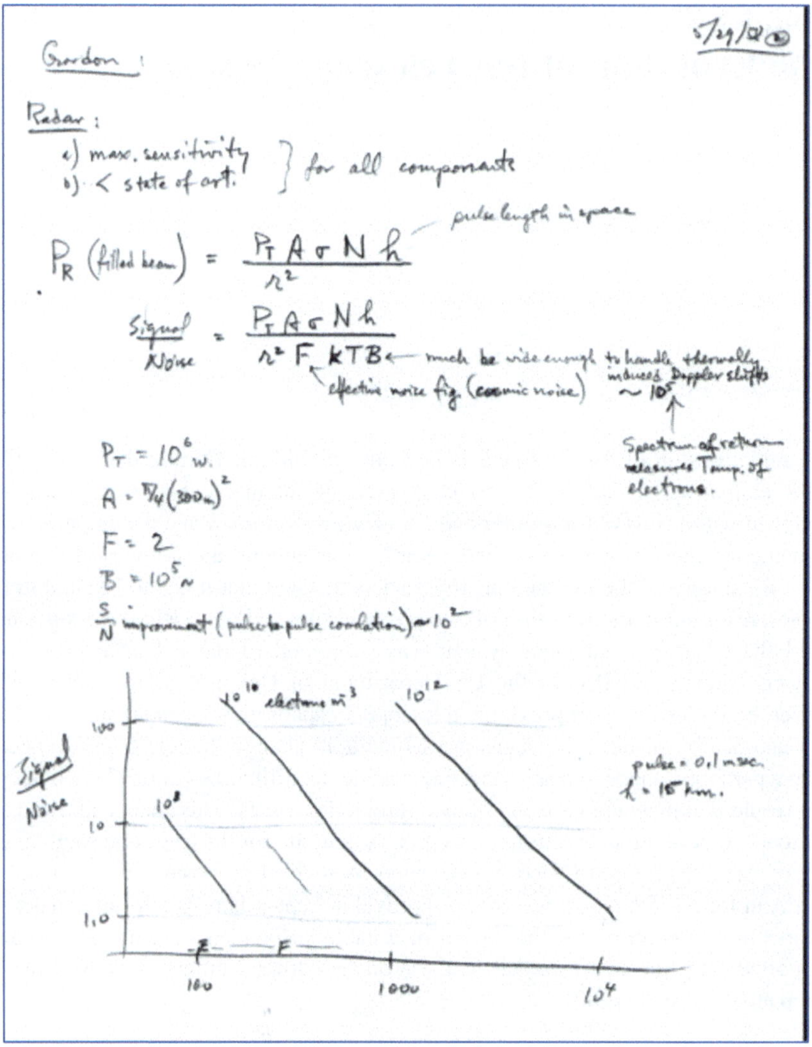

Fig. 2.1 The second of two pages of notes taken by Ralph Bolgiano at William E. Gordon's 29 May, 1958 seminar. The scale along the bottom of the plot is altitude in kilometers. Credit: Wesley Swartz for preserving the notes

determine whether the signal reflected from a layer of the ionosphere could be detected given an antenna with an area of A and a transmitter that generates pulses of radio waves with a power of P_T watts. A detection also depends on the altitude r and thickness h of the layer and the number of electrons per cubic meter at that height (see the Box for the details).

Box

Deciphering the notes of Fig. 2.1:

The first equation is the received radar echo power, P_r, from a layer with an electron density of N per cubic meter (about 1.3 cubic yards) and a layer thickness of h km at an average altitude of r km using an antenna with an area of A square meters and a single transmitter pulse with a power of P_T watts. The cross section or equivalent size of an individual electron is σ m^2. This is an incredibly small amount of reflected power, but detecting it would not be a problem except for the presence of radio noise so the signal/noise, the ratio of the received power P_r to the noise power given in the second equation, determines if the received power can be detected in the presence of the noise power. The noise power is given by F k T B, where k is a physical constant called Boltzmann's constant, T is the ambient temperature in the absolute scale of Kelvins, 300 K which is 27 °C or 81 °F, and B (10^5 Hz) is the estimated Doppler spread of the received echo power due, in Gordon's analysis, to the Doppler shift (see below) of the signal by the thermal motions of the electrons. The two major sources of noise are radio emission from the Milky Way, cosmic noise, and noise in the system that amplifies P_r. F is a multiplying factor called the Noise Figure and accounts for the cosmic and amplifier noise. Gordon assumed that it would be close to 2, a very conservative number even then. By averaging the radar echo from many transmitted pulses, the signal/noise can be improved, with the improvement going as the square root of the number of pulses averaged. Gordon assumed 10^4 pulses giving a signal/noise improvement of 10^2.

Assuming an antenna with a diameter of 305 m (1000 ft) and a transmitter power P_T of 10^6 watts (1 MW), the plot at the bottom of Fig. 2.1 gives the calculated signal/noise ratio on a logarithmic scale for three different electron densities (electrons per m^3) and altitudes r in kilometers—the scale along the bottom of the plot—assuming a layer thickness (the altitude resolution) h of 15 km. If the S/N ratio is greater than 5, the signal should be detectable.

The Doppler shift is the small change in frequency of the reflected radar signal due, in this case, to the velocity of the electron towards or away from the radar. A typical example is the sudden change in the tone of a train whistle from higher to lower as it passes through a station. The electrons will have a random distribution of velocities, so the signal echo will be spread out over a range of frequencies, the Doppler spread, B. The average velocity goes as the square root of the temperature T so B is dependent on T and a measurement of B allows the temperature to be determined. However, as was discussed in Chap. 1, it turned out that the B is determined by the velocities of the ions and, hence, by the temperature of the ions. As also discussed in Chap. 1, the smaller Doppler spread effectively increases the sensitivity of the radar system. However, the Doppler spread is also dependent on the mass of the ions allowing the composition of the ionosphere at different altitudes to be determined, a distinct plus.

Gordon knew that transmitters existed at that time that could provide at least 1 megawatt (MW) of power, so assuming this number, it was relatively easy to calculate the needed antenna size, A, given an electron density N, a height r and a layer thickness h (15 km or about 9 miles) determined by the transmitter pulse length used, 0.1 ms. Since the Doppler shift is dependent on frequency and there was some uncertainty as to the temperature at high altitudes, Gordon assumed a Doppler spread of 10^5 Hz. An antenna diameter of 305 m (1000 ft) would give the signal/noise results shown in the rough sketches at the bottom of Fig. 2.1. Assuming an electron density of 10^{10} electrons per m^3, the signal/noise ratio in the F region at about 300 km height would be close to 100, making the signal easily detectable, hence, as recounted in Chap. 1, Bowles ability to detect an echo from the F-region with a much less powerful radar system than what Gordon was proposing. At much higher heights, where the electron densities are much lower, increasing the transmitter pulse length to 1.0 ms would change the layer thickness to 150 km and increase the signal/noise ratio by a factor of 10 also making these heights detectable.

The received power is independent of the transmitter frequency, but the cosmic noise power is not; it progressively increases as the frequency decreases below about 1000 MHz, increasing the cosmic noise F factor in the signal/noise calculation of Fig. 2.1 and, hence, decreasing the sensitivity. However, at low frequencies it is much easier and cheaper to build large radio telescopes as the requirement for the accuracy of the surface is much more relaxed. The sweet spot was 200 MHz to 300 MHz and the decision as to the exact frequency was not made until much later. A somewhat higher frequency, 430 MHz, was finally settled upon as there was no government frequency allocation available near 200 MHz that would allow transmitting at this frequency and transmitters with high powered amplifying tubes called klystrons had been developed near 400 MHz for the three cold war Ballistic Missile Early Warning System (BMEWS) stations in Alaska, Greenland and Great Britain.

2.2 Evolution of the Telescope Design

Gordon was incredibly excited by the results of his calculations, a radar system to investigate the ionosphere looked feasible, and this excitement infected many of his colleagues at Cornell. Sometime prior to his May 29 seminar, Gordon had approached George Winter, chair of the Civil Engineering Department at Cornell and William (Bill) McGuire, another faculty member in Civil Engineering, for help in determining if a 1000 ft diameter telescope could be built. Bill McGuire immediately set to work to see what it would take to build a parabolic reflector sitting on level ground. How high would the edges be and could they be supported? An important consideration was how smooth the reflector surface had to be and whether it could be made of wire mesh or had to be a solid surface. These requirements are dictated by the frequency of the planned transmitter, which was initially between 200 MHz and 300 MHz but, as noted above, was extended to 400 MHz due to the existence of high power amplifying tubes, klystrons, at about this frequency. The

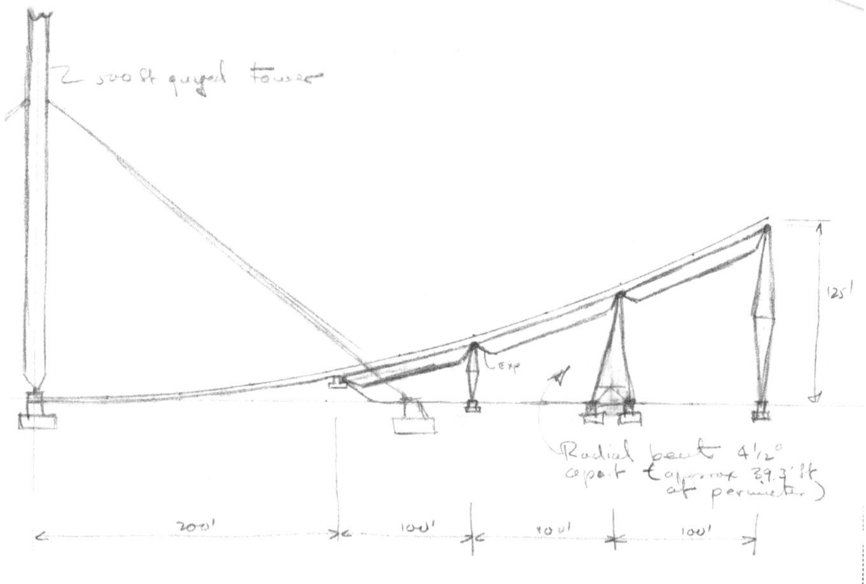

Fig. 2.2 One of several concept drawings by William McGuire made on 16 May 1958 showing that a 1000 ft diameter radio telescope could be built on flat land. William McGuire papers, Cornell University archives

wavelengths, the distance between the peaks of the propagating electromagnetic wave, corresponding to this frequency range are from 1.5 m (~5 ft) to 0.75 m (~2.5 ft). Using the rule that the reflecting surface should not vary on average by more than 5% of the wavelength from the desired shape, the reflector surface needed to be set to an accuracy of between 3.75 and 7.5 cm (1.5 and 3 inches) depending on the frequency. Similarly, a lightweight wire mesh surface could be used with spacing between the wires of about 1.5 inches—chicken wire would work reasonably well.

Between 16 and 22 May, McGuire produced a number of possible designs. Figure 2.2 shows his first rough concept drawing from 16 May 1958. The edges of the reflector are 38 m (125 ft) high supported by three progressively higher towers starting at 91 m (300 ft) from the center and connected by radial trusses similar to the type that hold up supermarket roofs. There would need to be three concentric rings of 80 towers spaced, as indicated on McGuire's drawing, 12.5 m (40 ft) apart on the outer 1 km (~3300 ft) perimeter of the reflector connected by the radial trusses and with cables between them to support the wire mesh. The central tower, which would hold the transmitting and receiving feed at the focal point, would be 150 m (500 ft) high supported by guy wires back to the ground. McGuire's initial work clearly showed that building a 1000 ft diameter parabolic "chicken wire" reflector was feasible with commonly used construction techniques; however, it would clearly be a massive structure. At that time, 500 ft guyed radio towers were relatively common so posed no obvious difficulty. McGuire made no cost estimates at this time.

An alternative to a reflector supported by towers, trusses and cables was one where the mesh was laid on ground that had been excavated into the required bowl shape. If the original terrain was flat, this would mean balancing the amount of material excavated for the central area with what would be needed to build up the edges to the right height. Clearly, building the telescope in a natural depression would save a lot of excavation and, hence, money.

Gordon also discussed the project with Donald Belcher, another Professor of Civil Engineering at Cornell and a recognized expert, perhaps the expert, on the analysis of aerial reconnaissance photography. During World War II he assisted General Douglas McArthur's Pacific campaign by analyzing island invasion sites and is credited with locating the site of Brazil's then new capital, Brasilia. It was, again, Gordon's good fortune to have just the right person on the Cornell faculty to give him advice on where to find a suitable "hole in the ground" in which to site his planned telescope. It would be very expensive to move enough soil and rock to make one. In a 27 May memorandum to Gordon, Belcher suggested that there were possible sites for the telescope in Kauai, Hawaii, in northern Mexico and in Cuba. Belcher chose these locations because Gordon had specified a latitude, preferably in the 22 °N to 23.5 °N range but not less than 20°, to maximize the number of days on which the Sun and planets could be observed with a telescope that would only be able to steer up to 3° from the zenith. On 20 or 21 June the Sun is at its most northerly excursion, the summer solstice in the northern hemisphere, and is overhead at midday at the latitude of 23.5 °N. As it moves through its maximum excursion, the latitude at which it is overhead at midday is changing very slowly, maximizing the number of days for which it is in view for a zenith pointing telescope.

On 9 June, Gordon held a meeting to review the specifications for the telescope with his selected team, all of whom were from Cornell. Bill Gordon was extremely fortunate that people with the needed talents for the initial design work and telescope site selection were at the university. Present in addition to Gordon were Donald Belcher (site selection), Henry Booker, then the Chair of the School of Electrical Engineering (antenna specifications), Mike Colbert, a technician with the radio astronomy project, Richard Mason, Assistant Professor of Civil Engineering (antenna structure), Bill McGuire (antenna structure), Ben Nichols, Professor of Electrical Engineering (radar system), and George Winter (antenna structure), where the brackets give their responsibilities. For unknown reasons, Marshall Cohen, Assistant Professor of Electrical Engineering since 1954 and involved with Cornell's Radio Astronomy Project, was not included although he was specifically tasked with investigating the possibility of using the telescope to detect radio emission from deuterium, an isotope of hydrogen with one neutron commonly called heavy hydrogen, in the Milky Way at a predicted frequency of 327 MHz.

As laid out by Gordon, the immediate objective was a "feasibility study and preliminary cost estimate of a radar capable of exploring space." "Specifically (1) the radar should have the sensitivity to detect an electron density of 1000 per cubic centimeter (10^9 per cubic meter, see Fig. 2.1) at an altitude of 1000 km using a pulse length of 1 ms, and (2) the radar should use techniques presently within the state of the art."[1] This led to a set of updated specifications:

- Transmitter peak power of 1 MW based on existing technology
- Telescope with a parabolic reflector of 300 m (1000 ft) diameter with a 60% aperture efficiency (effective area used) and a central tower of about 152 m (500 ft)
- Reflector surface accuracy of 10% (later corrected to 5%) of the wavelength chosen
- A noise fig. (F) of a factor of 2
- Wavelength of 1 to 1.5 m (corresponding to a frequency of 300 to 200 MHz, later changed to 400 MHz to 200 MHz)—the longest consistent with a noise figure of 2
- Bandwidth of approximately 10^5 Hz[2]—the expected Doppler spread at heights of a few hundred kilometers.
- Antenna able to be pointed up to 3 degrees from the zenith to allow tracking the Sun, planets and radio sources
- Location at latitude 20 °N so that the Sun and planets could occasionally be viewed with the vertically pointing telescope. (This latitude is not consistent with building the telescope in Puerto Rico, which has a maximum latitude of about 18.5°.)

An outcome of the 9 June meeting was a series of questions and tasks:

- The 1000 ft parabolic reflector would consist of square wire mesh. What size and spacing of the mesh wires was required? Booker to investigate.
- The wire mesh would come in panels. Did the edges of the separate panels have to be welded together for electrical continuity? Booker to investigate.
- Did it make sense to have the inner part of the reflector surface more precisely set than the outer part to allow radio astronomy measurements at higher frequencies?
- Would an array of smaller antennas that could be individually steered make sense? This should be explored but a large, fixed single parabolic antenna would remain the standard against which other systems would be compared.
- The feed at the top of the 500 ft tower must hold steady in wind gusts to within 10% of the wavelength, about 10 cm, to maintain pointing at the Sun, planets and radio sources.
- The transmitter power has to be taken to the feed at the top of the 500 ft tower via a large (22.5 × 45 inches) rectangular waveguide or a 40 inch diameter circular one. To avoid all this waveguide, would it be better to have a secondary Cassegrain reflector on the tower that would allow the transmitter radiating feed to be on the ground at the apex of the reflector?
- How many radar pulses can be integrated in the receiver? Nichols to investigate.

---

[1] Quote from "Design Study of a Radar to Explore Space", 9 June 1958; William McGuire papers, Cornell Archives.

[2] Note that in 1958 cycles per second (cps) was still the unit of frequency; the modern unit, Hertz (Hz), was introduced in 1960. For simplicity Hz is used throughout this account.

- What are the characteristics of the Air Force radars (the Ballistic Missile Early Warning System (BMEWS) radars)? Nichols to investigate.
- Are there any biological hazards associated with the transmitter and its radiation? Nichols to investigate.
- Natural radio sources can be used to calibrate the antenna and receiver.
- Low frequency coherent sounder measurements of the ionosphere up to the peak electron density of the F2-region can be used to calibrate the radar data from higher heights. (This may have been the first time that this calibration technique was proposed and it continued to be used during the life of the telescope.)

The issue of where to site the telescope was discussed. Belcher now suggested Puerto Rico as it would likely have suitable landforms, it was a U.S. Territory inhabited by about 3 million U.S. citizens, and would be reasonably easy to access from Ithaca. Its latitude of about 18 °N was a little south of the desired 20° which would reduce the number of times the Sun could be viewed per year from about 85 to 50, but otherwise it seemed ideal. Belcher was asked to investigate further.

There is a comment in the meeting notes that there was "Some pressures for Texas and/or a missile range for the site of the telescope." Given how early this was in the project, it is hard to imagine where this pressure came from unless it was the Office of Naval Research (ONR) or Wright Patterson Air Force Base since Gordon had, or was, submitting proposals for design support to them. Gordon had already been in touch with ONR because he needed money. Large private research universities such as Cornell normally only pay their faculty for the 9 months of the academic year. For the three summer months the faculty can earn additional income from research grants, consulting, etc. It was May 1958 and Gordon needed funds to pay at least some of the summer salary and expenses of a number of faculty such as McGuire, Belcher and himself to keep the project moving. He was again the head of the Radio Astronomy Project and had a contract with ONR for this work. As he recounted in his 1979 self interview, he contacted his contract manager at ONR, Arnold Shostak, and explained the situation to him, probably infused with the excitement that he felt. Shostak gave him permission to spend funds from the existing ONR contract, which certainly saved the project in the near term. This was another example of Gordon's good fortune but it is possible that Shostak then, or soon after, suggested Texas or the missile range. Certainly, the Texas site was kept as a possibility until it was dropped based on costs and, possibly on Cornell's part, a reluctance to put the proposed facility on a military site.

The idea was there, build a 305 m (1000 ft) diameter radio telescope, equip it with a one-megawatt transmitter and it would be possible to detect radar echoes from the ionosphere up to heights of 1000 km (600 miles). One problem was that no one had ever built a 305 m (1000 ft) diameter telescope or anything close to it. What initially made it appear quite feasible was that the ionosphere is overhead no matter where you are on the surface of the Earth, so it was only necessary to build a telescope that could look vertically. There was no real need to be able to point it in any other direction. This meant that the telescope's 1000 ft reflector could be fixed on or in the ground and its parabolic shape[3] would both direct the transmitted radar signal towards the zenith, and focus the radar echo on a receiver about 500 ft (150 m) above the center of the reflector. It would also be necessary to transmit from this focal point, so some means had to be devised to get the 1-MW of power 500 ft up to the focal point.

However, as discussed in Chap. 1, there was a strong desire to use the radar system to also study the Sun and terrestrial planets, Mercury, Venus and Mars. This meant that the telescope would have to be steerable at least over a small range of angles from the zenith, 3° was set as the specification in the June 1958 team meeting and, preferably, be located in the tropics. Meeting the 3° requirement (later set to 2.3°) led to a very complicated feed arrangement at the top of the 150 m (500 ft) tower. It also limited the number of opportunities for observations of the Sun and, for the planets, even fewer productive opportunities as a planet needed to be relatively close to Earth in its orbit. The strength of the reflected radar signal is strongly dependent on the distance to the planet[4] so the opportunities to do detailed studies would be rare. These considerations clearly weighed heavily with Gordon's targeted funding agency, the Advanced Research Projects Agency (ARPA) of the U.-S. Department of Defense, and they encouraged Gordon to look at the possibilities of a spherical antenna that could point up to 20° from the zenith and, hence, track a celestial object for over 2 h as well as being able to point at an object over a large area of the sky. This led to the final design of the telescope but, as detailed in this chapter, the path was far from straight.

2.3 The Parabolic Antenna Phase

Helped by being able to re-direct funding from the Office of Naval Research for the Radio Astronomy Project to pay for salaries and other expenses over the coming summer, Bill Gordon moved quickly to finalize a design, select a site and develop a cost and schedule for the project. The 9 June 1958 meeting at Cornell on a "Design Study of a Radar to Explore Space" set the specifications for the telescope and assigned tasks to specific individuals. The most critical issues to settle as soon as possible were the basic design of the telescope, feed location at the top of the mast or

---

[3] Almost all radio antennas such as satellite TV dishes have a parabolic shape. This means that the entire antenna has to be pointed in the direction of the incoming radio waves as there is only one axis of symmetry, the principal axis. The big advantage of the parabolic shape is that it focuses an incoming radio wave to a point where a suitable device, usually what is called a feed horn, can collect the signal and send it to an amplifier. If the telescope's geometry is suitable (i.e. it has what is called a high focal length to diameter ratio) it can be sensitive to signals coming from small angles relative to the direction the antenna is pointed. This is the principle behind a number of "hole in the ground" fixed parabolic radio telescopes built during the early days of radio astronomy.

[4] The strength of the received radar echo decreases as the fourth power of the distance to the planet, so twice as far away means a drop of a factor of 16 in the strength of the radar echo.

on the ground using a secondary reflector, how to make it steerable over a small number of degrees, flat ground versus a natural depression and where to build it. The decisions would be strongly influenced by cost with the location decision also strongly influenced by the desire to use the radar to study the Sun and planets. Bill McGuire and Richard Mason, a young faculty member in Civil Engineering, took on this task with their results summarized in a report "The Fixed Antenna for a large Radio Telescope—Feasibility Study & Preliminary Cost Estimate" by R. E. Mason and W. McGuire dated November 1958.[5]

As recounted in their report, Mason and McGuire looked at five options for the telescope with the two that looked viable being:

A. A fixed parabolic reflector with a transmitting/receiving feed at the top of a tower with heights between 250 ft and 1500 ft.
B. The same parabolic reflector but with a Cassegrain sub-reflector near the top of the tower that would allow the feed to be close to the ground at the center of the reflector.

The other three, an array of small reflectors, a Fresnel reflector where the surface is divided into separate concentric rings, each of which is at ground level, and a Mills Cross type telescope consisting of two orthogonal long cylindrical parabolas, were not viable due to cost, lack of frequency flexibility and/or difficulty in steering the telescope to track the Sun and planets. All efforts were concentrated on assessing options A and B. A consideration common to both these options was a flat site versus using a pre-existing depression. A flat site also had two options: have the reflector close to the ground by excavating a central depression and using the excavated material to build an elevated rim, or build the telescope based on a tower, truss and cable system as illustrated in Fig. 2.2. It was quickly determined that using an excavated depression would be much the cheaper and no effort was made to do a detailed cost estimate for the truss and tower design.

Mason and McGuire moved very quickly and by 16 July 1958 had preliminary estimates of the amount of material that had to be moved for tower heights from 250 ft to 1500 ft. The costs for the single towers of option A and the three towers required to support the sub-reflector for option B were obtained in August from a tower manufacturing company, Blaw Knox of Pittsburgh. Figure 2.3 shows Mason's 16 July drawing for option A giving the range of tower heights considered and the resultant height difference between the center and rim of the reflector. The additional cost of a higher mast is offset by lower earth moving costs due to a smaller center to rim height as the reflector would have less curvature. A 500 ft tower would require moving about 585,000 cubic yards of material to form a central depression and the raised rim. The center would be 66 ft below ground level and the rim 59 ft above ground level, a huge structure with water drainage being a significant challenge. Clearly, a natural depression in the tropics of about the right size and with natural

[5] Available in the Cornell University archives.

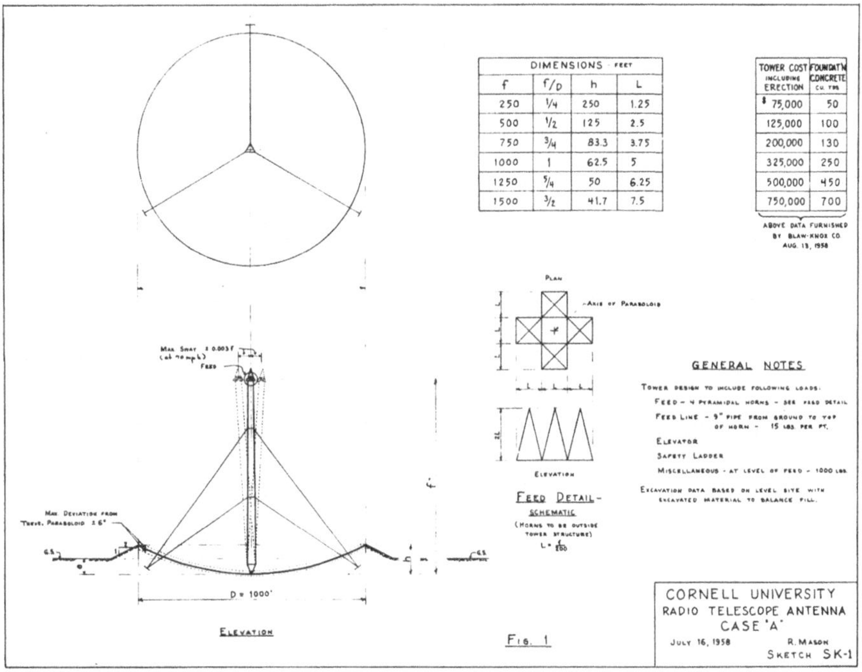

Fig. 2.3 Sketch of the telescope built in a depression created on flat ground by excavating material at the center and piling it up at the edges. For a 500 ft tower, 585,000 cu yds of material would need to be moved. Several tower options were considered and the tower in the drawing shows one that is about 750 ft high. Four radiating/receiving horns are clustered about the top of the tower. Drawing by R. E. Mason, 16 July 1958, with 13 August tower cost table added later. The excavation volumes on the original drawing were deleted apparently because these estimates were low by about 12%

drainage, Puerto Rico as suggested by Belcher, would lower the amount of material to be moved and, hence, the cost.

Probably during the summer or fall of 1958, a model of the option A telescope was made and Fig. 2.4 is a photo showing Bill Gordon and Bill McGuire discussing its merits. Not obvious in the photo are the four feed horns clustered around the top of the tower as shown in Mason's sketch drawing. There is no provision in this model or the drawing for a tracking capability.

Gordon wrote a memo on 11 July 1958, discussing option B, the use of a simple Cassegrain sub-reflector (sub-dish in Gordon's hand drawing of Fig. 2.5) that would redirect the rays to a feed near the apex of the parabola. While having the feed—the "radiator" in Gordon's drawing—close to the ground is an obvious benefit, the feed in this location is quite complicated in comparison to the simple horn type feeds at the top of the tower for option A. To direct all the transmitter power to a 70.7 ft diameter sub-reflector 470 ft above the center of the primary reflector would require a device that is about 7 wavelengths in diameter, about 33 ft (10 m) for a frequency of 200 MHz and half that for 400 MHz, very large devices. In order to obtain firmer

Fig. 2.4 William E. Gordon (left) and William McGuire discussing a model of the telescope with a central tower built into an artificial or natural depression probably in the late summer or early fall of 1958. Credit: Cornell University

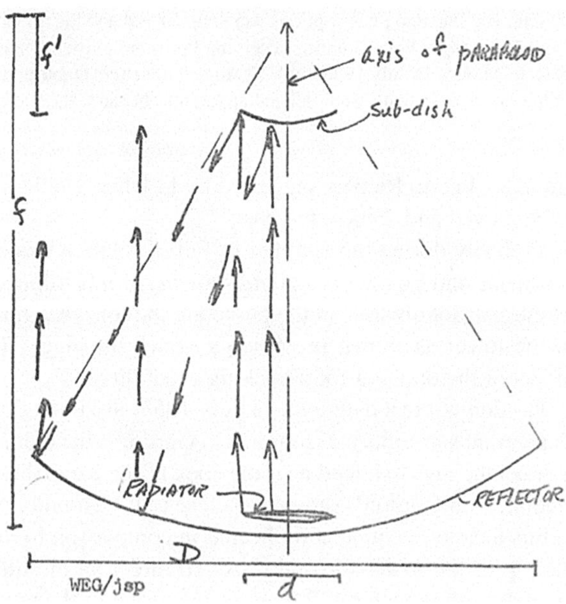

Fig. 2.5 A sketch by William Gordon of a possible sub-reflector based design where the transmitting and receiving systems would be near the center of the reflector and a Cassegrain sub-dish would redirect the transmitted power (see arrows) on to the 1000 ft main reflector and, hence, to space. The received signal would follow the reverse path. (W. E. Gordon memo, 11 July 1958, Cornell University archives)

ideas on how to implement option B, Cornell let a contract to Wheeler Laboratories of Long Island to study the possibility of using a sub-reflector and how it might impact the need to steer the telescope over the desired ±3 degrees in order to track the Sun and planets. The conclusions of the study were that the sensitivity of the sub-reflector system would be slightly better than having the feeds at the top of the tower due, mainly, to reduced losses in transmitter power in going up the 500 ft tower, but this would be offset by considerable complications in supporting a large semi-steerable sub-reflector near the tower top and the steering mechanism for the large feed needed near the ground.[6] A sub-reflector based design continued to be studied but its cost turned out to be higher than that for feeds at the top of the tower and it was eventually dropped.

Tracking: As discussed in Chap. 1, the desire to use the telescope and radar system to study the Sun and planets significantly complicated the selection of the location of the telescope and its design. Assuming that you transmit for the 16 minutes it takes for a radar signal to go to the Sun and back and receive the radar echo for at least a few minutes, then you have to be able to track the Sun as the Earth rotates for about 20 minutes. Since the Earth rotates 1° every 4 min, this tracking time corresponds to being able to point the telescope up to 2.5° from the zenith. The tracking requirement for radar observations of Mercury, Venus and Mars is less as they come closer to the Earth than the distance to the Sun. Given the complexity of the tracking system, it was designed to allow the feed horns to move up to 20 ft (6 m) from the center of the 500 ft tower (Fig. 2.6) allowing tracking only up to 2.3° from the zenith or for 2.4 min more than the echo round trip time to the Sun. However, it was more than adequate to observe the terrestrial planets provided that they were close to the zenith at the telescope's location and closer to the Earth than the Sun in their orbit about the Sun.

2.4 Site Selection

A suitably shaped bowl-like formation in which to place the telescope would be much cheaper than moving 585,000 cubic yards of rock and soil that would be necessary if the site was flat ground. As recounted earlier, because of the desire to observe the Sun and planets as often as possible given the planned 2.3° steerability, a site in the latitude range 22° to 23.5° was best with 20° being the lowest desirable latitude. However, this limited the site choices to Hawaii, Mexico and Cuba, all with significant disadvantages related to distance or bureaucratic issues such as customs. At the 9 June meeting, Donald Belcher suggested Puerto Rico as a possible site because of the presence of a limestone karst region in the northwest of the Island. The latitude would be about 18° reducing the number of days that the Sun and

---

[6] Study of Cassegrain antenna for radar telescope, Peter Hannan, 24 Oct 1958. Cornell University archives.

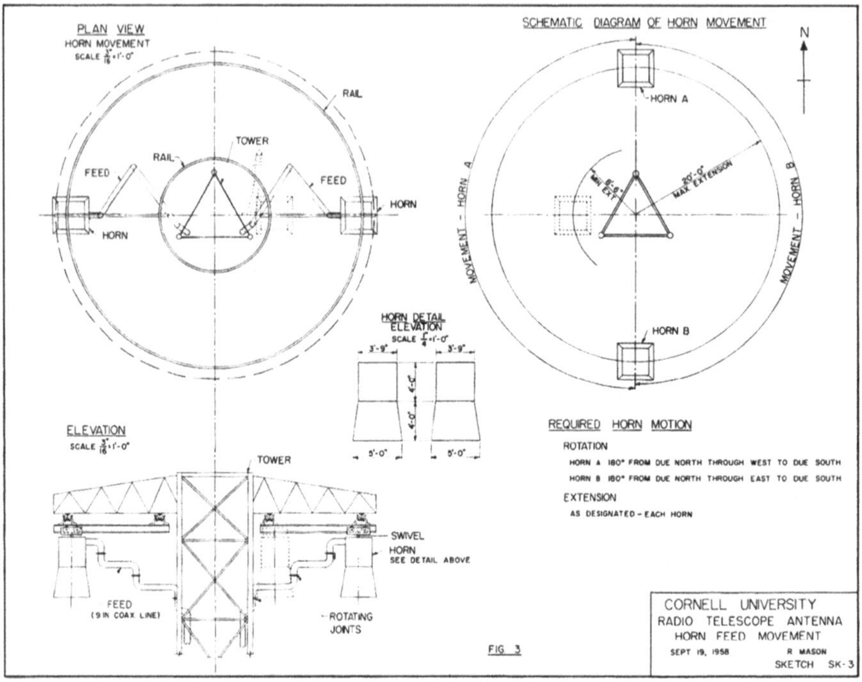

Fig. 2.6 The proposed tracking mechanism at the top of the tower. Two horns fed by co-axial lines that can concertina to allow radial motion along a rail and azimuthal motion over ±90°. The rail is supported from two tracks concentric with the tower center at radii of 8.5 ft and 20 ft

planets could be observed, but this was offset by the obvious advantages of Puerto Rico, a politically stable U.S. territory inhabited by about 3 million U. S. citizens and reasonably accessible from the Cornell campus in Ithaca, New York. Belcher was tasked through his Ithaca based company, Donald J. Belcher and Associates, to investigate possible sites in the Karst region that would accommodate the antenna with a minimum of excavation and fill. This type of terrain consists of limestone in which large caverns, formed by the limestone rock being dissolved by water, collapse creating what are known as sinkholes. Some have vertical sides but others are more rounded, perfect for the telescope. For several years, Belcher Associates had been attempting to generate interest in mineral exploration in Puerto Rico utilizing aerial photography, so Belcher had considerable knowledge of the Island and knew of the existence of aerial reconnaissance photography from 1951.

Belcher contacted the Puerto Rico government via an 11 June letter to the then governor, Luis Munoz Marin, seeking their cooperation. The governor's office referred him to the Secretary of Public Works, Roberto Sanchez Villella (the successor in the 1964 election to Luis Munoz Marin as governor) and, after being contacted, the Public Works Department offered access to the 1:15,000 stereoscopic aerial reconnaissance photography of the Island. Based on a \$3087 proposal to Cornell from Belcher Associates—the cost of a round-trip "tourist" class ticket

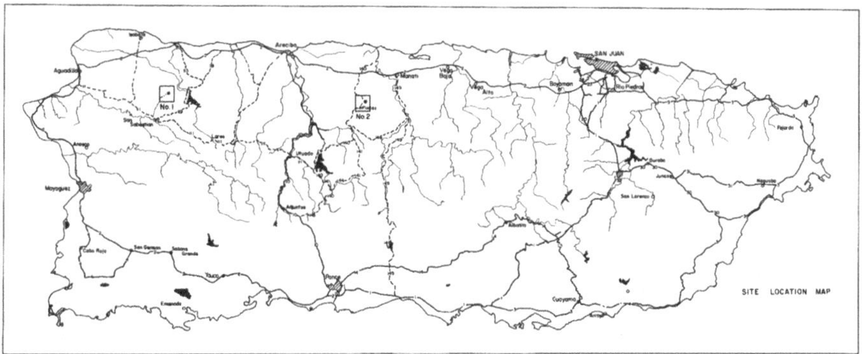

Fig. 2.7 Map of Puerto Rico showing the locations of the initial sites selected by Don Belcher as having sinkholes suitable for the telescope, San Sebastian (No. 1), Florida (No. 2) and Vega Baja. The final site is about 8 miles (13 km) south of the city of Arecibo just to the west of the road from Arecibo to Utuado (Route 10). Map from Belcher and Associates' report dated August, 1958. The island is ~100 miles (160 km) by ~35 miles (56 km)

from New York to Puerto Rico was only \$135—several Belcher staff members were sent to Puerto Rico with 3-D equipment to study the photos looking for possible sites, depressions with diameters of about 1000 ft (300 m) and rim heights of about 125 ft. A number of sites were found and then reduced to 14, all of which were visited and inspected. This led to three primary sites that are described in Belcher's 7 August 1958 memo "Site in Puerto Rico for Radar Telescope" to Gordon and McGuire and in Donald J. Belcher's and Associates formal report "Site Locations for a Proposed Radio Telescope" later in August.[7] This may be the first time that the term "Radio Telescope" was used for the project.

In his memo and report, Belcher describes three possible sites near San Sebastian, Florida and Vega Baja (see map of Puerto Rico, Fig. 2.7). The site near Vega Baja on the north coast was very close to the coast and the main highway from San Juan to Arecibo so was not pursued, probably due to concerns over salty air and radio interference. Both of the other two sites were close to the right size to hold a 1000 ft reflector. At that time, the reflector was still a parabola with a 500 or 750 ft central mast, so there was little interest in the surrounding topography, except as a shield against wind and radio interference, as long as the bowl shape was about right. The San Sebastian site near the western end of the Island seemed to be preferred as the terrain was the best fit for the telescope and it had good drainage and access via an asphalt road and an existing "tractor" road into the bowl. The land was relatively cheap ranging from \$50 to \$100 per acre for the hillsides and about \$350 to \$500 per acre for the central area that was planted in sugar cane. San Sebastian's local government people were very interested in hosting the telescope, even offering the use of municipal bulldozers, a not unusual offer in Puerto Rico. The next best site

[7] These reports are available in the Cornell University archives.

near the town of Florida was also close to the road from the coastal highway to Florida, a possible radio interference concern, and had similar land costs.

Two other sites in the continental Unites States were considered, one at a government installation in the southwest and the other at Forrestport in upstate New York, where there was an unused 1218 ft radio tower. Both of these would have required that the telescope axis be tilted to the south to observe the Sun and planets. This plus cost studies for the southwest site that were higher than for the Puerto Rican sites resulted in them being excluded.

Funding: On 17 September 1958, a meeting was held at Cornell of 27 representatives from General Electric, the Air Force Rome Air Development Center (RADC) in upstate New York, Wheeler Laboratories of Long Island, the Air Force's Wright Air Development Center (WADC), the newly formed Advanced Research Projects Agency (ARPA) of the Department of Defense, the Office of Naval Research (ONR), the Army Signal Corps, and Cornell University. The design study was now being supported by ONR, RADC and WADC, and the objective of the meeting was to present the results of the study to the sponsors and other interested groups. Depending on funding, the objective was to start work on the telescope in January 1959 and have it completed and operating in time for the June 1961 close approach of Venus to Earth. Arnold Shostak of ONR offered to coordinate funding for the project between ONR, RADC and ARPA. ONR and RADC remained interested at least through November 1958, but Cornell eventually submitted the first proposal that month directly to ARPA.

Frequency and the transmitter: At his 29 May seminar, Gordon proposed a frequency between 200 and 300 MHz (1.5 to 1.0 m wavelength) for the incoherent scatter radar. His 10 June memo setting the specifications for the telescope also specified this frequency range with the lower one set by the cosmic radio noise from the Milky Way, which gets stronger as the frequency is lowered, and the upper one by considerations of reflector surface setting accuracy and the mesh size. However, in practical terms, the frequencies of available high powered amplifying tubes for the transmitter and available government frequency assignments would dictate the possible frequencies, especially as the cost of developing new tubes at different frequencies would be large.

As discussed briefly in the 1 December design study report, there were two options for the transmitter, one with an upper frequency limit of about 200 MHz using an RCA A2342 triode tube giving an output power of about 1.0 MW peak pulse power and an average power of up to 90 kW, and one utilizing two Varian VA 842 klystrons[8] operating in the frequency range of 400 to 450 MHz giving a peak

[8]Triodes and klystrons both amplify a radio frequency signal. Both devices rely on a hot cathode to emit an electron beam. The triode amplifies by modulating the electron beam at the desired frequency using a grid that is placed between the cathode and the anode. Klystrons amplify by using a series of cavities that resonate at the frequency of the radio signal. A low level signal is fed into the first cavity and the electron beam interacts with that transforming some of the beam power into radio frequency energy and the proportion converted increases as the beam passes through the successive cavities.

power of 2.5 MW and an average power of 150 kW. These klystrons were similar to the ones used in the Ballistic Missile Early Warning System (BMEWS) radars. The 200 MHz transmitter option appears to have been an existing U.S. government owned one that could be modified at a cost of about $497,000. Cost estimates for a VA 842 klystron based one were received from Continental Electronics ($560,000) and Levinthal Electronics Products, Inc., later renamed Radiation Systems, ($500,000).

The 200 MHz transmitter barely met the power requirement laid down in the 10 June memo and it is not clear that a frequency assignment would have been possible since the TV channels 11 to 13 filled the frequency range from 192 MHz to 210 MHz. This meant that a frequency in the 400 to 450 MHz range would be preferable, 430 MHz was the eventual choice. This had good and bad consequences. The good news was that new low noise parametric amplifiers had been recently developed that contributed significantly less noise than the ones Gordon specified in his May 1958 seminar. This meant that the noise figure, which combines the cosmic noise, primarily from the Milky Way, and the noise introduced by the amplifier, would be less than 1 rather than the value of 2 that Gordon used for his original sensitivity calculations. In contrast, at 200 MHz, the noise figure would exceed 1 when the Milky Way was overhead at Arecibo and be less than 1 at other times. These new values, especially at 400 MHz, effectively doubled the original calculated sensitivity but, again, this did not affect the decision to build a 1000 ft antenna.

The negative consequence of choosing 430 MHz was additional loss of sensitivity as the telescope tracked to 2.3° from the zenith. The 1 December report puts this as a factor of 5. Despite this, the decision was made to go with a frequency close to 400 MHz; there really was no other option.

2.5 The First Proposal

These technical details and cost estimates formed the basis of a 13 November 1958 proposal to ARPA.[9] The proposal was for a 1000 ft diameter fixed parabolic reflector built in the San Sebastian sinkhole in Puerto Rico. Two by two-inch mesh was to be laid directly on the carefully graded parabolic surface with support posts where this was not feasible. A central 500 ft tower supported the ±2.3° from the zenith beam swinging system of Fig. 2.5. The transmitter cost estimate used was that from Levinthal Electronics Products, Inc., which utilized two 1.25 MW klystrons at a frequency close to 400 MHz.

The total cost of the proposal was $2,367,000. This included $1,205,000 for land acquisition, site development, an operations building, the reflector, the tower, feed and transmission lines, surveying and engineering. The Levinthal transmitter had a

---

[9]Research Report EE 395 dated 1 December 1958. Cornell University archives under Arecibo Ionospheric Observatory.

cost of $500,000, and $300,000 was assigned for the receiving and data acquisition systems. The first year's operating costs were $200,000. The technical staff was projected to consist of a supervisor, an engineer and two technicians at the telescope site plus two technicians in Ithaca. Four Cornell faculty and six research assistants, primarily located in Ithaca, were to be associated with the project.

Construction was projected to begin in January 1959, and to be completed by July 1960, a total of 18 months. Design work would take the first 3 months, contracting the following month and actual construction 9 months. There was a five-month contingency. A shakedown research period of 9 months was envisaged ending just in time for radar observations of Venus during its close approach to Earth in April, 1961. This was the close approach of Venus during which the first radar detections were made. Unfortunately, Arecibo was not ready in time and unable to take part in the first radar detections of Venus. It was not until early 1964 that the newly completed Arecibo telescope made radar observations of the planet.

In retrospect, both the cost estimate and proposed schedule were very optimistic even for the relatively simple fixed parabolic reflector and central tower design. It was not clear from the proposal how the project was to be organized, who would do the detailed design and whether Cornell would act as its own general contractor or hire a single firm to manage the entire project. The answers would have a significant impact on the budget, a budget that seemed to have no provision for incidental expenses related to Cornell's oversight, construction supervision, and other non-direct costs. The on-site staff during the first year of operation appears inadequate for maintaining the telescope, transmitter and receiving systems. No provision seems to have been made for general maintenance of the site and operations building. However, events were rapidly making this proposal moot.

2.6 Changing to a Spherical Reflector

It was clear that the tracking capability of a fixed parabolic reflector was unsatisfactory in terms of meeting the secondary objectives of the telescope, radar studies of the Sun and planets and, to a lesser degree, studies of cosmic radio signals. This was especially true once the decision was made to use a frequency close to 400 MHz as the ability of a parabola to focus radiation arriving at a small angle to the principal axis scales directly as the wavelength; the shorter the wavelength (higher the frequency) the smaller the tracking capability. This deficiency was also apparent to Ward Low, who worked at the Institute for Defense Analyses (IDA), a non-profit civilian institution established in 1956 to provide advice to the Department of Defense on weapons research. In 1958 the IDA established a division to support the Department of Defense's newly formed Advanced Research Projects Agency (ARPA). ARPA was the agency that Gordon was pinning his hopes on for obtaining funding for the telescope and Ward Low was his main contact (see Chap. 3 for a detailed discussion). Low clearly had a broad vision for the possible uses of the future telescope perhaps driven by the instinct that it might not be possible to get

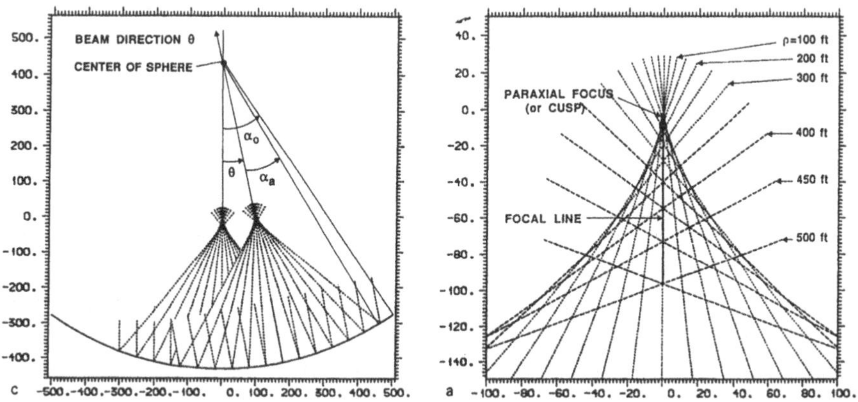

Fig. 2.8 Left: A spherical reflector with radio waves impinging on it from the zenith and at an angle of θ as depicted by the short parallel rays. The reflected radio waves almost come to a point focus but, as can be seen in the "blow up" of the focal area in the right drawing, the focus is not perfect. For the Arecibo design shown, all of the rays cross a 96 ft (30 m) focal line and a device, a line feed, needed to be designed that would pick up all the waves and bring them to an effective focus at the top. In late 1958 there were only rudimentary ideas as to how to design such a device but if it could be built, the telescope could be "pointed" by just moving the correcting line feed through different θ (zenith) angles. An effective focus could also be obtained using a carefully shaped radio mirror that intercepted all the rays and reflected them to a true focus. Such a "Gregorian" reflector system was implemented for the Arecibo telescope in the 1990s

funding for the limited instrument that Gordon first proposed in November, 1958. Low knew that there was ongoing development work on spherical reflectors at the Air Force's Cambridge Research Laboratories (AFCRL) in Massachusetts.

Because of the symmetry properties of a sphere—it looks the same no matter in which direction you look from its center—an antenna made from a portion of a sphere could, in principle, track a celestial object over much larger angles than a fixed parabolic reflector which is only symmetric about its principle axis (Fig. 2.8). The disadvantage is that a radio wave impinging on a spherical surface is not focused to a point, as can be seen from Fig. 2.8 for the 1000 ft diameter Arecibo spherical reflector. All the rays reflected from the spherical reflector cross a line, the focal line in Fig. 2.8, with the ray from the edge of the reflector crossing the focal line 96 ft (30 m) below the ray from the center of the reflector. The problem comes down to designing a device, a line (source) feed, that lies along the focal line and picks up the signals arriving along the ray directions and propagates them to the top of the line feed with just the right velocity that an artificial focus is formed. This problem was first addressed by R. C. Spencer, C. J. Sletten and J. E. Walsh (1949) of the Antenna laboratory at AFCRL. They published an internal AFCRL report "Correction of spherical aberration by a phased line source" in May 1951, that formed the basis for future work at AFCRL and elsewhere.

An alternative to a line feed is a concave Gregorian sub-reflector that collects all the signals along the rays shown in Fig. 2.8. Its shape can be adjusted to focus the incoming radiation to a point. AFCRL investigated Gregorian reflectors for spherical antennas as did Cornell for the proposed Arecibo telescope. However, the sub-reflector would have had to be extremely large and the focal point would be in a very inaccessible location. The big advantage of Gregorian sub-reflectors for spherical antennas is that they can operate over a wide band of frequencies as opposed to the very frequency-dependent line feeds; every operating frequency requires its own line feed. It was not until the 1990s that a successful Gregorian sub-reflector was installed on the Arecibo telescope with a third reflector solving the problems of accessibility and efficient utilization of the primary reflector.

According to Philipp Blacksmith in a recorded interview in 1979, Ward Low met with Carlyle (Carl) Sletten, the head of the Microwave Physics Laboratory at AFCRL, and Blacksmith at AFCRL resulting in Sletten requesting Blacksmith to visit Cornell. Blacksmith had been leading the group working on spherical antennas and monitoring work that was being done under contract by the Technical Research Group (TRG) of Somerville, Massachusetts, and the Wiley Electronics Co. of Phoenix, Arizona. Bill Gordon in his self-interview from the same period, remembered visiting AFCRL and talking with Blacksmith about their work on a small spherical antenna. Blacksmith and, possibly, other AFCRL staff visited Cornell, probably on 19 December 1958, to discuss line feeds for spherical antennas. According to Blacksmith there were other consultants and ARPA representatives present.

By coincidence, Carl Wiley and Allan Love from Wiley Electronics Co. had visited Cornell on 16 December. Carl Wiley was a brilliant, versatile engineer who invented Synthetic Aperture Radar in 1951 while working for the Goodyear Aircraft Company and introduced the idea of solar sails for spacecraft. Wiley formed his own firm, Wiley Electronics, and Allan Love, who later played a critical role in developing line feeds for Arecibo, joined the firm in 1957 to work, initially, on an airborne imaging radiometer for the U.S. Air Force under a contract with the Wright-Patterson Air Force base. It turned out that this system would work most effectively if the antenna had a spherical shape rather than having the usual parabolic one. This launched them into a project to design a line feed for the required 42 inch (107 cm) antenna and, in 1958 they came up with an open waveguide channel design. This led to a contract with AFCRL to develop a line feed for a larger spherical reflector. As recounted by Love (1989), they had just attended a conference on "Large Antennas for Radio Astronomy" in New York, where they had given a talk on line feeds for large spherical reflectors, and met Henry Booker and Marshall Cohen. Booker and Cohen encouraged them to come to Cornell to talk with Bill Gordon to discuss the idea they had been working on for a line feed, a three sided waveguide channel that radiated from the open face.

As discussed in a memo[10] written after the meeting, the major issues were handling the transmitter power, and radiating circular polarization since the open channel waveguide radiated linear polarization. Possible solutions to the polarization

[10] 16 December conference with Carl Wiley and Allan Love of Wiley Electronics with Cohen, McGuire, and Gordon. Cornell University archives.

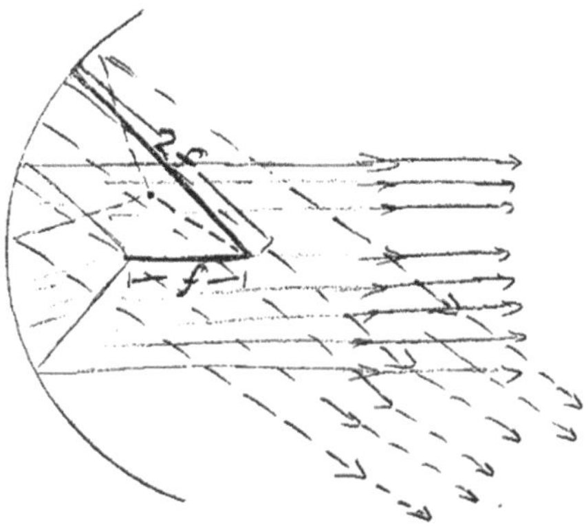

Fig. 2.9 Sketch of the rays in two directions for a spherical reflector fed by a line feed and an explanation of how a spherical reflector can be used for beam swinging. From the 10 December 1958, memo by William E. Gordon. Cornell University archives

problem included using four open waveguide channels facing 90 degrees from each other, or a closed waveguide with radiating slots. In the words of Merle LaLonde, who worked closely with Bill Gordon during the construction of Arecibo and, later, with Allan Love on designing and building line feeds, "The decision to build the Arecibo reflector in the form of a sphere and thus enable pointing of the 1000' antenna by moving the feed (instead of the reflector!) was influenced greatly by the development of a linearly polarized line feed in the late 1950s by Wiley and Love, under the sponsorship of AFCRL." The fortuitous meeting with Wiley and Love gave Gordon the confidence to move ahead with a spherical antenna. However, given the uncertainties regarding a dual-polarization line feed capable of handling several megawatts of RF power, it was a significant gamble to go for a spherical reflector design. If it paid off, it would result in a wonderful instrument for studies of the ionosphere, planets and cosmic radio emission. It was this prospect that must have motivated ARPA and Cornell to take the gamble.

On 10 December 1958, Gordon sent a memo "Increased beam swing capability" to Belcher, Booker, Cohen, McGuire and Nichols outlining ideas for expanding the tracking range of the telescope with emphasis on changing from a parabolic to a spherical reflector (Fig. 2.9).

A suitably designed line feed will illuminate the 1000 ft spherical reflector when it is pointing towards the zenith but, as Figs. 2.8 and 2.9 make clear, once you steer away from the zenith, the feed is still illuminating 1000 ft but not all of this is on the actual 1000 ft reflector. The illuminated area spills over its edge by about 30% when pointing 20° from the zenith effectively reducing the size of the reflector and lowering the telescope's sensitivity. As Gordon pointed out in his memo, solving this would require a reflector 1285 ft in diameter for steering out to 10° from the

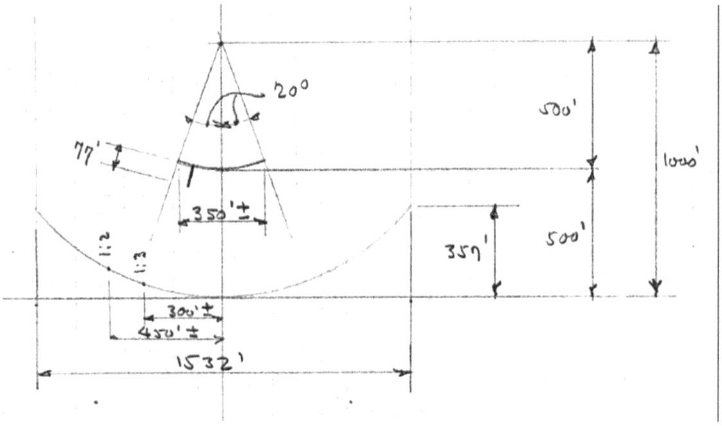

Fig. 2.10 16 December 1958 sketch by Bill McGuire of the basic geometry of the spherical reflector telescope. The sketch assumes a radius of curvature of the reflector of 1000 ft. This was later lowered to 870 ft to reduce costs by shortening the feed arm (the arc over which the line feed has to travel) from 350 ft to 304 ft but increasing the length of the required line feed from 77 ft to 96 ft. The reflector is shown with the diameter of 1532 ft required to avoid any spillover as the feed moves to 20° zenith angle. The final telescope was built with a 1000 ft diameter reflector

zenith and 1532 ft for an angle of 20°. Increasing the reflector diameter by 50% more than doubling its area would be highly desirable but extremely costly. Attempts to keep this option open complicated planning for over a year. The memo opened with Gordon stating that ARPA wanted Cornell to estimate what it would cost to increase the beam swinging capability and expressed willingness to consider a proposal that would increase the cost by a factor of two or less. The November 1958 proposal to ARPA had construction costs for the simple parabolic reflector with a central tower and a beam swinging capability of 2.3°, of $2,367,000. At a meeting in Washington that Gordon had with Low and others on 2 March 1959, Low suggested a budget of $4,500,000 to allow for beam swinging. This may have seemed more than adequate, but the complexities of going from the simple parabolic reflector system to a spherical reflector with a beam swinging capability of 20° were large. There seemed to have been little consideration of limiting the steerability to 10°.

The initial geometry assumed a radius for the spherical reflector of 1000 ft, which placed the top of the line feed tracing out an arc, the paraxial surface, at a distance of 500 ft from the reflector. Bill McGuire's 16 December 1958 drawing (Fig. 2.10) using these dimensions shows that the arc over which the line feed must move extends to 175 ft from the center. This arc must also rotate 360° about the vertical axis in order to be able to point the telescope to any location up to 20° from the zenith, clearly a much more formidable structure in terms of size and weight than the 20 ft radius arms at the top of the tower for the parabolic reflector (Fig. 2.6). Added to this was the need to design and construct a 77 ft long dual polarization line feed when no such feed had ever been built before. Based on the work of Wiley and Love, it was believed that a line feed could be built that would illuminate a 60° to 70° sector of a sphere. To reduce costs, a 70° sector illumination was chosen. This lowered the radius of curvature of the sphere to 870 ft and the length of the arc for the line feed to 152 ft from the center. The length of the line feed increased to 96 ft. Surprisingly, these dimensions stayed fixed through the design and construction of the telescope.

As usual, Bill McGuire jumped in with some preliminary designs. At this time, McGuire and everyone else still thought in terms of a central tower or towers to support the structure needed to allow the line feed to point in any direction within 20° of the zenith. Figures 2.11, 2.12, and 2.13 show McGuire's three central tower based spherical reflector schemes from December 1958. He also developed two schemes based on multiple towers supporting a ring girder, on which two 150 ft long trusses with rails for the line feed would rotate in azimuth. There is no evidence that these multiple tower schemes were ever pursued. The central tower scheme remained a serious possibility until the final design decision in early 1960.

In a telephone conversation with Bill McGuire on 21 January 1959, Gordon indicated that he was expecting proposals by 1 February or soon after from Blaw-Knox for a central tower supported cantilevered feed system and one from D. S. Kennedy of Cohasset, Massachusetts. D. S. Kennedy was a major antenna technology and manufacturing company in the 1950s. He must have also requested one from the Allison Steel Manufacturing Co. since they submitted a cost estimate for constructing a central tower with the cantilevered feed system on 14 February 1959. Gordon also mentioned that Kendrick Lippitt of D. S. Kennedy had suggested an arched tower system to support the moving feed with bases around the perimeter of the primary reflector (Fig. 2.14) and that they were investigating an alternative to a line feed, a 10 to 15 ft diameter lens with a point feed behind it. This was, in fact, not a practical idea, as the lens has to intersect the ray paths from the primary reflector as shown in Fig. 2.8. A 15 ft diameter lens clearly does not do this.

Alison steel provided a cost estimate of $936,000 for the central tower with the 150 ft cantilevered feed arm rotating on a bearing at the top of the tower partially counterweighted by a shorter arm with a large fixed mass at its end. Blaw-Knox, a leading supplier of microwave towers, sent a letter on 18 March saying that a tower supporting a cantilevered feed 150 ft long was not feasible at any reasonable cost and certainly not within the $1.5M cost cap that Gordon had set. D. S. Kennedy provided cost estimates on 13 and 27 January for the entire telescope including both arched and central tower designs. Both designs assumed that a lens could be used instead of a line feed, although it is not clear how this affected the cost estimates. The 13 January estimate assumed a four arched tower system ($2.241M), the reflector assuming the San Sebastian site or equivalent ($1.239M), R. F. system ($301,000) and drive and control systems ($93,050) for a total of $3,874,050. On 27 January they submitted a cost for a central tower of $0.847M reducing the overall cost to $2,480,000, just within Gordon's budget of $2.5M. Later in the year, perhaps about 30 June after the Arecibo sink hole had been selected, they submitted another estimate of $2,838,500 based on a three arched tower design and, from the list of costs, assuming that the reflector mesh could be laid on the surface of the contoured cavity.

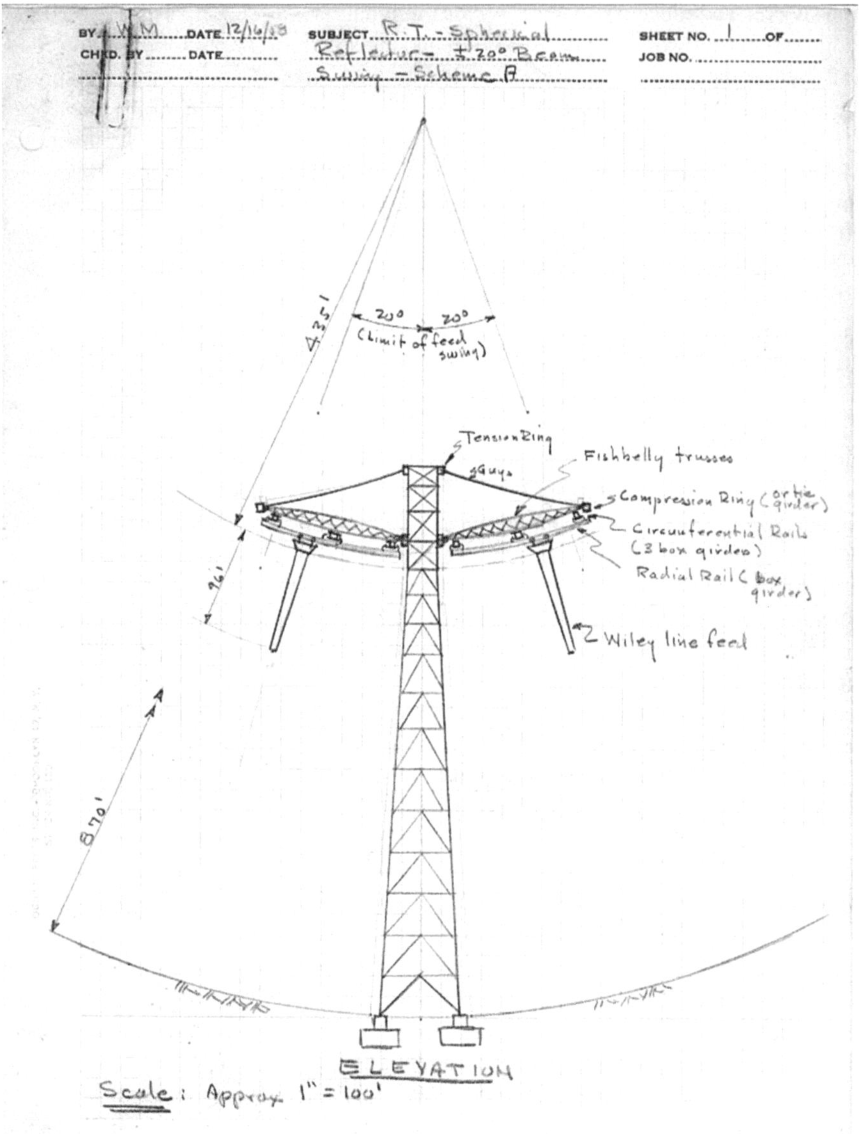

Fig. 2.11 Beam swing scheme A: 16 December 1958 sketch by Bill McGuire of a central self-supporting tower system for steering up to 20° from the zenith. Multiple cables from the top of the tower support a ring girder with a rail at a radius of about 150 ft. The two trusses in the figure can rotate on this rail and the "Wiley" line feed can move on the rail supported by the truss. One disadvantage of the central tower is that the telescope cannot look to the zenith. Note that the radius of curvature is now 870 ft, the final value, as opposed to the 1000 ft of Fig. 2.9. The line feed length is 96 ft. Papers of William McGuire, Cornell University archives

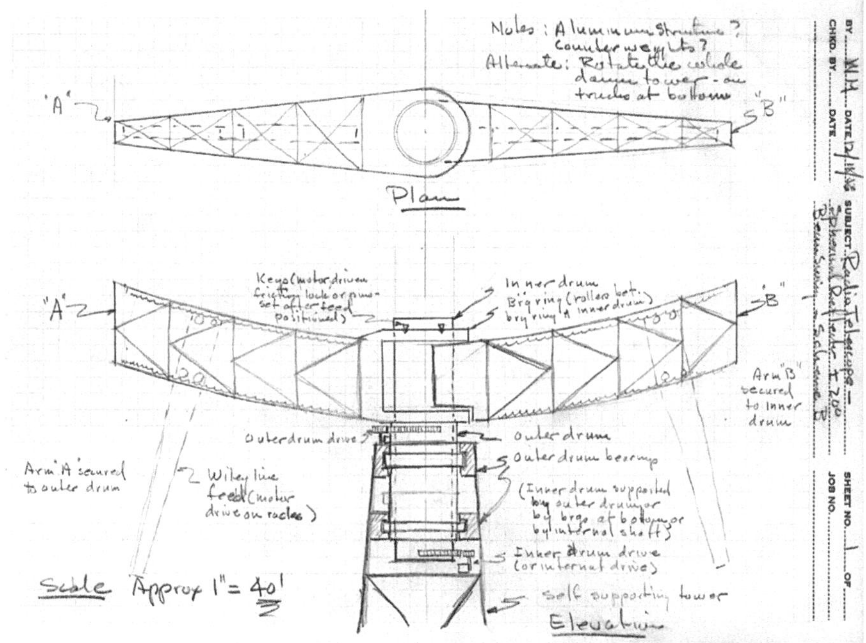

Fig. 2.12 Beam swinging scheme B: An 18 December 1958 central self-supporting tower concept by Bill McGuire, this time with cantilevered trusses rotating on bearings at the top of the tower. Each truss is 150 ft long supporting the moving "Wiley" line feed. The two bearings would allow the trusses to rotate independently, but this could result in a large, unbalanced load and it is not clear what the advantage was. The top note comments that it may be easier to just rotate the whole tower on trucks at its base. A variant of this concept was presented to potential design contractors a year later. Papers of William McGuire, Cornell University archives

2.7 Back to Puerto Rico

One outcome of the change to a spherical reflector was the requirement for a sinkhole that would allow a 1500 ft diameter reflector as an option. Neither the San Sebastian nor the Florida sites satisfied this need. Donald Belcher and his associates again pored over aerial reconnaissance maps of Puerto Rico and came up with one good candidate sinkhole about 8 miles south of the city of Arecibo (Fig. 2.14). It looked perfect but was located in relatively rugged terrain and lacked an access road for the last mile. It could accommodate either a 1000 ft or a 1500 ft reflector although the latter would require a considerable amount of excavation. Given how suitable it appears, it is hard to think that it was not included in the 14 sites that Belcher considered during his first survey. It is possible that the difficult access precluded it from consideration when the San Sebastian and Florida sites were selected.

It is not clear whether any alternatives to the Arecibo sinkhole were considered. The mystery, if one could call it that, is that the concept for the telescope still

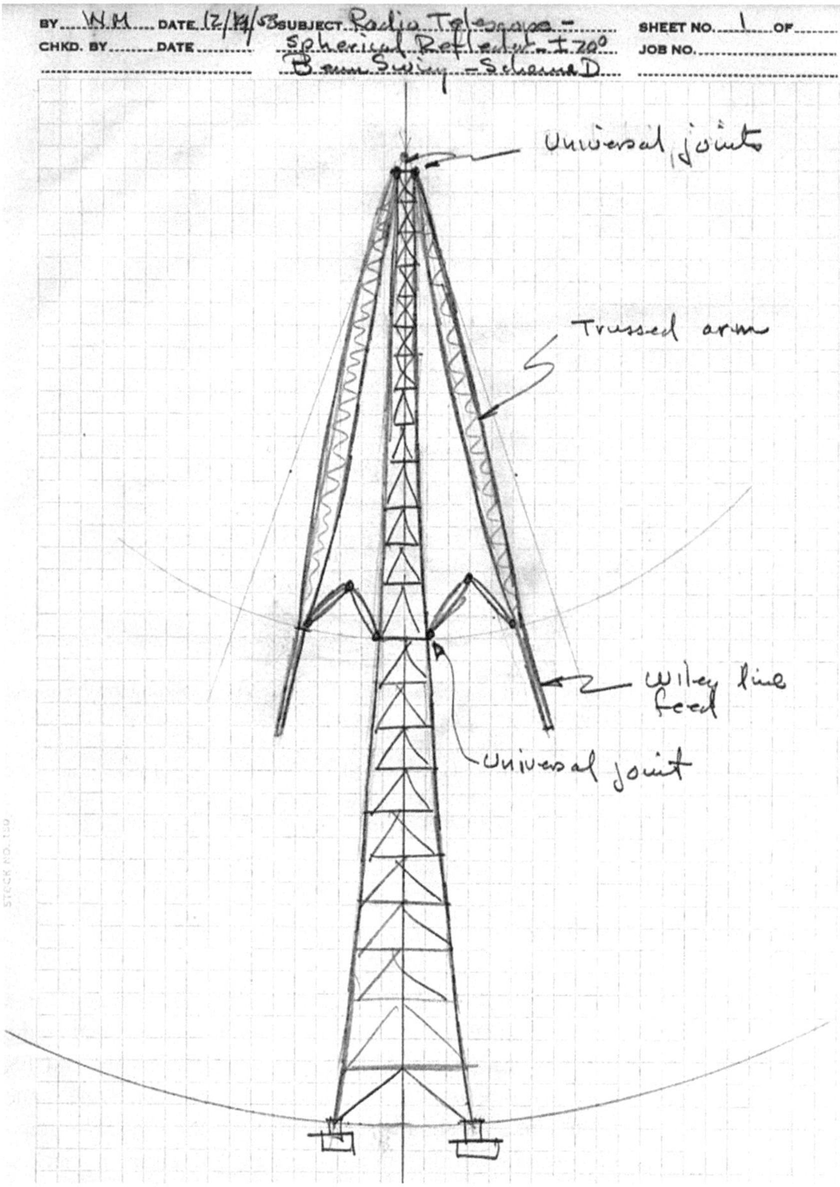

Fig. 2.13 Beam swinging scheme D: 19 December 1958 sketch by Bill McGuire of perhaps the most novel approach to solving the beam swinging requirement. With a self-supporting tower about 870 ft in height the 435 ft long trusses carrying the "Wiley" feed could rotate about the top and about a ring 435 ft from the bottom. Jointed arms could move the top of the feed over a 150 ft arc. Despite its elegant simplicity, this scheme was not pursued. Papers of William McGuire, Cornell University archives

Fig. 2.14 An aerial view of the Arecibo sink hole selected as the site for the 1000 ft telescope. The Rio Tanamá flows in the valley just south of the sink hole at the top of the photo going underground for a short distance at top center. Credit: Cornell University

included a central tower not a cable supported structure; yet the selected sinkhole has two hills spaced about 120° apart suitable for the towers needed for a cable supported structure. This greatly facilitated the construction of the eventual cable supported structure.

2.8 Converging on the Final Design and Obtaining Funding

On a 2 March 1959 visit to Washington, Gordon was told by Ward Low that the IDA had recommended to ARPA that the first proposal be funded at a level of about \$2.5M. However, Low indicated that he would like to see a revised proposal submitted that would include a 20° beam swinging capability estimated to cost an additional \$2M for a total of \$4.5M. Shortly after, Gordon received an oral message that the director of ARPA, Roy Johnson, had given "scientific approval" for the project at a level of \$2.5M but that it still had to go through the bureaucratic process at ARPA. Puzzlingly, on 17 April 1959, Low phoned Gordon to say that the project had been approved by the ARPA Council at a cost of \$4,500,000; \$1.3M would be made available out of FY1959 funds and the remainder added later. However, as discussed in Chap. 3, by early April ARPA had not made up its mind to support the project and at their meeting on 24 April the Council recommended delaying any decision until the Federal Fiscal Year 1960 budget was clear, so this must have just been Low's general opinion of what was likely to happen. A formal revised proposal for $4.5M was submitted by Cornell to ARPA on 16 June 1959, for construction of the Arecibo Radio Telescope. ARPA thanked Cornell for the proposal on 22 June but what surprised Cornell was a thank you received from AFCRL on the 19th. It is not clear why Cornell was surprised since they had sent a draft of the proposal to AFCRL a few days before the formal submission and the thank you may have been for this submission. By this time Cornell should have been aware that ARPA intended to have AFCRL do the contracting with Cornell for the construction of the telescope and with the (army) Corps of Engineers for the roads, buildings and other infrastructure.[11]

As recounted in Chap. 3, the summer of 1959 was primarily taken up with discussions about funding for the telescope. The amended proposal submitted to ARPA on 16 June was to "design, build and operate the radar in Puerto Rico" and covered the period 1 July 1959 to 30 June 1961, at a projected cost of $4.5M.

Success came on 30 September 1959, when ARPA Director Roy Johnson wrote the Commander, Air Research and Development Command (ARDC), Andrews Air Force Base (ARPA Order No. 106-60) with the request to proceed immediately with "design, construction and installation of a vertically directed ionospheric radar probe to be located in Puerto Rico." The design and construction shall generally conform to the Cornell University proposals dated 1 November 1958, and 1 June 1959, which resulted from Cornell studies undertaken under Contract No. AF 33(616)-5547. $1M was made available to the Air Force (for AFCRL) initially in order to commence design work immediately.

The letter instructed the Air Force to work with the Corps of Engineers to "arrange for the site acquisition, site development and radar installation in Puerto Rico." The letter went on to state that the Air Force must coordinate between the radar designer and builder and the Corps of Engineers to ensure timely information exchange between contractor(s) and the Corps. This is a somewhat confusing section of the letter since it seems to place responsibility for the "radar installation" on the Corps of Engineers while at the same time talks about the "radar designer and builder", that is, Cornell University. In practice the Corps was made responsible for site acquisition, access roads, primary power and water and all buildings with the specifications for the buildings to be provided by Cornell. Cornell was to be responsible for the actual design and construction of the telescope. The order specified "completion of the project for operational use within two years" in agreement with Cornell's schedule.

The instructions to the Air Force included a "vertically directed ionospheric probe" and a later sentence reinforces this: "Prior to undertaking any procurement for equipment which tends to commit the radar to any given beam-swinging capability, written approval shall be obtained from ARPA." Since the entire effort since December 1958 culminating in the 1 June 1959, proposal had concentrated on

[11]This sequence of events is based on memos and a one page hand written sequence covering 13 August 1958 to 30 September 1959, in the files of the Research Office of Cornell University, Cornell University archives.

Fig. 2.15 Philipp Blacksmith in the early 1960s. Blacksmith and Bill Gordon shared a penchant for bow ties. Photo courtesy of the Blacksmith family

designing a system to allow beam swinging up to 20° from the zenith, this injunction seemed to negate the whole program. However, from future events it seems possible that what was intended was a restriction on implementing a 1500 ft diameter reflector since the beam swinging design went ahead, and it is not clear if ARPA ever gave specific permission.

The ARPA order stated that as part of the design work, consideration had to be given to the scientific objectives related to the ionosphere, studies of the solar atmosphere, possible contributions to the study of the Moon and the possibility of "carrying out work on a laboratory scale." It is not clear whether this last referred to theoretical studies or actual laboratory studies but its inclusion is a puzzle since it was not mentioned in the Cornell proposal as reflected in the 1 August Cornell EE Research Report 435.

One additional result of the approach to AFCRL was the designation of AFCRL by the Air Research and Development Command as the responsible unit for the Arecibo project and, in turn, Philipp Blacksmith (Fig. 2.15) was asked to take on the task of Project Officer. Blacksmith recounted that he only agreed to do this if he was given "Command Line Authority" meaning that he could contact anyone, including ARPA officials, directly without having to go through the chain of command. This was approved. Blacksmith considered the $4.5M budget to be too low, $6.5M would be more realistic, but he indicated that he was not particularly concerned at the time; the money would be there if it was needed and justified.

The Air Force Cambridge Research Laboratories (AFCRL) grew out of the World War II Radiation Laboratory at the Massachusetts Institute of Technology, the MIT Rad Lab, which was famous for the development of radar systems during the war, and a lesser known laboratory at Harvard, the Radio Research Laboratory (Altschuler 2013). It was initially called the Cambridge Field Station and was located in that city in Massachusetts. In 1949 the name was changed to AFCRL then, in 1951, to Air Force Cambridge Research Center (AFCRC). In 1960 it reverted to AFCRL, keeping that name until it was closed in 1976. For simplicity, this history will only use the AFCRL name. Until a reorganization in 1961, the laboratories had two main divisions, the Electronics Research Directorate and the Geophysics Directorate. By the 1950s the laboratories had outgrown their Cambridge home and moved progressively to Hanscom (Air) Field in Bedford, about 20 miles west of Boston, where the Air Force established a base. In 1960 Brigadier General Bernard Holtzman was appointed Commander of AFCRL staying in this position through the completion of the Arecibo telescope. In 1961 he completely reorganized the Laboratories replacing the two directorates with five staff and support offices. However, the 14 technical divisions remained largely unchanged.

The Air Force levels above AFCRL underwent continual reorganizations and name changes. In 1950, the Air Force Research and Development Command (ARDC) was established with electronic research concentrated at the Rome Air Development Center (RADC) in upstate New York, and at AFCRL. In 1951 the Air Force Office of Scientific Research (AFOSR) was created as part of ARDC intended to mimic the Office of Naval Research (ONR), which was established in 1946 to support basic research at universities and other civilian institutions. In April 1959, Major General Bernard Schriever was appointed Commander of ARDC and, in 1961, ARDC was merged with the Air Material Command (AMC) to form the Air Force Systems Command (AFSC). However, due to concerns about mixing basic research and development, the Office of Aerospace Research (OAR) was created also in 1961, which reported directly to the Air Force Chief of Staff, and included the now renamed AFCRL.

Key people for the Arecibo project at AFCRL were Philipp Blacksmith as the Project Officer, his deputy, initially Edward Bouché and then James Murphy, and the contracts monitor Thomas Walsh. When he was given the job, Blacksmith was an engineer in charge of a group in the Microwave Physics Division within the Electronics Research Directorate working on spherical antennas. This was positive in that he probably knew more about antennas than anyone at Cornell and he had access to a pool of expertise at AFCRL and the Massachusetts Institute of Technology's (MIT's) Lincoln Laboratory that, like AFCRL, was situated at Hanscom Field. This resulted in a deep involvement in the project that led to some improvements in the telescope's design but was not always appreciated by the Cornell technical people. Blacksmith stated in his 1979 oral interview[12] that he spent 1 week a month in Arecibo and significantly impacted the design effort. Unfortunately, there is virtually no extant correspondence from or to Blacksmith related to the project apart from the monthly reports that he submitted to ARPA, which also contained a status report that Bill Gordon submitted to him. Without the 1979 interview he would be a very enigmatic figure in the project. However, it is clear that he and his successive deputies, Edward Bouché and James Murphy, were heavily involved in reviewing draft contract documents, approving the subsequent

[12] A transcript of Blacksmith's oral interview is available in the Cornell University archives under Arecibo Ionospheric Observatory.

sub-contracts, overseeing the construction work by both Cornell and the Corps of Engineers, and interacting with ARPA especially, as we will see, in supporting the never ending increases in costs and explaining the continual delays in the construction.

Relations between AFCRL and the Cornell administration were difficult throughout the construction of the telescope. Basically, the involved AFCRL contracting personnel, especially Thomas Walsh, thought that the Cornell people they dealt with, especially the ones in Ithaca, were not very competent. The problem seemed to be one of attitude to the construction of the telescope. The Cornell people thought of themselves as building a complex research instrument where small design changes were to be expected. At AFCRL this was likely appreciated at one level but, basically, their attitude was that Cornell was just a contractor responsible for the construction of the telescope.

On 3 November 1959, Cornell received a Statement of Work for a contract between AFCRL and Cornell University for a "Design study of a vertically directed ionospheric radar probe" under which Cornell was to provide personnel, services and facilities to "conduct design studies and design a vertically directed ionospheric radar probe" which included a beam swinging capability to 20°. Provision was to be made for a later extension of the antenna from 1000 ft to 1500 ft. The design was to give specific consideration and priority to the list of scientific experiments that was included in the ARPA Order to the Air Force, but it also included "a classified project." In his 1979 interview Blacksmith said that this was related to the detection of nuclear bursts in space and, as of about 1962, it was declassified. However, it did mean that key Cornell personnel had to have security clearances.

Cornell also had to conduct a study of "additional scientific uses to which an instrument of this nature can be put" including "Radio Astronomy, Radar Astronomy, Astrophysics, and related propagation phenomena" and specifying the instrumentation that may be needed. The study should result in a "proposed scientific program, which shall indicate both the military and scientific importance of the data that is to be acquired as a result of each experiment."

Finally, there was a requirement to provide design criteria for the facilities—roads, buildings, etc.—needed for the initial scientific program and to develop a master plan for the entire Arecibo Radio Observatory. The plan was to include future extensions of the antenna, auxiliary antennas and allied equipment, expanded operations, housing and service facilities, additional land acquisition, roads, landscaping, etc. It is clear that ARPA and AFCRL had a long-term view of the observatory that does not seem consistent with the nominal 10-year life that Gordon and others always mentioned.

Short monthly letter reports were to be sent to AFCRL and semi-annual Technical Summary Reports were to be submitted within 15 days of the end of the reporting period. Technical reports were to be submitted as required and in multiple copies except for classified reports, which were to be submitted only on a reproducible master copy (made in those days on a mimeograph machine, one of the most common being a Gestetner machine).

Fig. 2.16 A view of the farmed section of the sink hole from late 1959. Credit: Philipp Blacksmith

About this time Blacksmith and Gordon made a trip to Puerto Rico to show Blacksmith the sink hole and overall site. Figs. 2.16 and 2.17 are photos taken by Blacksmith during that visit showing that sculpting the sink hole into a spherical shape would be the significant task it turned out to be.

The initial design contract between Cornell University and AFCRL was signed by Cornell on 19 November with an effective start date of 6 November 1959, the official beginning of a project that Bill Gordon had initiated with his May 1958 seminar in the School of Electrical Engineering. It was for $377,655 and covered the period 1 December 1959 to 31 May 1960. Total expected cost of the project was $4.5M including initial operation, of which $2.82M was for the antenna reflector, feed support structure, drives and transmission lines. Site development costs, land acquisition, roads, buildings, power and water had been estimated by the Cornell group to be $280,000 but this activity was to be the responsibility of the Corps of Engineers under a separate contract with AFCRL. The Corps would have their own view of the costs, about $624,000 for the same activities not including contingency or their overhead charge. An 18 November schedule for the construction showed a completion date of March 1961, only 17 months away with no overall design concept yet established. Optimism about schedule and costs permeated the project from the very beginning, perhaps related to the inexperience of the Cornell group

Fig. 2.17 Bill Gordon at the tobacco drying shed in the Arecibo sink hole during a visit with Phil Blacksmith in late 1959 after ARPA had approved the funding. Credit: Philipp Blacksmith

with large construction projects in remote locations and to the "can do" outlook of the times.

2.9 Settling on a Final Overall Design of the Telescope

However, Cornell and AFCRL were clearly intent on moving rapidly. In anticipation of the 19 November signing of the contract, Gordon and McGuire arranged for a briefing conference for a design competition for the reflector, feed support structure and transmission lines to be held at Cornell on 19 November. It was attended by participants from 40 companies plus Cornell, AFCRL and the Corps of Engineers personnel. The plan was for each organization interested in submitting a proposal for the design to present its qualifications, design approach and cost estimate by appointment about 3 weeks after the briefing conference. Cornell would then choose the firm that had the most viable design coupled with overall engineering competency with the intent of issuing a design contract by 20 December. The central tower and tripod designs of Figs. 2.14 were provided to the participants with the statement that "Neither scheme represents a finished design and it is hoped that these two views will not prejudice your thinking." Fig. 2.18, a publicity photo, shows an artist's conception of a tripod design nestled in what is clearly meant to represent the

Fig. 2.18 The caption on the image says in part that it "is the artists drawing on one possible conception of the powerful radar to be designed and constructed by Cornell University's Center for Radiophysics and Space Research for the Defense Department." "In this approach, a 600 ft high tripod is used to support a rotatable arm carrying the antenna feed on a radial track to provide scanning the radar beam 20 degrees off vertical." "The primary purpose of the instrument is to obtain information about the structure and composition of the upper atmosphere. (Rendering by Eric Svenson, student in the School of Architecture, Cornell University.)." It acknowledges ARPA and AFCRC (AFCRL) and states that the project is under the direction of Professor William E. Gordon. Credit: Cornell University

Arecibo sinkhole. This image was also provided to the participants. Note that a line feed is used, not a lens.

It was emphasized that the tripod design was preferable because a single, central tower prevented the telescope looking to the zenith and would also shadow the feed when it was close to the tower, but that cost would probably be the deciding factor. It was also emphasized that alternative design concepts were very welcome.

Eleven proposals were received plus three letters that were not considered formal proposals. Interviews with 12 engineering groups (one of the non-responsive firms was included) took place between 10 and 14 December. Several proposals for a suspended feed structure and a cable supported reflector were submitted.[13] The

[13] One of these proposals by GB Electronics, a subsidiary of the General Bronze Corporation, was presented by George and Helias Doundoulakis at the event. On 11 September 1961 a patent application was filed in the name of H. Doundoulakis for a "Radio telescope having a scanning feed supported by a cable suspension over a stationary reflector". The patent was awarded on 13 September 1966. In the early 1970s a claim for compensation was made under the patent against the

upshot of this process as described by Bill McGuire in a one-page document was that "no one had unique ideas for the construction of these structures which were obviously superior to all the others and acceptable in all respects." There was no uniformity of opinion as to whether tower or arched feed support or suspended systems were preferable, and this also applied to ground or cable support for the reflector. McGuire was clearly not convinced as to the viability of a cable supported feed structure due to tolerance and cost issues but felt that it should not be dismissed. Blacksmith stated in his interview that the Cornell professors could not agree on a concept, and that he suggested that Cornell let three $25k design study contracts and select the best one for the final design work. However, notes by R. W. Robinson, Cornell's deputy Coordinator for Research, on a meeting on 18 December at AFCRL attended also by Gordon, McGuire and Richard Hufnail, the Administrative Director of the Center for Radiophysics and Space Research at Cornell (see next chapter), clearly indicate that Cornell recommended selecting the proposal by a consortium that included Von Seb, the Development Engineering Corp., Praeger & Kavanaugh, Engineers, and Severud-Elstad-Krueger-Associates (hereafter called Von Seb or the Joint Venture). Bill McGuire claimed in a September 1961 letter to Robinson, that Cornell had selected an engineering group for the design work, the Joint Venture, and planned to give them a $75,000 contract to do an initial, relatively detailed, design but that the Air Force (i.e. AFCRL) overruled them and insisted on the three design studies as the next step.

Consequently, a decision was made to issue the $25k design study contracts to firms that had good track records in engineering design work. The firms selected were Burns & Roe of Chicago, the Von Seb Joint Venture and the Research Corporation of America's (RCA's) Moorestown Division. The first two were selected by Cornell while AFCRL selected RCA. Contracts were awarded on 7 January 1960, following a meeting with all three firms at Cornell on the same day to discuss the project. Proposals were due 20 February. Studies for both tower and cable supported feed structures were to be conducted and for what they considered the best method to implement the reflector. Representatives of the three firms visited Puerto Rico and the Arecibo site between 10 and 14 January accompanied by Bill Gordon and Ed Bouché, the assistant project officer from AFCRL.

Bill McGuire did a detailed review of the three proposals that covered the preferred approach of each of them.[14] Both suspended and tower-based options were investigated by Burns & Roe and von Seb but a suspended structure was the preferred one for both firms. RCA submitted a proposal for a feed structure supported by multiple arched towers and a cable supported reflector and Burns &

U.S. Government leading to a challenge to the validity of the patent. A settlement was reached in 1975 resulting in the patent being validated in the names of Helias Doundoulakis, Constantine A. Michalos and William J. Casey. As part of the settlement, the U.S. government was given a license make use of the ideas covered by the patent.

[14] 10 March 1960 letter to W. E. Gordon from William McGuire. It includes some detailed calculations primarily of the Burns and Roe proposal. William McGuire papers, Cornell University archives.

Roe's proposal was for a cable supported feed structure and a reflector supported on wooden posts. Von Seb proposed a three-tower cable supported feed structure and two cable supported reflector options. After a detailed review, the von Seb proposal was chosen despite continued misgivings about suspending the feed structure from cables. It was felt that there were too many unanswered questions about RCA's feed structure design and that its method for supporting the reflector mesh was not workable in practice. The Burns & Roe design had the feed structure supported by cables from three steel towers around the perimeter of the reflector that were up to 500 ft high. However, it was the over 4000 wooden posts required to support the reflector mesh that was thought to be impracticable. The von Seb design[15] used concrete towers about 265 ft high placed on the hills surrounding the sinkhole at a uniform radius of 710 ft from the reflector center (Figs. 2.19 and 2.20) allowing an expansion of the reflector to 1500 ft only over a fraction of the circumference. Of their two designs for a cable supported reflector one was regarded by McGuire to be much preferable. Despite his misgivings about a cable suspended feed structure, McGuire recommended that the Von Seb Joint Venture proposal be selected.

While the final design differed from the one in the von Seb proposal in a number of the details, the two are remarkably similar. Most notably, in the final design the towers have step backs rather than being tapered, the curved feed arm is 12 ft wide instead of 9 ft and there is a central pintle bearing for the feed arm to rotate about. The four cables from each of the corners of the suspended structure, which passed over the tops of the towers and down to the anchor blocks, were replaced with four cables to the top of each tower and five separate cables to the anchor block. From a personnel standpoint, the very scary elevator up Tower 8 and a cable car riding on one of the cables to the suspended structure was replaced by a simple cable car from the ground plus a catwalk that also carried the waveguide needed to bring the transmitter power to the suspended structure plus power lines, control and monitor signals, and radio frequency transmission lines. Both the catwalk and the cable car are supported at the feed structure by separate cables from the tops of the three towers to avoid any asymmetric forces on the suspended feed structure.

The firms forming the Joint Venture were Von Seb, the Development Engineering Corp., Praeger & Kavanaugh, Engineers, and Severud-Elstad-Krueger-Associates. Thomas Kavanagh of Praeger Kavanagh, the chief design engineer, was one of the most respected structural engineers in the United States at the time (Fig. 2.21). Both Bill Gordon and Phil Blacksmith had great confidence in him.

The Joint Venture submitted three proposals, one solely for the design of the suspended feed support structure and the reflector plus the drive system and transmission lines, one for the design work plus supervision of the construction work and one that also included acting as general contractor for the construction work. There was some disagreement between Cornell and AFCRL over which proposal to accept,

[15] Preliminary Report "Design of Ionospheric Probe for Arecibo Radio Observatory", Von Seb Inc., Development Engineering Corp., Severud-Elstad-Kreuger Associates, Praeger Kavanagh, 1960. William McGuire papers, Cornell University archives.

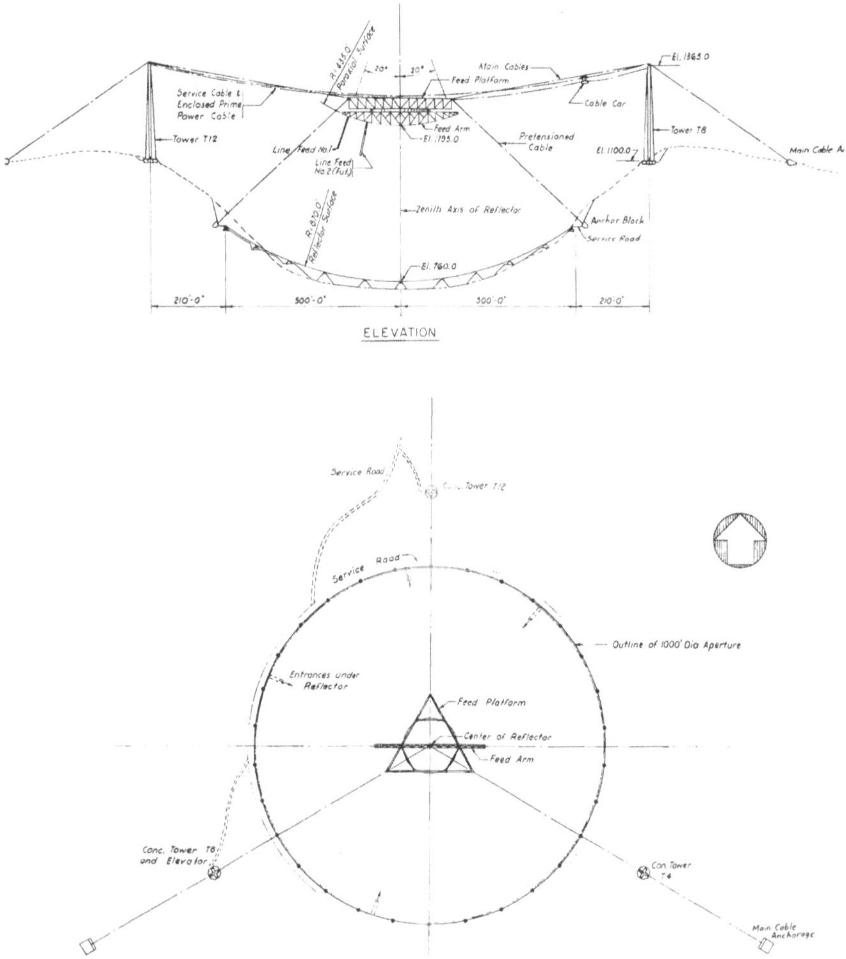

Fig. 2.19 The Von Seb Joint Venture proposed design for the Arecibo telescope. It is very close to the final design

but it was finally agreed that the Joint Venture should be contracted for the design and supervision of the construction work. The decision to not use the Joint Venture as a general contractor appears to have been driven by the desired schedule; contracts for individual components of the telescope could be let progressively by Cornell acting as the general contractor without having to wait for the completed final design. Cornell later claimed that it would have preferred to have the von Seb Joint Venture act as the General Contractor but the Air Force (AFCRL) insisted

Fig. 2.20 Artists drawing of The Von Seb Joint Venture's 1960 proposal for the Arecibo Radio Telescope

Fig. 2.21 Thomas C. Kavanagh (1912 to 1978) of the then engineering firm Praeger-Kavanagh, who was the chief designer of the Arecibo telescope. NASEM Archives: Deceased Member Files

that Cornell act in this capacity. The value of the contract signed with von Seb on 21 April 1960, was $341,000 ending in July 1961, the expected completion date of the construction.

A major technical challenge was the line feed with the success of the whole project resting on its successful implementation. Specifications were prepared and distributed on 16 December to 35 companies with the potential capability to design the feed. Thirteen proposals including one from Wiley were received by the 25 January 1960 deadline, all but three for line source feeds. However, as recounted in the 30 June 1960 report to AFCRL, the line feeds had a wide variety of design elements. The Technical Research Group Inc. (TRG) of Somerville, Massachusetts was awarded the contract for the design and fabrication of the feed on 3 March 1960, for a total of $258,096. Cornell and AFCRL were in agreement on the selection. As recounted by Allan Love in his 1989 Reminiscences (Love 1989), Carl Wiley was furious and complained to Congress, ARPA and DoD but to no avail. TRG proposed a 96 ft tapered square waveguide with radiating slots on each of its four faces that would transmit a circularly or linearly polarized wave and receive two circular or orthogonal linear polarizations. TRG assigned the project to Dr. Alan Kay.

The stage was now set for the final design, contracting and construction phases of the project. However, it would be useful to first review the funding efforts and the management organization of the project.

References

Altschuler, Edward C. 2013. *The Rise and Fall of Air Force Cambridge Research Laboratories*. Create Space Independent Publishing Platform.

Love, Alan W. 1989. *Reminiscences of a Career in Electrical Engineering*. IEEE Antennas and Propagation Society Newsletter, October.

Spencer, R.C., C.J. Sletten, and J.E. Walsh. 1949. Correction of Spherical Aberration by a Phased Line Source. *Proceedings of the N.E.C.* 5: 320–330.

Chapter 3
The View from Washington

The Advanced Research Projects Agency (ARPA) of the US Department of Defense was Bill Gordon's target organization for funding the construction of the telescope. ARPA was particularly interested in the cold war issue of ballistic missile defense (Project Defender) so clearly had an interest in the ionosphere through which ICBMs had to pass. ARPA, as a new agency, also had a significant budget, much larger than the National Science Foundation. After a year's effort including shifting from a parabolic antenna to a spherical one, ARPA approved funding for the construction of the telescope in September, 1959. The Air Force Cambridge Research Laboratories (AFCRL), acting for ARPA, subcontracted Cornell to design and build the telescope and the US Army Corps of Engineers to construct the roads, buildings and utilities infrastructure. Within Cornell, the hiring of Thomas Gold as Chair of the Department of Astronomy led to the formation of the Center for Radiophysics and Space Research, which became the management organization at Cornell for the construction of the telescope. A major issue arose in early 1960; AFCRL made it clear that they intended to control the science program of the future telescope. This led to an extended tussle between AFCRL and Cornell which was finally settled in Cornell's favor by ARPA and resulted in responsibility for the future operations funding for the Arecibo Ionospheric Observatory being assigned to the Air Force Office of Scientific Research (AFOSR).

3.1 Convincing ARPA to Fund the Telescope Construction

Gordon recounted in his 1979 self-interview that he started "selling" the telescope project in the fall of 1958 and this led him to Ward C. Low. Low worked for the Institute for Defense Analyses (IDA), a non-profit civilian institution set up to advise the Department of Defense and with a section devoted to advising the Advanced Research Projects Agency (ARPA). ARPA was founded in 1958 at least partly due to the perception that the United States was falling behind in space and, more

D. Campbell, *The Arecibo Observatory: A History of Innovation and Discovery*,
Historical & Cultural Astronomy, https://doi.org/10.1007/978-3-031-75225-4_3

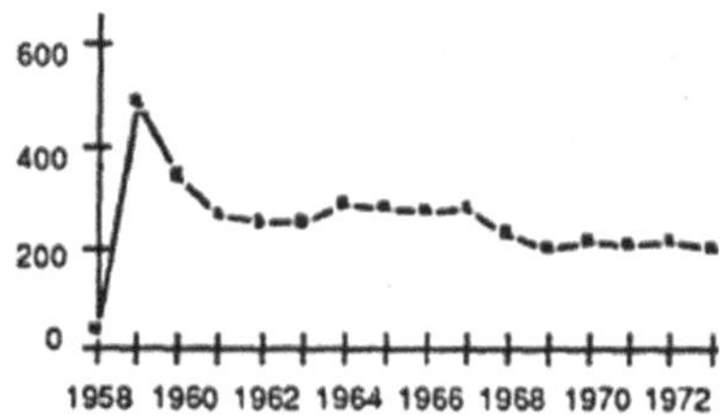

Fig. 3.1 ARPA's budget in then year dollars in millions from 1958 to 1973. The sharp decline after 1959 represented the transfer of civilian space related programs to the recently formed NASA. Van Atta et al. (1991a section of Fig. II-1)

generally, in science and technology and specifically because Sputnik's launch made "President Dwight Eisenhower anxious about U.S. military capability and high-tech research, and eager to reorganize the lumbering and diffuse U.S. missile and space effort" (Herzfeld 2008). Its name was changed to DARPA (Defense Advanced Research Projects Agency) in 1972. As related in a publication on DARPA's technical accomplishments (Van Atta et al. 1991a) "By 1960, ARPA focused on long-term, basic research emphasizing two specific Presidential issues, ballistic missile defense (Project Defender) and nuclear test detection (VELA), that were in the forefront of national security concerns." Gordon realized that ARPA would have an interest in the ionosphere since missiles and spacecraft had to pass through this relatively unknown part of Earth's atmosphere and, importantly, ARPA had adequate funds to support the construction of the telescope. In his search for someone to talk with at ARPA, Gordon was likely directed to Low, who was the Chairman of the Upper Atmosphere Physics Panel of the Ballistic Missile Defense Group (BMD) of the Advanced Research Projects Division (ARPD) of the IDA, and linked studies of the upper atmosphere, the ionosphere, with issues related to research on the detection of ballistic missiles, ARPA's Project Defender. Low obtained his PhD from Boston University in 1955 with a thesis entitled "Analysis of Experiments on Aerodynamic Heating of High-Velocity Missiles in Accelerated Flight" so, while relatively young, he had a background in issues related to Intercontinental Ballistic Missiles (ICBMs).

In 1959 ARPA's budget was about $460M (Fig. 3.1). It was a very new agency looking for projects to fund that would advance its mission. By comparison, the National Science Foundation's 1958 budget was $40M, and it was already funding the construction of the 140 ft telescope of the National Radio Astronomy Observatory. This makes it clear why Bill Gordon set his sights on ARPA as most likely to fund his telescope, especially as it was possible to persuasively argue that knowledge of the ionosphere was related to the research goals of Project Defender. At that time, universities working on defense related projects was perfectly acceptable. It was only during the Vietnam War in the late 1960s and early 1970s that many universities withdrew from most such work especially if it was classified.

Gordon recounts that it took numerous visits to Washington to keep Ward Low convinced that his telescope project was a sound one. Low would raise issues and Gordon would come up with answers to reassure him. Major concerns were related to the concept of incoherent scatter, whether the proposed radar system would achieve its objectives and, probably, whether it would be relevant to Project Defender's objectives. Ken Bowles' first detection in October 1958 of incoherent scatter from the ionosphere using the National Bureau of Standards (NBS) 41 MHz radar near Long Reach, Illinois (Bowles 1958) and subsequent measurements[1] clearly demonstrated that the concept of incoherent scatter was correct.

In June of 1958, the U.S. Navy through the Naval Research Laboratory (NRL) broke ground at Sugar Grove, West Virginia, for a 600 ft (183 m) fully steerable antenna nominally intended to pursue the new field of radio astronomy.[2] However, it also had a national security mission related to long distance communications via bouncing signals from the Moon and to monitoring Soviet communications, including those related to missile launches, also via reflections from the Moon. It also had the potential to do radar studies of the Sun and terrestrial planets, Mercury, Venus and Mars. Its initial estimated cost was $20 million, which rapidly escalated as the complexity of the project became more apparent. A continuing concern at ARPA as they considered funding the Arecibo project was duplication of effort: why build Gordon's telescope when a more capable, although smaller, telescope was under construction by the Navy?

Two months after the start of construction of the Sugar Grove telescope, the National Radio Astronomy Observatory (NRAO), a newly formed astronomical center supported by the National Science Foundation (NSF), broke ground for a 140 ft radio telescope at Green Bank, West Virginia, only about thirty miles from the Sugar Grove site. Both projects were within a radio quiet zone that had been established by the U.S. congress in 1958 to protect the telescopes from locally generated radio frequency interference. During 1960 it became apparent that both projects were in serious trouble with the NRAO 140 ft telescope put on hold due to engineering problems related to its equatorial mount. The Sugar Grove project developed more serious problems related to the plan to start construction of the foundations and mount before there was a final design. The weight of the moving structure kept increasing due to the need to maintain the precise parabolic shape of the reflector as it moved and under changing wind and temperature conditions. The weight finally exceeded the design load of the foundation and mount and, rather than expending the huge sum needed to correct for the problem, the project was cancelled in 1962. The Green Bank 140 ft telescope was finally completed in 1965. Fortunately, these problems arose after ARPA had agreed to fund Arecibo and construction had started in September of 1960 although it was still in the very early stages. Otherwise, it is likely that ARPA may have had second thoughts about funding Arecibo; with the two large antenna projects in trouble, would ARPA have wished to

[1] Kenneth L. Bowles, "Incoherent scattering by free electrons as a technique for studying the ionosphere and exosphere: Observations and theoretical considerations", NBS Report #6070, 18 September 1959.

[2] A detailed account of the Sugar Grove telescope saga and of the construction of NRAO's 140 ft telescope can be found in Kellermann et al. (2020).

risk the possible problems and resultant public scrutiny arising from building an even larger telescope?

Ward Low had concerns about the very limited steerability of the parabolic reflector telescope initially proposed by Bill Gordon; it would not have been very productive for radar studies of the terrestrial planets[3] or for radio astronomical observations. However, Low's primary concern was with its capabilities for studying the ionosphere and for this a vertically looking parabolic antenna was just fine. The desire on the part of Low and, perhaps, ARPA for an antenna that could point up to 20° from the zenith and, hence, be capable of exploring the terrestrial planets and studying cosmic radio waves was indicative of a broad perspective.

As recounted in Chap. 2, in late 1958 Low put Gordon in touch with AFCRL, where there was a group working on spherical antennas leading to a rethink of the design of the telescope. However, Gordon was already preparing a proposal for a fixed, parabolic antenna with a small beam swinging capability, and Cornell submitted this proposal to ARPA on 13 November (it was dated 1 November 1958). It was for $2,367,000 including the first year of operations. As recounted in DARPA Technical Accomplishments Vol. II (Van Atta et al. 1991b), "The proposal was assigned to project DEFENDER, which was concerned with phenomenology of missile flight, part of which would take place through the ionosphere. ARPA did not respond to this Cornell proposal with its then characteristic speed. This was due to several factors: the controversy within ARPA over the proposal's relevance to DEFENDER; the difficulties that were being experienced at the time by other ambitious, large antenna projects; and also because of a positive suggestion made by ARPA staff to use a spherical rather than a parabolic dish. This technical suggestion would not make a big difference in ionospheric research, which was ARPA's main stated justification for support, but could help a lot in radio and radar astronomy, and so added to the attractiveness of the facility for a wider range of investigators." (This statement is not entirely accurate as the two large antenna projects at Green Bank and Sugar Grove were not obviously in trouble until sometime in 1960.)

From ARPA's, and the IDA's, viewpoint funding Cornell's proposed radar depended entirely on its prospects for assisting with ballistic missile defense and, possibly, its potential for detecting nuclear explosions in space. At the time a considerable effort was underway, the Guide Line Identification Program for Anti-Missile Research (GLIPAR) project led by ARPA, to identify ways to counter Intercontinental Ballistic Missiles (ICBM's). However, as the desire for a larger beam-swinging capability indicates, this does not mean that individuals at ARPA and the IDA were not appreciative of the significant ancillary benefits that the telescope would provide, especially the capability for radar studies of the Sun and planets.

[3]The sensitivity of a radar observation of a planet declines as the fourth power of the distance so that it is normally critical that the observation be made when the planet is closest to the Earth in its orbit. For a telescope with a very limited range of viewing angles this is rarely possible.

On 14 January 1959, the Ballistic Missile Defense (BMD) group of the IDA issued a two-page memorandum "Recommendation for Support of Cornell University Proposal for Radar and 1000 ft Reflector for Ionospheric Measurements"[4] addressed to the Assistant Director, ARPA and signed by W.R. Hutchins, BMD Group Coordinator, and Ward Low as Chairman of the Upper Atmosphere Physics Panel. The memo stated that the principal reason for the recommendation of support for the radar probe "is to provide information on detailed features of earth's atmosphere in the midcourse region of ICBM's." By detailed features it meant the electron density and temperature. The memo referred to Bowles's successful measurement of incoherent scatter but did not point out that this measurement suggested that a 1000 ft antenna was not really necessary. The memo went on to point out that the radar could be used for studies of the planets and the Sun, that the 2° of beam-swinging for the proposed design was inadequate and that "the added advantages of a beam-swinging capability of, say, ±20° would be very great, and preliminary investigation suggests that the additional costs would be approximately $2 million." Specific recommendations were that Cornell, as soon as it had a contract for the design and construction of the facility, should undertake a detailed study of the costs and technical implications of a beam-swinging design. However, no steps were to be made to implement a beam-swinging design without ARPA approval. The contract should be administered through ARDC/AFCRL and expenditures should not exceed $2.5 million for two years. Curiously, the memo ends by recommending that ARPA make funds available to ARDC for the project without additional formal approval by ARPA suggesting that the project had already gone through an ARPA approval stage.

On 10 April Ward Low wrote a memo, "Descriptive Applications of Cornell University Electromagnetic Sounder (1000 ft dish)", in which he points out that the Cornell University dish could provide information about the ionosphere that was important to the objectives of GLIPAR. The memo also pointed out the use of the dish for lunar, solar and planetary radar studies with emphasis on how easily Mars and Venus would be detected.

This memo was followed on 16 April by a "Research paper prepared by the policy and planning division to provide answers to pertinent questions raised by the ARPA Program Council" presumably about the Cornell Ionospheric Sounder as the project began to be called. The Council was apparently asking whether there were alternatives sources of support or whether the objectives of the project could be met by other instruments such as the 600 ft Sugar Grove antenna then under construction. The answers were no, at least on the time scales of interest for ballistic missile defense.

ARPA apparently did not take the IDA's advice, at least not immediately. A 28 April 1959 memo from ARPA Director Johnson to the Director, ARPD of the IDA, states that at its 24 April 1959 meeting the ARPA Program Council

[4]This memo and others referred to in this chapter can be found in the Cornell University archives under Arecibo Ionospheric Observatory.

recommended deferring consideration of support for the project until after 1 July 1959, "when its utility can be weighed against available FY 1960 funds."

On 16 June 1959, Cornell submitted a revised proposal based on a spherical 1000 ft reflector to be built in a newly selected sinkhole south of Arecibo, Puerto Rico. Total projected cost was $4.5 million including $200,000 for the first year of operation. The scheduled completion date is not specified but the proposal was for the period 1 July 1959 to 30 June 1961, which may have included the first year of operation. Cornell would design the transmitter, receiver, antenna feed, transmission lines and radio frequency switching; and an engineering firm would design the reflector, feed support structure and the development of the site infrastructure, buildings, roads and primary power. Clearly, Cornell was not aware that ARPA was already considering having the Corps of Engineers do the infrastructure work to Cornell's specifications.

The proposal makes clear that Cornell "proposes to design, build and operate the radar in Puerto Rico," and ends with the statement that "The radar will be operated as a facility of the School of Electrical Engineering and the sponsors." By 'operate' Cornell clearly meant controlling the scientific program in conformance with its obligations to ARPA. AFCRL either already, or soon after, had other plans.

Ward Low again recommended on 5 August 1959 that the Cornell Ionospheric Sounder be funded at the level of $2.5M in FY1960 with an additional $2.1M in FY1961. He argued that the Sounder provides "the only apparent means whereby reasonably small regions of the ionosphere can be examined essentially continuously" and that "local values of the electron density and electron temperature in each such parcel can be observed and the kinetics of electron motion deduced" finishing with "disturbances produced by vehicles passing through such a region can be evaluated". He makes the point that "the greatest possible sensitivity is required" justifying the need for a 1000 ft reflector and high peak power transmitter and pointing out its applicability to studies of deep space.

On 12 August Low directed comments on the Cornell Sounder to the Skifter group. Hector R. Skifter was Assistant Director of Defense Research and Engineering within the Department of Defense. He was chairman of the Ballistic Missile Defense Steering Group and coordinated planning and development for the Nike-Zeus anti-ballistic missile system and for ARPA's Defender program.[5] The Cornell sounder came under its aegis and Low provided input. This was originally a classified document that was declassified in 1978. Low starts with a general discussion of Ballistic Missile Defense and then points out the importance of the "atmospheric medium where mid-course flight occurs. It is somewhat of a jolt to realize how far that medium is from being 'empty' space. The proposed C. U. sounder represents a bold attempt to capitalize on a rarely observed physical phenomena—incoherent scatter by electrons—in gaining data about that medium". Low goes on to reiterate his previous arguments and to discuss using the Navy's Sugar Grove facility- will not be ready soon enough, available observing—time limited and not

[5]New York Times obituary for Hector R. Skifter, 17 July 1964.

sensible to use a $90M instrument when a $5M one will do the research—and phased arrays—not well enough developed. While no explicit recommendation is made, the document is clearly very supportive of the Cornell proposal.

A final review of the underlying physics of the proposal was conducted by Professors Marvin Goldberger and Richard Blankenbecler of Princeton University, Professors Gordon and Booker of Cornell and William Bradley and John Bond of the IDA. A report, written by John Bond dated 21 September 1959, states that "the two main questions that needed to be resolved concerned (1) the theory of scattering of electromagnetic waves by a plasma and (2) the magnitude of back-scatter in the side lobes." Both of these questions were resolved satisfactorily and a list of additional, more minor issues are discussed and there is a list of additional work that should be done. This latter includes the possibility of carrying out work on a laboratory scale and studying the possibility of detecting nuclear bursts in space. Clearly, no major issues were uncovered and ARPA approved funding for the project on 30 September 1959 (see below).

While there was agreement between Cornell and ARPA that the Sounder was a research instrument intended to make basic measurements of the ionosphere, electron densities and temperatures, it is clear that Cornell's motivation for proposing the instrument and ARPA's for funding it were very different. Bill Gordon and his colleagues were interested in fundamental questions related to the then very poorly understood ionosphere. ARPA's interest was mission directed; they wanted to know if missiles passing through the ionosphere would leave a signature in the electron density that might point the way to some means to defend against ICBM's. There was also the possibility that the Sounder would be sensitive to nuclear explosions in space. None of the ARPA objectives in funding the project appear to have been realized. Curiously, Arecibo did detect a traveling ionospheric disturbance coming from the south-east, a very unusual direction, in the early morning hours of 22 September 1979. This was used as collaborative evidence for a low yield nuclear explosion near southern Africa that a VELA satellite may have detected via the characteristic light signal from such an explosion. The U. S. Administration organized a panel, chaired by Jack Ruina, the former Director of ARPA during most of the construction of Arecibo, to investigate. The panel concluded that the VELA signal "was probably not from a nuclear explosion. Although we cannot rule out that this signal was of nuclear origin." The issue remains controversial.

Why was ARPA so supportive of adding the beam-swinging capability with its larger cost and increased risk compared with a fixed parabolic reflector when the latter would meet the ionospheric related objectives? The answer may just be that it was much more technically interesting and the then anticipated additional cost of $2M or so was small compared with ARPA's roughly $350M budget in FY1960. However, the reason may also lie with ARPA's original charter in 1958, which included coordination of the military satellite program; there was no civilian program at that time. In response to a September 1957 directive from Assistant Secretary of Defense Holaday, what was called the Advisory Group on Special Capabilities submitted recommendations on 15 January 1958, which "noted that unmanned explorations of the moon, Venus and Mars appeared to be within the capabilities of presently planned systems and recommended that a lunar probe be part of the IRBM (Intermediate Range Ballistic Missile)-based satellite program." (Watson 1997). Interestingly, apart from military satellites, the army's long- range plan submitted in January 1958 included an unmanned Moon landing in 1958, a manned one in 1967 and a 500-man lunar expedition in 1971. With the formation of NASA in July 1958, responsibility for the civilian side of the space program was transferred to the new agency leaving ARPA to coordinate the purely military aspects. However, the awareness within ARPA of the plans and needs of the civilian program may have pre-disposed it to be supportive of the beam-swinging capability of the Arecibo telescope that would allow radar observations of the planets and the Sun. A measurement of the round-trip light time to, say, Venus using a powerful radar would have significantly improved the accuracy of our then knowledge of the Astronomical Unit, the mean distance from the Earth to the Sun, thus enabling accurate navigation for planetary spacecraft missions. This measurement was made in 1961, before the Arecibo radar was completed although later Arecibo radar time delay measurements to Venus further refined its value.

3.2 Setting Up the Project Organization

On 30 September 1959, Roy Johnson, Director of ARPA, wrote the Commander, Air Research and Development Command (ARDC), Andrews Air Force Base (ARPA Order No. 106-60) with the request to proceed immediately with "design, construction and installation of a vertically directed ionospheric radar probe to be located in Puerto Rico. The design and construction shall generally conform to the Cornell University proposals of 1 November 1958 and 1 June 1959, which resulted from Cornell studies undertaken under Contract No. AF 33(616)-5547." $1M was made available to the Air Force (AFCRL) initially in order to commence design work immediately.

ARPA issued an order to the Commander, Air Research and Development Command (ARDC) on 30 September 1959 authorizing ARDC to manage the construction of the 1000 ft telescope and radar system based on the Cornell proposals of 1 November 1958 and 1 June 1959. The (Army) Corps of Engineers was to be responsible for purchasing the land, making the access and site roads, providing primary power, and constructing the buildings to Cornell's specifications. Total cost was to be $4.6M and $1M was provided immediately for design and other initial work.

Within ARDC, the Electronics Research Directorate of the AFCRL was assigned responsibility for the construction of the now designated Arecibo Radio Observatory and Phillip Blacksmith was named Project Officer. A contract was signed by Cornell on 19 November 1959, initially for $377,555 and another was issued to the Corps of Engineers for over $700,000 for the purchase of the land and the infrastructure work.

3.3 Thomas Gold and the Center for Radiophysics and Space Research

At Cornell, the organization of the construction project and the longer-term direction of the Observatory was impacted by the arrival in July 1959 of Thomas Gold, the newly appointed chair of the Department of Astronomy. Since its inception in 1932, the department had had only once, very briefly, more than two faculty members. Its chair since 1942 was R. William Shaw, whose interest was primarily in teaching. Perhaps Shaw's most notable action as chair was the hiring in 1947 of Martha Stahr as an assistant professor, the first female faculty member in Cornell's College of Arts and Sciences (Rossiter, 1995). Stahr became involved with the Radio Astronomy Project—originally the Microwave Astronomy Project—established in 1946 by Cornell's School of Electrical Engineering and supervised by Bill Gordon for much of the time after his arrival at Cornell in July of 1948.

The astronomy department was clearly a poor match for the coming space age as represented by the Soviet Union's launch of Sputnik in 1957. Shortly after the launch of Sputnik, Dale Corson, Chair of the Cornell's very highly regarded physics department, and Solomon Hollister, Dean of Engineering, approached the then Cornell president, Deane Malott, to express concern about the state of the astronomy department and to suggest that it be completely revamped and expanded under a prominent astronomer to be hired.[6] Malott agreed and appointed a search committee headed by Corson. The outcome of the search was an offer to Thomas Gold, then a faculty member at Harvard University, to create a new Department of Astronomy. At 39, Gold (Fig. 3.2) was a relatively young but well-known English astrophysicist of Austrian heritage, best known at the time as part of the trio with Herman Bondi and Fred Hoyle which postulated the steady-state theory of the universe. However, he had a very broad view of astronomy and space sciences that boded well for the creation of a department with a faculty with diverse interests. Before joining the Harvard faculty, Gold had spent the fall (autumn) semester of 1956 at Cornell and was relatively familiar with the university and the faculty in astronomy and physics. Several physics faculty members were, or would become, very involved with astrophysics, most notably Hans Bethe and Edwin Salpeter.

Gold's appointment as chair of the Department of Astronomy, and as professor of both Physics and Electrical Engineering, included the right to appoint several new faculty members and to create and direct an interdisciplinary research center devoted to space sciences. He was well aware of the plans to build a very large telescope and radar system in Puerto Rico and saw its potential as a groundbreaking instrument for radar studies of the planets and for radio astronomy. As an inducement, Cornell's job offer included making the telescope part of the new center to be formed and, hence, under Gold as its director. However, by the time Gold arrived at Cornell in July 1959, the university had already told Bill Gordon that he would be the Project

[6]See the transcript of a 1980 interview with Dale Corson by Joanna Rankin in the Cornell University archives under Arecibo Ionospheric Observatory.

Fig. 3.2 Thomas Gold in November 1959 shortly after his arrival at Cornell University. Courtesy: Thomas Gold

Director for the construction of the telescope. Gordon did not welcome Gold's involvement since as project director Gordon, at least nominally, reported to Gold rather than to the Director of the School of Electrical Engineering, the Dean or the Vice President for Research. Cornell's actions had created a difficult situation that would lead to significant disagreements between the two over the next few years. Gold clearly felt that the future observatory's ultimate success rested with its potential for ground breaking discoveries related to radar studies of the planets and radio astronomy while Gordon was clearly most interested in the original objectives of the observatory, studies of the Earth's ionosphere. However, the underlying issue was the control of the facility, should it be run locally, initially by Bill Gordon, or should oversight rest with the director of the newly formed Center for Radiophysics and Space Research (CRSR) headquartered on the Cornell campus.

CRSR was founded in November, 1959 (Fig. 3.3). Thomas Gold was the director and Henry Booker, now the director of the School of Electrical Engineering, was its associate director. Since there had not been time to appoint any new faculty members in astronomy, all of the initial senior faculty members in the new center apart from Gold were from Cornell's College of Engineering, primarily from Electrical Engineering (Fig. 3.3). All of the contracting work for the telescope was done through the aegis of CRSR out of Phillips Hall on the Cornell campus, the home of the School of Electrical Engineering, under the administrative direction of Richard Hufnail.

In July 1959, Dale Corson had his first introduction to the project when he moved from his position as chair of the Department of Physics to become Dean of Engineering. He became a very strong supporter of both the project and Bill Gordon, support that was to prove crucial over the following years. Due to his wartime radar work, Corson was well connected in Washington. As he recounted in an interview in 1980, this did not help much in persuading ARPA to fund the project. By the

Fig. 3.3 The "Outer Space Team" at the time of the founding of CRSR; Left to right, Benjamin Nichols, Professor, School of Electrical Engineering; William Sears, Professor of Aeronautical Engineering; William Gordon, Professor, School of Electrical Engineering; Thomas Gold, chair of the Department of Astronomy and director of CRSR; and Henry Booker, director of the School of Electrical Engineering and associate director of CRSR. Credit: Cornell University

summer of 1959 ARPA was already convinced that it should support construction of the telescope, but, as we will see, Corson's support was very helpful with both the government and the Cornell administration, including the Board of Trustees, during some of the inevitable difficult stages of the construction and the issue of what institution would control the scientific program at the new observatory.

In late 1958 shortly after his acceptance of the offer to join Cornell and well before taking up his position there, Gold visited the Millstone Hill facility of MIT's Lincoln Laboratory outside of Boston to learn more about the new field of radar astronomy. In his autobiography Gold (2012) recounts that Gordon Pettengill, a staff member at Millstone Hill who was working on radar studies of the Moon, was assigned to show him around and explain the equipment. Gold was very impressed by Pettengill and, as remembered by Pettengill,[7] Gold indicated that he planned to offer him the position of Associate Director of the new observatory to be built in Puerto Rico with direct responsibility for the planetary astronomy program. The

[7]Gordon Pettengill, personal account to the author, June 2014.

appointment was initially complicated by disagreements with AFCRL over whether the title should be Assistant Director or Associate Director, but, eventually, Pettengill was appointed Associate Director and joined the observatory staff in Arecibo in 1963. Prior to this, he made numerous trips to Arecibo to assist primarily with telescope instrumentation issues and, despite having been offered the appointment by Gold, he and Bill Gordon had a very good relationship. However, Pettengill's appointment highlighted the complex relationship between Gold and Gordon.

3.4 Broken Promises: Project Budget and Schedule

Cornell's 13 November 1959 proposal to ARPA specified a project cost of \$4.5M, in line with the \$2M increase specified by Ward Low over the 13 June 1958 proposal cost of \$2.5M needed to cover the cost of the 20° beam swinging capability. Philipp Blacksmith, AFCRL's Project Officer, claimed that he thought that \$6.5M was a more realistic number but that he went along with the proposed cost knowing that additional funds would be available as needed and justified. William McGuire in a letter to Bill Gordon after a visit to the Arecibo site from 28 November to 2 December 1959 pointed out that either the cost or the schedule for the project was unrealistic. Essentially, expending unlimited funds via cost plus contracts might just get the project completed in the required fifteen months. Alternatively, staying within the cost estimate of \$4.5M may be possible, but it would take much longer. It is clear that seeing the sinkhole and the difficult access from the existing paved road made Bill McGuire appreciate the difficulty of the task that Cornell had taken on.

It was clear very quickly that the budget of \$4.5M was not going to hold. On 21 December 1959, ARPA was notified by ARDC (AFCRL) that the cost estimates were not firm or reliable. This was at least partly due to significant differences in the cost estimates by the Corps of Engineers for the work it had to perform compared with Cornell's original estimates for the same work. It had not been Cornell who decided to engage the Corps although Cornell's cost estimates were likely overly optimistic given the realities of the field conditions. A revised development and funding plan was sent to ARPA by RADC on 9 March 1960, for a total of \$6,007,200. ARPA approved the amendment on 25 April, although apparently not for the full amount. After negotiations with AFCRL, the Cornell contract was finally amended in July to a new total of \$4,698,000 with a completion date of 1 September 1961. The contract with the Corps of Engineers was for \$785,000 with completion on 1 June 1961. \$120,000 covered AFCRL's costs. Despite being increasingly unrealistic and, then, impossible, the Cornell contract completion date of 1 September 1961 was officially adhered to until January 1962 when it was changed to 31 October 1962.

3.5 Operational Plans: A Tussle Over Control

With completion of the observatory scheduled for mid-1961 the issue of the operation of the facility after its completion was of immediate importance. As early as March 1960, AFCRL made it clear that they planned to make the Arecibo Radio Observatory an Air Force institution with Cornell and/or the University of Puerto Rico contracted to "keep the lights on". A 9 March 1960 Revised Development and Funding Plan for ARPA Order 106-60 written by AFCRL states that "Although it is contemplated that Cornell University supplemented by the University of Puerto Rico will initially operate this site under contract to AFCRC,[8] it is planned that soon an Air Force civilian staff will also be assigned permanently to this installation. As the Air Force program expands the strength of this staff will grow. It is contemplated that there will always be a requirement for contractual help from Cornell or the University of Puerto Rico." In other words, they may need Cornell's or the University of Puerto Rico's help to hire local staff and maintain the facility but ARCRL would control the scientific program.

Did the people at AFCRL seriously believe that Cornell, after all the planning and effort that it had expended, would be content to just operate the facility while the Air Force controlled the management of the scientific program? Cornell did not see a copy of this plan until a meeting at AFCRL on 19 November 1960, but they were given an indication of the Air Force thinking via an 18 July 1960 letter to Bill Gordon from the AFCRL contracting officer saying that Cornell was "serving as the prime contractor for the design and construction of the installation and does not have a contract for the operational phase of the activity. It is anticipated that Cornell, together with other Universities or contractors, will assist the Air Force in the carrying out of experiments to meet ARPA and the Air Force missions for which the facility has been established."[9] This must have been a distinct surprise to Bill Gordon since it implied an involvement of the Air Force, clearly through AFCRL, in defining the scientific program of the future observatory. AFCRL did operate a small radio astronomy facility, the Sagamore Hill Radio Astronomy Observatory, at Hamilton, Massachusetts under the direction of Dr. Jules Aarons. It was primarily used for studies of solar radio emission. The plan was for a close association of the Arecibo Observatory with Sagamore Hill with Aarons playing a leading role in determining Arecibo's research program. However, in assigning responsibility for the construction of the Arecibo Observatory to the Air Force, ARPA made no mention of an "Air Force Mission" related to its future operation.

---

[8] As recounted in Chap. 2, until 1960 AFCRL was known as the Air Force Cambridge Research Center. For simplicity, AFCRL is used throughout this history.

[9] Letter to William E. Gordon from Carmin J. Iadonisi, Contracting Officer, dated 17 July 1960.

In his 4 August 1960, report on the project to ARPA, Phillip Blacksmith refers to the letter, which is included with his report, and states that it was sent at his request. Both documents are in the Cornell University archives under Arecibo Ionospheric Observatory.

It is not clear what prompted Tom Gold to arrange a discussion at the IDA on 15 September 1960, concerning the future plans for the operation of the Arecibo Observatory. Present were (Prof.) Charles Townes, then vice president and director of research at the IDA and inventor of the MASER (the radio wave equivalent of the LASER), Jack Ruina, DoD and future Director of ARPA as of 20 January 1961, Charles Cook of ARPA, Gordon Pettengill and Gold. Gold, in his 2.5-page 21 September memo recounting the discussion, states that "Everyone thought and expressed this view very strongly that it should be made quite clear that Cornell would be the authority running the Observatory." Tellingly for the future, Gold claimed that it was Jack Ruina who was "most vehemently and outspokenly on the side of the University and against the Armed Services for the running of research facilities." By charging the AFCRL contract $12 for a day of Gold's trip to Washington, Cornell gave an opening to Thomas Walsh, AFCRL's contract administrator, who was quoted in Cornell notes about two separate discussions as saying "Gold should not be talking with ARPA about future plans anyway. When the time comes, Bedford (i.e. AFCRL through the AFCCDD (Air Force Command and Control Development Division)) will talk with ARPA" and, even more adamantly, "Get Gold the h---- out of my contract. There is a strong feeling here about Gold. I'll do the talking to ARPA." The relationship between AFCRL and the Cornell administration was clearly not good.

The discussion about plans for the operation of the new Observatory at the 19 November meeting at AFCRL included Charles Cook of ARPA, Jules Aarons, R. Barrett, Deputy Director of the Electronics Research Directorate, and several other AFCRL representatives, Lt. Colonel Ashley from Air Force Headquarters in Washington, Bill Gordon and Dale Corson. According to the notes that Corson made after the meeting, there was general agreement that, at least for the first three years, ARPA would fund the operation of the facility for the ionospheric component of the science program and the Air Force through AFCRL would also provide some funding. Corson made detailed notes covering the meeting. However, it was still clear that despite Bill Gordon being the director of the observatory, the science program would be largely dictated by AFCRL. It was not until part way through the meeting that the Cornell people were finally given a copy of AFCRL's 17 June tentative plan for the observatory's management. At the end of his notes, Corson expressed confidence that it would all work out in Cornell's favor although it was not clear what the basis for his optimism was.

Over the next few months there was considerable discussion between ARPA, AFCRL and Cornell over the future management of the Observatory. On 20 November, AFCRL issued some modifications to its tentative plan for the management of the observatory. It still had the Air Force playing a leading role with Jules Aarons providing oversight and, at the same time, being the deputy director of the facility reporting to Bill Gordon. Since Cornell would be under contract to the Air Force to operate Arecibo, Gold certainly did not consider it acceptable for the Air Force to then appoint an Air Force person as deputy director. On 26 and 27 January 1961, Charles Cook visited Arecibo to talk with Bill Gordon and Gordon Pettengill, who happened to be there. This visit coincided with a change at ARPA that would have a large impact on the project and the future management of the observatory; Jack Ruina was its new director. In a 2 February letter to Booker, Gordon recounts that Cook felt that Ruina would be open to supporting a basic research program at Arecibo, not just the ARPA program in support of Project Defender. He also said that Ruina would stand no nonsense from the Air Force. Gordon and Cook discussed the possibility of ARPA directly monitoring the management contract although Cook had some concerns about the work load for ARPA. Alternatives apart from the Air Force were the Office of Naval Research (ONR), NSF and NASA. Gordon wrote to Cook after the visit with the information that it would take about $1M to support the entire research program (i.e. the yearly operating costs of the observatory) versus about $0.5M for the limited ARPA only program.

AFCRL issued another "Tentative Operational and Management Plan for the Department of Defense Ionospheric Research Facility (DODIRF)," the new name for the facility, dated 13 March 1961. While worded differently, the basic plan was the same as in the first operational and management plan but more explicit. The Electronics Research Division (ERD) of AFCRL would be the managing entity and would appoint a General Manager, Dr. Jules Aarons. A facility manager would be selected with ARPA's approval and Bill Gordon was designated for the position. There would be two Associate Directors, one, Gordon Pettengill, would be in charge of the ARPA program and the other, appointed by ERD (Electronics Research Division of AFCRL), would be responsible for the Air Force program. The entire scientific program would be done at the direction of an Arecibo Advisory Group whose members would be nominated by the General Manager. This still boiled down to Air Force control of the facility and its research program.

The revised AFCRL Tentative Management Plan was received at ARPA on 24 April. Dale Corson received a copy on 20 May and reacted very quickly. The now ARPA Director, Jack Ruina, met with Corson on 23 May 1961, to discuss the role that Cornell might play in managing the new facility. Colonel A. P. Gagge, who was on ARPA's staff and a former director of the Air Force Office of Scientific Research (AFOSR), was requested to review AFCRL's Tentative Plan. He sent a memo to Director Ruina on 23 May in which he pointed out that "As presented by the Cambridge (AFCRL) Group, Cornell will be placed essentially in a "service" position and be completely diluted in having any scientific control of the research program resulting from the use of the Dish in Puerto Rico." The memo continues, "The roles and responsibilities as outlined on pages 6 and 7 give ARPA a very superficial role. It gives the Air Force appointment control over every key position including membership of the Arecibo Advisory Group (AAG)."

Gagge's recommendations were (1) the contract for operating Arecibo be transferred from AFCRL to the Air Force Office of Scientific Research (AFOSR); (2) Cornell, as the prime contractor, be given the principal scientific responsibility for the program. There would be an Arecibo Advisory Group (AAG) composed of representatives of the government agencies contributing to the project such as ARPA, the Army, the Navy and the Air Force. It would be chaired by ARPA,

Fig. 3.4 Jack Ruina, Director of ARPA from 1961 to 1963. Credit: A 1957 photo courtesy of Andy Ruina

periodically review scientific progress and suggest research areas of specific interest to the sponsoring agency.

On 23 May, the same day, Director Ruina (Fig. 3.4) sent a memo to AFCRL telling them that before acting on their proposal, ARPA wished to consult with several groups including AFOSR and US Air Force Headquarters. He went on to say that they had direct discussions with Cornell and had asked Cornell to provide "additional information concerning the role they might play in the operation and management of this facility." This could not have been good news for AFCRL. A copy of this memo was sent to Dale Corson.

Cornell moved quickly and submitted to ARPA a "Draft Proposal for the Operation of the Arecibo Facility", plus supplementary information in a separate document, both dated 6 June 1961. It is clear that Cornell had talked with AFOSR with ARPA's permission since the initial proposed management actions were for ARPA to issue an order to AFOSR to establish a contract for scientific operation and management of the Arecibo Radio Observatory (ARO). AFOSR would then contract Cornell for these tasks. Bill Gordon would be appointed Director of the ARO and Gordon Pettengill Associate Director. Cornell would suggest a list of suitable members for the Arecibo Advisory Group (AAG) and the AAG would be tasked with reviewing Cornell's plans for the scientific operation of the observatory for the first two years. The Proposal included a detailed discussion of the anticipated scientific program. Operational budgets for fiscal years 1962, '63 and '64 were developed—$611k, $1.131M and $1.234M, respectively, plus provision for new instrumentation.

The Supplementary Information discussed Cornell's capabilities and resources and made it clear that the proposed management structure would be through CRSR with Thomas Gold as Director and Henry Booker as Associate Director. They would report to Dale Corson as Dean of Engineering and, in turn, Bill Gordon, as Observatory Director, would report to Gold and Booker. A discussion of the division of time on the facility stated that ARPA's Project Defender would utilize approximately 50% of the time for the first 3 years.

The Cornell proposal must have received a good review at ARPA and at AFOSR since on 16 June W. W. Bolton, Director of ARPA's Program Management Division, wrote Dean Corson to say that they had met with Major General Hooks, Director of the Office of Aerospace Research, which oversaw both AFCRL and AFOSR, and that they had agreed that "the operational phase of the DODIRF (Department of Defense Ionospheric Research Facility as it was designated at that time) would be contracted through OSR." Cornell had prevailed over AFCRL, but it must have resulted in some strain in their relationship especially as AFCRL still had responsibility for the continuing construction of the observatory. There was some concern within ARPA about potential difficulties with AFCRL over the transfer of the operations contract to AFOSR, but in the end responsibility for the construction work stayed with AFCRL.

The puzzling question is what made AFCRL think that they could relegate Cornell to a minor role at best in the operation of the observatory especially after all the discussion over the previous six months. The whole idea for the facility originated with Cornell. It appears to have been a complete misreading of the outlook of ARPA and, probably, Ward Low at the IDA. ARPA's attitude was summed up in a handwritten note of uncertain authorship "It is a Cornell facility for the nation, not an in house CRL facility." On 13 July 1961, ARPA issued an Order, No. 270-62, to AFOSR requesting that AFOSR provide assistance to ARPA by contracting with Cornell for the operation and management of the ionospheric research facility being constructed under ARPA orders Nos. 106 and 122. The yearly budget was to be approximately $1.2M, the name of the facility changed to the Arecibo Ionospheric Observatory (AIO) and the initial period of performance being 1 November 1961 until 30 September 1963. The Order was forwarded to Dale Corson on July 20th for comment. Corson's reply was highly supportive only expressing concern about any renewal of the agreement being well ahead of the 30 September 1963 date, preferably about one year ahead.

3.6 The Post-Construction Organization

To set the framework for a Cornell proposal to AFOSR for the operation of AIO, a meeting was held at Cornell on 25 July attended by Knox Millsaps, Executive Director of AFOSR, Lloyd Wood, AFOSR Director of Physical Sciences, and Captain John Reid also of AFOSR, Colonel J. R. Fowler of the Office of Aerospace Research, the parent organization of AFOSR, Charles Cook of ARPA, and Gordon, Corson, Hastie and Robinson of Cornell. John Hastie and Robert Robinson were the Head and Associate Head, respectively, of Cornell's Research Office. As discussed in a memorandum by John Hastie written after the meeting, the major issues were related to the desirability of an independent scientific advisory group to report to the director of ARPA, the priority of Project Defender for time on the telescope and, of course, the budget. Classified work would be kept to a minimum with any related data to be shipped off site for analysis. There was agreement on an estimated operational cost of about $1.2M per year after the start-up period and that the start date for operational funding would be 1 November 1961.

A few days later, Dale Corson passed along to Knox Millsaps Cornell's suggested names for the Arecibo Advisory Committee (AAC); Lloyd Berkner, President of Associated Universities Inc., Colin Hines of the University of Chicago, Ronald Bracewell of Stanford University, Roger Gallet of the National Bureau of Standards (NBS), Bruno Rossi of MIT and Otto Struve, Director of the National Radio Astronomy Observatory. Millsaps liked the selections but pointed out that Colin Hines was a Canadian citizen so was potentially a problem because of the possibility of classified projects. Based on past experience, Millsaps was also opposed to membership by employees of government agencies such as NBS preferring that they attend as observers. After discussions between AFOSR and Cornell the suggested names were changed to Ron Bracewell, Chairman, Lloyd Berkner, now at the Graduate Research Center Corporation, Fred Haddock, University of Michigan, Otto Struve and Arthur Waynick, Pennsylvania State University. Unfortunately, Lloyd Berkner indicated that he would not be able to serve on the AAC. In a turnaround in his attitude, Millsaps in a June 1962 letter to ARPA and the Office of Aerospace Research, recommended Herbert Friedman, a civilian scientist at the Naval Research Laboratories, and Jules Aarons of AFCRL. AARP agreed with the appointment of Herb Friedman and, by a process that is not clear, the final makeup of the AAC was Bracewell (Chair), Friedman, Waynick, Hines, Haddock and Struve. The Bracewell committee, as the AAC was commonly called, met for the first time on 27 and 28 November 1962, in Puerto Rico. Struve was unable to attend. Representatives of ARPA including Director Ruina, AFOSR including Marshall Harrington of the General Physics Division and the designated Project Monitor for the AIO operations contract, Bill Gordon, Dale Corson, Tom Gold and others from Cornell and a number of Arecibo staff including Phil Waldron, the then new Business Manager.

Cornell drafted a proposal "The Operation of the Arecibo Radio Observatory to 30 June 1964" to be submitted to AFOSR in late July 1961. This may have been a discussion document for the 25 July meeting at Cornell. A final and more detailed proposal was submitted to AFOSR probably in late September or October 1961, the date is unclear. It was for a total of $2,280,636 for the period 1 November 1961 to 30 September 1963. The proposal discusses Cornell's history with ionospheric and radio astronomy research, the involvement of faculty and graduate students through the Center for Radiophysics and Space Research (CRSR) and the expected research program at Cornell in support of Arecibo. It goes on to discuss the AAC and its charge—reviewing proposals, advising on scientific priorities and reviewing the operation of the facility, and Cornell's obligations—operate and staff the facility in accordance with the requirements of ARPA and publish results in scientific journals. The observatory director, initially Bill Gordon, would report to Tom Gold as Director of CRSR who, in turn, reported to the Dean of Engineering, Dale Corson. Corson reported to the Cornell Provost and the President.

In addition to the normal operating costs, staff, utilities, general maintenance items, library acquisitions, etc., the proposal also listed enhancements to the observatory's instrumentation that would not be covered under the construction contract. These included general items such as a telescope control computer, lease of a Control Data Corporation (CDC)160A data analysis computer, a 40 MHz transmitter and feed for ionospheric, solar and planetary radar observations, a test feed at 1420 MHz to see if the telescope could make useful observations of the 21 cm wavelength spectral line of atomic hydrogen and photogrammetric equipment for measuring the surface accuracy of the telescope's reflector.

Discussions with AFOSR on the operations contract dragged on for several months with the start date being progressively delayed to 1 December 1961, to 1 January 1962, and then to 1 March. This was undoubtedly due to the obvious delays in the construction program. A major issue was the budget; AFOSR insisted that it could not exceed $2,150,000 and emphasized the need to keep the construction and operations strictly separate with no overlap in charges. Other issues were overhead rates, the never-ending issue of area allowances and moving costs for those staff hired from outside of Puerto Rico, the need for the CDC 160A computer and antenna tracking computers, and general maintenance costs.

During the discussions on the operations budget in late 1961, Cornell requested an additional $559,000 for the construction work and an extension of the construction completion date from 31 March to 31 October 1962. To partially fund this increase ARPA reduced the operating contract funds by $315,000 to $1,835,000 while maintaining the two-year contract period.

On 27 January1962, Bill Gordon and Robert Robinson met in Washington with Captain John Reid of AFOSR to try to finalize the contract. There was general agreement on most issues including a 25% overhead rate for Cornell and provision for schooling and access to the facilities at Ramey Air Force Base for expatriate staff. The main sticking point was the area allowance of the expatriate staff. Cornell wanted 25% to a maximum of $3000 while Reid was offering 16% with a maximum of $3000. After conferring with AFOSR Director Knox Millsaps, Reid upped the offer to 20% with a maximum of $3000 and Cornell agreed to this. The final contract was for $1,501,740 with a start date of 1 February 1962, and completion date of 31 January 1964.

Cornell's understanding was that as each construction sub-contract was completed and signed off then the Operations Contract would take over for the maintenance—e.g. the buildings once the Corps of Engineers was finished, the service roads and bowl surface, which needed stabilizing vegetation once the excavation contract was completed.

3.7 What's in a Name

Once the Arecibo site was chosen, the project was often referred to at Cornell as the observatory in Arecibo, which quickly morphed into the Arecibo Observatory. In Washington it was commonly referred to as the Cornell University Sounder. With the signing of the contract with AFCRL, the name became the Arecibo Radio

Observatory. However, it was occasionally referred to as the Cornell University Arecibo Radio Observatory in publicity articles, which also mentioned Cornell's future operation of the facility. This clearly irritated AFCRL especially as at that time AFCRL had its own plan to manage the Observatory after its completion. The 18 July 1960 letter that Gordon received from the AFCRL contracts manager insisted that the observatory be referred to as the Air Force Arecibo Radio Observatory. Apparently, this name did not sit well with ARPA and on 6 October 1960, Cornell was instructed by AFCRL to change the name to Department of Defense Ionospheric Research Facility, DODIRF, a particularly ungainly name. These name changes were a significant irritant for the Cornell staff in Arecibo as they were organizing the start of the construction phase of the telescope. They required successive changes in the correspondence letterhead and other publicity material.

As discussed above, ARPA instructed AFOSR to change the name to the Arecibo Ionospheric Observatory (AIO) in their July 1961 order for the operation of the observatory. Cornell had requested Arecibo Radio Observatory and this was the original name in the ARPA order. However, in the Order to AFOSR, "Radio" was hand changed to "Ionosphere" probably to make the point that the observatory's primary mission was its work in support of Project Defender. The name remained AIO until 1971 when the observatory's management organization at Cornell changed from CRSR to the National Astronomy and Ionosphere Center (NAIC) under funding from the National Science Foundation, and the observatory's name reverted to the original one, Arecibo Observatory.

References

Bowles, K.L. 1958. Observations of Vertical Incidence Scatter from the Ionosphere at 41 Mc/sec. *Physical Review Letters* 1: 454–455.

Gold, Thomas. 2012. *Taking the Back Off the Watch: A Personal Memoir*, Astrophysics and Space Science Library. Vol. 381. Springer.

Herzfeld, Charles. 2008. How the Change Agent Has Changed. *Nature* 451: 403–404.

Kellermann, Kenneth I., Ellen N. Bouton, and Sierra S. Brandt. 2020. *Open Skies: The National Radio Astronomy Observatory and Its Impact on US Radio Astronomy, Chapter 9*. Springer.

Rossiter, Margaret. 1995. *Women Scientists in America: Before Affirmative Action, 1940-1972*. John Hopkins University Press.

Van Atta, R.H., S.J. Deitchman, and S.G. Reed. 1991a. *DARPA Technical Accomplishments*. Vol. III. Institute for Defense Analyses Paper P-2538.

———. 1991b. *DARPA Technical Accomplishments*. Vol. II. Institute for Defense Analyses Paper P-2429.

Watson, Robert J. 1997. *History of the Office of the Secretary of Defense, Vol. IV, Into the Missile Age, 1956-1960*. Washington, DC: U.S. Government Printing Office.

Chapter 4
Turning Soil at Last

In the summer of 1960, William Gordon and four other Cornell related families moved to Arecibo, Puerto Rico, to oversee the telescope construction work and build the needed receiver and data acquisition systems. The line feed and transmitter contracts had already been awarded and there were seven telescope construction sub-contracts that had to be negotiated. The grading and excavation contract was quickly let allowing work to start in September. By mid-1961 all of the contracts had been let. However, progress was very slow and it was soon clear that the very aggressive schedule, which had completion by September 1961, would not be close to being met. Solid progress was made, just much slower than scheduled. These delays had a real cost and 1961 saw several requests to ARPA for additions to the construction budget, all of which ARPA agreed to. The schedule and funding issues led to a major meeting of all the interested parties, ARPA, AFCRL and Cornell, in San Juan in October 1961. This meeting was highly constructive in getting everyone working together and resulted in a new completion date of October 1962.

4.1 Moving to Puerto Rico

In late July and August of 1960, five Cornell related families moved to Arecibo, Puerto Rico, Bill and Elva Gordon and their two teenage children, Merle and Kay LaLonde and their daughter, George and Gloria Peter and their four children, Al and Ruth Johnson, and Thomas and Elizabeth (Betty) Talpey and their two small children. Merle LaLonde was a recent, slightly older than normal, Cornell engineering graduate, who would become Gordon's right-hand person in dealing with the contractors and played a prominent role in the initial operational phase of the Observatory. Tom Talpey and George Peter would be developing the receiver and data acquisition systems and Al Johnson, who studied physics at Cornell and ended

D. Campbell, *The Arecibo Observatory: A History of Innovation and Discovery*, Historical & Cultural Astronomy, https://doi.org/10.1007/978-3-031-75225-4_4

up as an administrative assistant in the Physics Department,[1] became the business manager. Talpey was a Cornell electrical engineering PhD who worked for Bell Laboratories. He made recruiting visits to Cornell and, on one of these visits, Bill Gordon recruited him to join the Arecibo project while on leave from Bell Labs. George Peter was a senior technician, who had worked on the Radio Astronomy Project at Cornell, and brought great organizational ability to the project. He remained a Cornell employee associated with the Arecibo Observatory for all his working life eventually becoming one of the Cornell employee representatives on the University's Board of Trustees.

In mid-August the group was augmented by Donald Yabsley, a long-time researcher with the Radiophysics Laboratory of the Commonwealth Scientific and Industrial Organization in Australia. Yabsley, who was primarily stationed in Ithaca at Cornell but spent significant time in Arecibo, would be working on issues related to the transmitter and the waveguide system needed to get the transmitter power to the line feed. George Sigal, the on-site supervising engineer for the Von Seb Joint Venture, arrived in early September completing the small initial expatriate group.

Cornell's decision to act as its own General Contractor meant that the University had to have a strong presence at the Arecibo site to supervise the work of the individual contractors and to mediate between them when they inevitably interfered with each other's work schedule. However, this did not mean that Gordon himself had to move to Puerto Rico much less Peter and Talpey and certainly not so early in the construction. The initial receiver work could probably have been done better on the Cornell campus given the logistical difficulties in obtaining components in Arecibo. Gordon's decision to move so early to Arecibo probably reflected his feeling of intense personal involvement with the project, a depth of involvement that is possibly unique. His presence at the work site was certainly helpful as everyone, including the contractors and Air Force Cambridge Research Laboratory (AFCRL) staff, liked and respected Gordon. However, the problems of communicating from Arecibo were significant, the US postal service worked well, as little as 2 days to Ithaca, but the telephone system was highly unreliable. This meant that Gordon and Al Johnson were at least partially cut off from day-to-day interaction with the Cornell Campus to the detriment of the project management there, which was handled by Richard Hufnail, the Center for Radiophysics and Space Research (CRSR) administrator, who oversaw the contracting, and John Hastie and Robert Robinson, the Head and Associate Head, respectively, of Cornell's Research Office. Others in Ithaca who provided assistance were Dale Corson, the Dean of Engineering, Thomas Gold as Director of CRSR, Bill McGuire, Henry Booker, Ben Nichols and several other faculty members in Electrical Engineering.

Arecibo, 50 miles (80 km) west of San Juan on the north coast of Puerto Rico, is a very old city first settled by Europeans in the mid sixteenth century (Fig. 4.1). When the five families arrived in 1960, Puerto Rico was in its eighth year of self-rule as an

[1] See Paul Hartman's "The Cornell Physics Department, Recollections and a History of Sorts", page 261, Cornell e-Commons, 1984, revised 1993.

Fig. 4.1 Aerial view of the older part of the city of Arecibo in September 1961. The tree lined main plaza in front of the cathedral can be seen above and to the right of center. The Cornell office was in the street to the left of the cathedral in this view down towards the ocean, probably the second last building on the right. The Mir hotel was to the right of the plaza. The photo was taken by the prominent Australian radio astronomer, Joseph Pawsey, during a visit to Arecibo in 1961. Credit: Joseph Pawsey papers, Australian CSIRO archives, and Miller Goss for providing it to the author

unincorporated territory of the United States and known as the Commonwealth of Puerto Rico or, in Spanish, the Free Associated State of Puerto Rico. The first elected governor, Luis Munoz Marin, had surrounded himself with an extraordinarily able group of people in the government and was working hard to improve the economy of Puerto Rico and the living standards of its roughly 2.4 million people.

The American military had a very strong presence on the Island with a large naval base, Roosevelt Roads, at the eastern end, Fort Buchanan in San Juan, and Ramey Air Force base, a Strategic Air Command base, on the northwestern corner, plus a number of smaller bases. Ramey started off as the Borinquen Army Air Field in 1939. In 1948 it was renamed the Ramey Air Force base after Brigadier General Howard K. Ramey, who was killed during the Second World War, and in 1950 it was designated as a Strategic Air Command base housing the 72nd Bombardment Wing. Until its closing in the early 1970s, Ramey was an important resource for the observatory staff hired from outside of Puerto Rico, mainly because of the English language school, but also due to access to the relatively cheap food and other items such as sound equipment at the base Commissary and Exchange (PX). Curiously, despite the possible presence of nuclear weapons, expatriate staff who were not U.S. citizens were equally welcome to use the base facilities. However, use of the base was a divisive issue with the locally hired observatory staff, who were also U.S. citizens and somewhat resentful of the privileges of the expatriate staff.

Arecibo, with a population of about 29,000 in the city in 1960 and about 69,000 in the municipality including roughly 2000 in Barrio Esperanza near the telescope site about 10 miles (16 km) south of the city, was noted for its Ron Rico rum plant, the Cambalache sugar mill, the Pan American paper plant, which used the left over bagasse from the sugar mill to make cardboard, and a number of small factories sponsored by the government's industrial development corporation, Formento. The only hotel was the small Mir Hotel on the central plaza run by the Bernat family, welcoming but offering rather basic accommodations except for the large central common room on the second floor that had a good view of the plaza, the place to be on festival nights. The local public schools were, of course, all in Spanish and had a different system of organization and expectations than the Ithaca schools. The local professional class and others who could afford it, sent their children to the San Felipe Catholic school. Instruction was largely in Spanish but teaching the students to communicate in English was taken seriously. English was spoken either well or to some degree by many local people, the older ones because Puerto Rican schools were taught in English until 1930, a large number of Puerto Ricans served in the military, and there was a significant number of young people who had grown up in the United States and returned to Puerto Rico with their parents, who had gone to the mainland looking for jobs. At that time, there were no real estate agents and houses were rented based on word-of-mouth, the technique was to look for empty houses and enquire of the neighbors. There was one small supermarket in town, Tuyo's, which had a good supply of food items with the exception of fresh vegetables. Most food was imported so prices were high. The Festival Patronales de San Felipe in May was a time of steel bands in the Arecibo plaza and a parade in which the Observatory staff participated for many years. There were three movie houses, the Alameida, Oliver and, mildly X-rated, the San Luis. The movies had Spanish sub-titles; dubbed ones were not popular in Puerto Rico at the time.

Cultural events were few with one option being to drive to San Juan, a 4 h round trip on a very busy two lane road. The cultural highlight of the year was the Casals festival held in the theatre of the University of Puerto Rico in Rio Piedras, a suburb of San Juan. Pablo Casals had moved to Puerto Rico in 1956 refusing to live in his homeland Spain while it was under the Franco regime. He either played the cello or conducted the orchestra for the festival through the mid-1960s. The very positive aspects of life in Arecibo were the friendly people—if you had a problem there was always someone wanting to help—the wonderful climate, the summers were not particularly hot due to the trade winds especially if you lived near the ocean or on the ridge above the city, and the great beaches. It became a tradition for the expatriate observatory related community and others to meet at Caracoles beach east of the city on Sunday mornings.

This was the environment that the five families were dropped into in the summer of 1960, a large change from Ithaca, New York. Some of the local leading families "adopted" them including José Victor Oliver, a 1936 engineering graduate of Cornell, whose family owned the rum plant, and his wife Blanca. The families were invited to join the Country Club, a social club, and this, in turn, introduced them to a lot of people. However, schooling was an immediate problem. After one semester at San Felipe, the Gordon and Peter families decided to send their children to the English language school at the Ramey Air Force base. The disadvantage of the Ramey School was its distance; the one-way drive from Arecibo took about 45 min along a busy road.

4.2 Getting Started

Once they were settled in Arecibo, their first objective was to establish an office and workspace in downtown Arecibo to be used until buildings were available at the telescope site. A suitable place was found at Calle Gonzalo Marin #6, behind the cathedral away from the main plaza (Fig. 4.1). It had been the home of the local land registry, which was moving to the new Court House, and had room for offices and enough additional area to build the receiver system. However, it took some time to get possession, make necessary modifications to the space and purchase work benches, desks and chairs. Until the move in in September 1960, everyone worked in Al Johnson's rented house close to the Country Club.

Facing reality: By the summer of 1960 and the move to Arecibo, considerable progress had been made in many aspects of the project. The contract with the Joint Venture for the design and supervisory work had been signed on 21 April for $341,000 and design work was underway. The feed contract had been let to TRG in April for $258,096 and the transmitter design and construction contract let in June to Levinthal Electronics Products (later renamed Radiation at Stanford) for $580,000. Since Cornell was still operating under the initial $377,555 contract with AFCRL, there were provisions in these three contracts that limited Cornell's obligations to within this amount. Cornell's initial construction contract with AFCRL for $4,698,000 was finally signed in July 1960. There were seven more construction sub-contracts that had to be let corresponding to the progression of the design work by the Joint Venture: excavation and grading, towers and anchorages, the suspended feed structure and its support cables, the reflector, mechanical systems, electrical systems and the waveguide system for the transmission of the transmitter R.F. power to the line feed from the operations building. The receiver and data analysis systems were not contracted as the decision had been made to build them "in house" in Arecibo. There was good news that the Zenith Radio Corporation was developing an electron beam amplifier, known as a parametric amplifier, that would potentially significantly lower the noise contribution from the first amplifier at 430 MHz in the receiving system, increasing the overall sensitivity of the radar.

In Arecibo, the Corps of Engineers was working on the acquisition of the land around the sinkhole and had been asked to arrange for the bulldozing of a rough track to the sinkhole suitable for jeep access so that drilling equipment could be used to examine the rock under the tower and anchorage locations. The Joint Venture had Norman Porter Associates of New York carry out a detailed survey of the bowl in order to plan for the excavation and grading work, the objective of the first contract to be let. This included the precise positioning of the reflector in order to minimize the amount of excavation and fill material required and to optimize the positions of the towers and anchors. It was at this time that provision for increasing the diameter of the reflector to 1500 ft to prevent spillover of the feed pattern onto the ground was abandoned. This provision would have meant considerable additional rock removal and the raising of the center of the reflector by 18 ft (5.5 m) compared to its optimum location for a 1000 ft reflector. The additional cost was large and costs were already

Fig. 4.2 George Peter (left), Merle LaLonde and Don Yabsley examining reality in the bottom of the sinkhole on 26 August 1960, asking themselves "What do we do now?" Credit: Cornell University

rising significantly compared with the $4.5M budget. Invitations to bid for the excavation and grading work were sent to 35 prospective bidders on 1 July 1960. The 430 MHz frequency assignment for the transmitter was approved by the U.-S. government's Interdepartmental Radio Advisory Committee (IRAC) and discussions were underway with the government of Puerto Rico for a radio quiet zone stretching 4 miles from the Observatory to minimize radio frequency interference from new installations. A bill to this effect was passed by the Puerto Rican House of Representatives in May and later by the Senate. It became law on 15 June 1960.

Despite all the progress, with the move to Arecibo reality must have looked a lot different to the issues faced in Ithaca of organizing the funding and design details of the telescope. Once in Arecibo and faced with the complexities of what they had to do, the whole enterprise must have seemed extremely audacious. The small group had the responsibility of converting a rugged sinkhole (Fig. 4.2) into the world's largest radio telescope, while at the same time constructing the complex receiving system and supervising the procurement of the feed and transmitter. All this while working from a small office cum laboratory in Arecibo without good communications with the outside world. The telescope site was 10 miles away along a narrow and hilly winding road with the last mile being just a bulldozed track with steep grades that could only be negotiated by a jeep (Fig. 4.3). There was a significant support group in Ithaca but, as Project Director, the responsibility to move the project forward lay with Bill Gordon.

The Corps of Engineers: Placing the Corps of Engineers in charge of the purchase of the needed land around the sinkhole, building the access road, constructing the buildings and organizing the utilities must have seemed like a good idea to both

Fig. 4.3 The environs of Barrio San Rafael in June 1960 showing paved Rt 325 coming from the lower left with the intersection with the one-mile bulldozed track to the sinkhole at the center. The sinkhole can be seen just left of top center with the end of the bulldozed track at the future site of the operations building in the saddle between the hills. The bar at the intersection was a favorite meeting place especially as, for the first year, the road to the site was normally only passable for four wheeled vehicles so cars had to be left at the intersection (Fig. 4.4). Credit: Cornell University

AFCRL and, later, probably Cornell. However, it had the disadvantage that Cornell had no control over the Corps' work, especially the completion of the access road.

The Corps' first task was arranging for the purchase of the land. Inheritance laws in Puerto Rico require that some or, in the absence of a will, all of the estate is jointly inherited by the existing children. This means that purchasing land often requires obtaining agreement from multiple joint owners. If one of them is a minor, a sale becomes even more complicated. To get around this problem, the land for the observatory was acquired by the U.S. government through the Corps by eminent domain. Except for the cultivatable few acres on the floor of the sinkhole, the land had little agricultural value and most, if not all, of the owners apparently welcomed the purchase by the government as it paid relatively generously for the land. A total of 11 parcels covering a little over 118 acres (Fig. 4.5) plus another 8 parcels for the 7 acres required for the easement for the access road (Fig. 4.3) were purchased. This process was finally completed in early September 1960, just in time to allow the first contractor to begin work.

Fig. 4.4 The local bar at the beginning of the track to the site in 1960. Before the road was paved in September 1961, the wooden building was removed and a concrete one built further to the right. Credit: Cornell University

It was obvious to everyone that the roughly one-mile access road to the site was critical for the construction activities. On his first visit to the sinkhole in early December 1959, Bill McGuire recounted[2] that he took Bob Findlay and an engineer from the Raymond Construction Co. of Puerto Rico to the site. The issue of the access road was discussed, and Findlay offered to construct the road for $125,000. However, at a meeting between AFCRL and the Corps in Jacksonville on 24 November 1959, a tentative date of 1 September 1960 was set for completion of the Corps' work, a satisfactory timescale so Findlay's offer was not acted upon.

On a 25 to 27 April 1960 visit to the site with a representative of the Joint Venture, McGuire found that the Corps had bulldozed a preliminary road suitable for "jeeps, weapons carriers and tracked vehicles" but no other progress had been made making it clear that the 1 September date would not be met. It was not until October 1960 that the Corps let a contract to the Samford Overseas Construction Company for $555,121 covering all of the Corps work, access road, buildings and utilities. It was more convenient and, possibly, cheaper to have a single contractor but developing the specifications for the buildings and utilities must have taken time delaying the contract and the start of the access road work.

Meanwhile, the terrible condition of the access road (Fig. 4.6) probably increased the value of the bids for the tower and anchor construction and was certainly a major

[2]William McGuire 27 September 1961, summary of issues related to the project sent to R. W. Robinson. Folder 5 Arecibo Observatory/Research Office Documents, Cornell University archives.

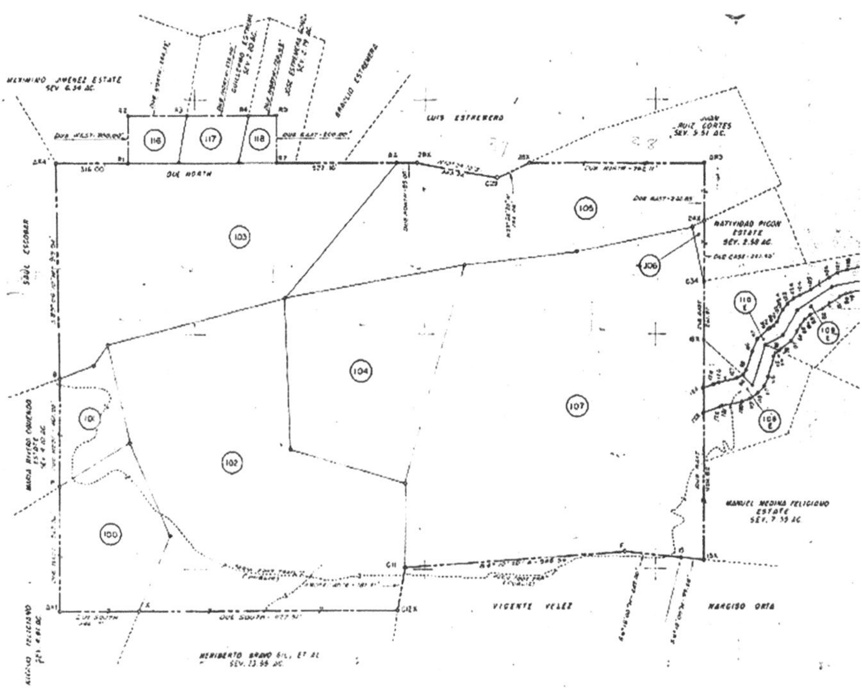

Fig. 4.5 The land parcels that were acquired by the Corps of Engineers under eminent domain for the telescope site totaling 118 acres (48 hectares). Including the access road, 19 parcels were acquired of which 8 were for the access road. Only 3 of these are shown. Section of Corps of Engineers drawing number D-15-02-424 dated 1 August 1960

impediment to getting the contractors' heavy equipment to the work site. Bulldozers had to be used to pull the equipment through the mud. In letting the contract to Samford, the Corps had made no provision for the use of the access road by Cornell's contractors while it was being built. The road was not completed and paved until September 1961, a year late. Samford's work was further delayed by a strike starting on 27 February 1961 related to a unionization drive. It was not settled until 31 March.

4.3 Letting the Contracts

The grading and excavation contract for the bowl was let to the low bidders, a joint venture of Raymond Construction Co. of Puerto Rico and Alpha-Burke-Wilhelm, for $449,600. As a Puerto Rican based firm, Raymond was familiar with the terrain and the hiring of local personnel. Formal notice to begin work was given on 14 September and work did begin later that month. Completion was expected by 12 April 1961. In retrospect, allowing less than 7 months to complete the work

Fig. 4.6 The beginning of the access road in December 1960. The tree is the same one as in Fig. 4.4. Credit: Cornell University

seems extremely optimistic especially given the state of the access road. Nearly 300,000 cubic yards of rock and soil had to be excavated and about 250,000 cubic yards of fill material emplaced to convert the sinkhole into a graded bowl suitable for the reflector (Fig. 4.7). Considerable excavation and grading were also needed for the hilltop locations of the towers and anchors and for the access roads to these locations. High ground had to be cut back to form steep cliffs to the NW, N, SW and SE, while major areas needing fill were to the NNE, SSE, and W (Fig. 4.7). Most of the limestone rock was "hard" requiring blasting although some could be excavated with a bulldozer.

Request for bids packages for the towers and anchors were released in August and seven bids received in September. They were all over the estimated cost. After discussions with the four lowest bidders as to how to reduce the cost, these firms were asked to bid to new specifications. A major cost driver was the tapering of the towers (see Fig. 2.23). A change was made to a constant width cruciform shape with step backs at regular intervals (Fig. 4.8). This would allow slip forms to be used, a technique whereby the forms are continuously moved higher as the concrete is poured with the rate set by the initial setting time of the concrete. Tower T-12, close to the north and Tower T-4, 120° to the south-east, were both 265 ft tall while Tower T-8, 120° south-west from T-12, was 390 ft tall due to the absence of a suitable hill (Fig. 4.7). The bases of the shorter towers were 36 by 38 ft and 15 ft deep while that of the 390 ft tower was larger in area and 25 ft deep. These were very large concrete columns.

The Joint Venture's original design had four cables, which were continuous from the anchor blocks over the top of the towers and to one of the apexes of the triangular

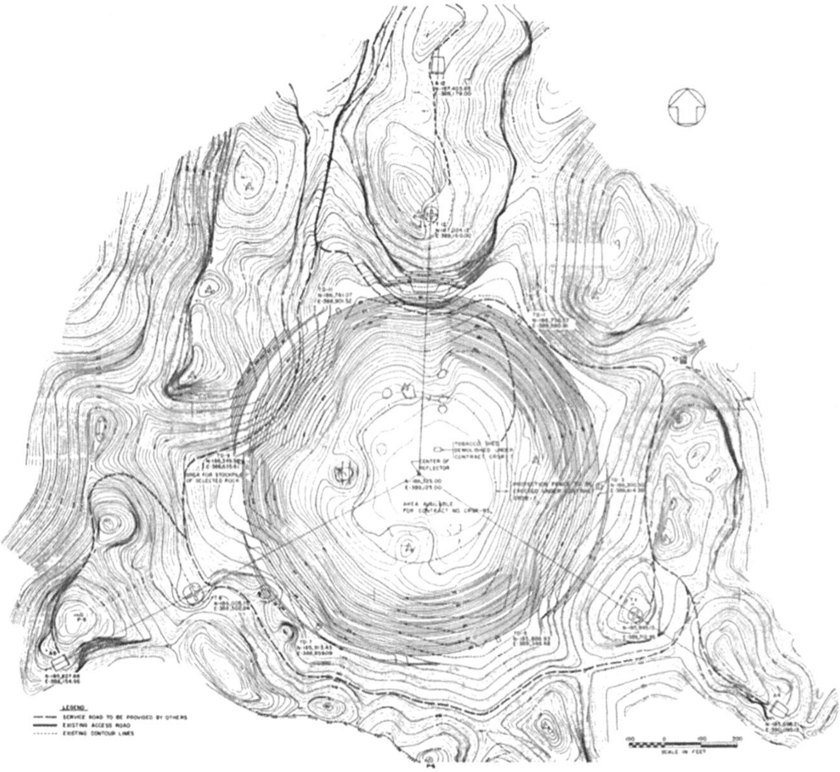

Fig. 4.7 Height contours of the desired reflector shape superposed on the contours of the original sinkhole. These latter contours are dotted where material needs to be removed or fill emplaced to obtain a smooth bowl with the desired shape. The solid lines from the center cross the locations of the towers (T) and anchors (A), T-12/A-12 at the top, T-4/A-4 at the lower right and T-8/A-8 at the lower left. The solid black line from top, just left of center, is the access road to the saddle between two hills where the operations building will sit. The dashed lines are the required service roads to the towers, anchors and center of the bowl. The survey was conducted by Norman Porter Associates of New York under a contract with the Von Seb Joint Venture

feed platform. This was modified to have four 3-inch cables from the top of the tower to the feed platform with the horizontal load on the tower tops being compensated by five 3.25-inch cables from the tower top to very large concrete anchors (Fig. 4.9). Each of the cables would carry about 250 tons of load, about half their breaking strain. This meant that the force on the anchor blocks would normally be about 1250 tons. The vertical component of this force, about 700 tons, had to be resisted by the weight of the concrete anchor block requiring a minimum of about 10,000 cubic feet of concrete (300 cubic meters) or about 21 ft on a side. In practice the concrete anchor blocks were designed to support the vertical component of the full possible load on the cables, about 2500 tons. The horizontal component of the force on the

Fig. 4.8 The completed Tower T-12 in 1962 showing the cruciform design with step backs. This photo was taken in late 1962 during preparations for the lift of the platform triangular structure. The two cables at upper right are temporary cables involved with the lift. Three of the eventual five anchor cables are at upper left. Two of the remaining cables are attached to the nearest apex of the triangle and will be pulled up using the remaining two cables as the triangle is lifted. The jin pole supported from the top step back is used to lift cable ends and other items to the tower top. The waveguide run from the control building to the future catwalk can be seen at lower left. Credit: Philipp Blacksmith

Fig. 4.9 Left: The Tower T-8 anchor block showing the internal construction. Five pairs of 2-inch cables are run internally through corrugated pipes to heavy steel plates close to the back of the block so that the loads of the five backstay cables from the tower top are transferred through these to the entire block. The assembly is encased in concrete as shown in the right hand image of the T-12 anchor. After pre-tensioning of these internal cables, the pipes were filled with cement grout under pressure to protect them from corrosion. Credit: Cornell University/ARPA

anchor blocks is resisted by the rock in front of the anchor blocks requiring that the blocks be sunk deep in the ground (Fig. 4.9).

In October the contract for the towers and anchorages was given to A. S. Wikstrom Inc. of Skaneateles, New York operating under Caribbean-Wikstrom at a cost of $925,935 and a completion date of 25 May 1961. This again seemed highly optimistic especially as the pace of work on the towers and anchors was dependent on the completion of the access roads and grading for the tower bases and anchors.

Mounting problems: Starting in October 1960 there were three contractors on site, Samford, RABW (short for the excavation and grading consortium), and Wikstrom, vying for access and space. And it started to rain heavily. RABW's first task was to finish the service roads needed to access the tower and anchorage sites and the center of the bowl. This work went very slowly. Quoting Gordon's 30 November report to AFCRL[3] "Work on service roads 7 days a week, slow progress. Difficult terrain, heavy rains, frequent blasting required. Behind schedule, had meeting with contractor on 18 Nov, service roads to be completed in two weeks." The 30 December report had the service roads still only 50% complete. However, the road to T-12 and its anchorage, A-12, was completed (Fig. 4.10) so that Wikstrom could start work. Wikstrom was having its own problems. They needed to assemble a concrete plant near the service road to T-12 but their suppliers refused to deliver concrete and aggregate over the access road as it was in such bad condition. This was resolved and by the end of January 1961 the concrete mixing plant was set up and the forms were in place for the base of T-12. However, this was only 4 months before all three anchors and towers were to be completed, clearly the schedule was slipping badly.

The next contract to be let was for the cables and suspended steel feed structure. This contract included the fabrication and erection of the suspended structure consisting of everything that can be seen in the photo of Fig. 4.11 plus the fabrication and installation of the main and backstay cables, the six platform tie down cables and the service cables that support the platform ends of the catwalk and cable car. The three main components (see Fig. 4.11, a photo from March 1963) are: (1) the triangular feed platform that is 216 ft long on each side and 30 ft high, (2) the azimuth ring girder with rails on the inside and outside on which trolleys run to move the feed arm in azimuth, (3) the 304 ft long, the length of a football field, banana shaped feed arm that has rails along its bottom on which the two carriages houses seen in the photo move in elevation.

One of the carriage houses will support the 96 ft long 430 MHz line feed under construction by TRG while the other will have feeds at different frequencies for radio astronomical observations. In the original design, this second carriage house was a weight that moved in and out on the feed arm to counterbalance the weight of the single carriage house holding the line feed as it moved. Phil Blacksmith suggested changing the weight to another carriage house to hold additional feeds and raised the $30,000 cost to implement the idea. A central bearing would keep the

[3]Gordon's monthly reports to AFCRL were combined with AFCRL's monthly reports to ARPA and both are available in the Cornell archives under Arecibo Ionospheric Observatory.

Fig. 4.10 The excavation and grading work progress in December 1960. The last section of the access road is to the lower right, the service road to the T-12 site (center) is completed and the cliff face trimming for the operations building in the saddle to its right is making progress. The service roads to the Tower T-4 site (upper left) and its anchor are also making progress. At lower center is the winding track which allowed foot and donkey access to the farm that can be seen at upper center on the other side of the Rio Tanamá. Credit: Cornell University/ARPA

feed arm centered as it moves around the azimuth track. The entire structure was stabilized against oscillation and rotation by the splayed pairs of 1.5 inch cables in the photo that tie the upper corners of the triangular platform to concrete blocks located near the periphery of the reflector. The total weight of the structure including feeds and utilities but not including the cables was about 575 tons. The partially completed catwalk in the photo would carry the 430 MHz waveguide, power, radio frequency signal cables and control cables as well as providing people access.

Bids were received on 18 January 1961. After negotiations the contract was let to a joint venture of the Harada Corporation of San Juan and Nat Harrison Associates Inc. of Miami for $1,052,040, about $100,000 over the original estimated price. Notice by Cornell to start work was given on 24 February 1961, with the completion date of 1 February 1962. This joint venture decided to go by the name Harada Corporation of Miami using the offices of Nat Harrison Associates. At the strongly put suggestion by Praeger-Kavanagh, the design firm of the von Seb Joint Venture, Bethlehem Steel was chosen to provide the cables and the steel for the main

Fig. 4.11 Photo from March 1963 showing the details of the suspended feed structure, catwalk, main cables, platform corner tie-down cables, and service cables that support the end of the catwalk, that were all included in the Harada contract. The platform is not yet at its design height so the upper end of the catwalk, which is supported by the three service cables from the tops of the towers, is much higher than the platform at this time. Not shown are the backstay cables from the tower tops to the anchor blocks. In this photo the hoisting of the feed arm for attachment to the trolleys running on the azimuth track is almost completed. The carriage houses, which move on the elevation rails and hold the line feed and other feeds for radio astronomy, can be seen without their aluminum siding hanging at the bottom center of the feed arm. The concrete rim wall for the reflector is visible at the bottom of the photo. Credit: Cornell University/ARPA

structural members. It was this contract that effectively moved the completion date of the project from 1 September 1961 to a date in 1962. There would soon be four contactors on the site vying for resources and access.

The design and specifications for the reflector system were completed in early April and sent out for bids. The photo of Fig. 4.12 shows the details of the work under this contract. Approximately 18 acres (7 hectares) of welded wire mesh with 0.5 inch (13 mm) square openings was to be supported by two sets of cables at right angles to each other. The overall spherical shape was set by ten 1.25 inch galvanized cables spanning the bowl at 100 ft spacing in the north-south direction (cables left to right in the photo). They are supported by large concrete blocks at the rim wall and

Fig. 4.12 Photo showing the major components of the reflector system. The concrete rim wall has the attachment points for the approximately 3 ft (1 m) spaced 3/8 inch (9 mm) upper cables with gaps for the large concrete anchor blocks for the orthogonal 100 ft (30 m) spaced 1.25 inch (31 mm) lower cables. These lower cables are forced into the correct curvature for the spherical reflector via tie downs at about 16 ft (5 m) spacing to concrete blocks in the ground. The cable system supports the 18 acres of ½ inch by ½ inch opening galvanized welded wire mesh. Credit: Thomas Talpey and Cornell University/ARPA

forced into a spherical shape by tie downs to concrete blocks set into the ground at approximately 16 ft spacing. At right angles to these large cables are 0.375 inch galvanized cables at about 3 ft spacing connected to the concrete rim wall via a vertically adjustable fitting and tied to the large cables by U-bolts (see photo). Weights and/or crimping of the cables were needed to bring the 100 ft spans of these light cables into the right shape. The rim wall is almost continuous around the 3140 ft (~1 km) circumference of the reflector and is ballasted by a cement shelf on the outside to withstand the horizontal load of the ~300 light cables.

Bids were due 3 May. The low bidder was R. P. Farnsworth & Co. and notice to begin work was given on 21 May 1961, at a total cost of $753,000 and a completion date of 7 January 1962.

The contract for the RF Transmission system was also sent to the bidders in early April also with a 3 May deadline for bids. This contract covered the procurement and installation of the waveguide system to take the output power from the 430 MHz transmitter in the operations building to the line feed attached to the bottom of one of the carriage houses on the feed arm, a distance of about 1400 ft (Fig. 4.13). The contract was let to I.T.E. Circuit Breaker Co. in early June at a cost of $238,071. There were objections from Harada, which had combined with D. S. Kennedy Co. to

Fig. 4.13 A portion of the waveguide run installed in June 1963. The catwalk with the waveguide above the walkway is at lower left with the waveguide then passing over the steel block that is held in place by the three service cables coming from the towers. The ends of the cables supporting the catwalk and cable car are attached to this block independently of the feed platform so that they do not place any lateral load on the platform. Credit: Cornell University/ARPA

bid, claiming that they had not been given adequate notice of the bid opportunity. However, despite the sensitivity of the issue given the need for a good working relationship between Harada and the waveguide contractor, I.T.E. was given the contract. Completion was clearly dependent on when Harada finished installing the suspended feed structure and the catwalk.

In May bids were also requested for the mechanical and electrical work on the suspended structure. These contracts included the hydraulic drive motors to move the feed arm in azimuth and the carriage houses in elevation, the cable car drive system, the electrical lines to the platform to provide power for the drive motors, the receiving equipment in the carriage houses, the feed structure and catwalk lights and the lights on the top of the towers. Both of these contracts were let to the Harada Corporation of Miami also in early June saving future complications due to contractor interactions if they had been let to other firms. These two contracts totaled \$582,550 with, again, completion depending primarily on the pace of Harada's work on the suspended structure. At this point all the contracts for the telescope had been let and the task came down to keeping the contractors working with a minimum slippage of the schedule.

4.4 Making Progress

By the end of June 1961, work under both the excavation and grading contract and the towers and anchors contract should have been completed. Progress had certainly been made but the work on both contracts was far from complete and, due to the insistence of AFCRL, both contractors were sent delinquency notices on 27 June. Service roads to the three towers and anchors were complete allowing the tower and anchor contractor, Wikstrom, to proceed. The road to the bottom of the bowl was in progress but hampered by significant quantities of water in the bottom of the bowl, what became known to the Cornell contingent as Lake Cayugita after Lake Cayuga, the subject of Cornell's anthem "High above Cayuga's waters." However, the cutting of the cliff faces and the needed fill work were far from complete (Fig. 4.14). Pouring concrete for tower T-12 was proceeding as was excavation of anchorage A-12. The foundation of T-4 was ready for the concrete pour (Fig. 4.14). Both contractors had a lot of work remaining and their delays would delay Harada's work on the suspended feed structure in the bottom of the bowl which, in turn, would delay the reflector installation. Curiously, the monthly progress reports from AFCRL to ARPA still had a completion date for the project of 1 September 1961, only 3 months away.

Fig. 4.14 An aerial view of the bowl and work on the operations building from June 1961. Tower T-12 at center left is being poured as is the foundation for T-4 at upper left. The bowl has partly taken shape but there is clearly a great deal of excavation and fill work still to be done. At lower right, the operations building is under construction by the Corps of Engineers. Credit: Cornell University/ARPA

Fig. 4.15 The state of the work on December 15, 1961. Lake Cayugita is still covering the center of the bowl. Tower T-12 (center) is completed as is its anchor block, which can be seen just to the right of the small building on the road to T-12. Tower T-4 is being poured again after the repair of the collapsed hoist tower and the foundation for Tower T-8 is being readied for the concrete pour (upper right). The difficult road to the location of T-8's anchor block can also be seen at upper right. Just to the right of T-12 is the concrete anchor block for the two catwalk support cables. The operations building and the Visiting Scientist Quarters and cafeteria located below the operations building have been completed by the Corps of Engineers. Credit: Cornell University/ARPA

It rained, often heavily, on 16 days between 10 and 29 September 1961, usually associated with thunderstorms. Lightning destroyed the steel tower on Tower T-4 used to hoist concrete and reinforcing steel, delaying this work for about 5 weeks until a new tower could be built. Lake Cayugita (Fig. 4.15) became a real problem and this plus the continuing delays prompted Tom Talpey to write a ditty to the tune of Cornell's Alma Mater:

Far above Lake Cayugita
With its muddy slime
Stands our laboratory building
Not quite done on time
See the towers, reaching upward
Numb'ring one point five
Can we finish next summer?
Come, aw come, let's strive!
Thomas Talpey

Prior to the construction activity, the sink hole was naturally drained through at least one known cave. The excavation contractor was instructed to keep this open but, nonetheless, it became clogged with fine limestone rock powder resulting in Lake Cayugita. After investigating a number of ideas such as pouring acid down the drainage cave to dissolve the clog, a solution was devised in late 1961 that involved lifting the center of the bowl several feet with fill, constructing a moat along the eastern and southern edges of this area, and installing a diesel pump to send the water through a pipe up and over the side of the sinkhole and pour it into the Rio Tanamá running in a valley just behind the sink hole. This finally solved the problem although not until early 1962.

Tom Talpey's ditty highlighted the frustrations of the slow progress. The Corps of Engineers finally finished paving the access road in September 1961, a year late. The operations building, Visiting Science Quarters and utilities including power and a 900 ft deep water well were finally finished in October. However, a serious problem with the buried part of the high voltage line to the site—the line came from a sub-station to the south-east—delayed the move of the Cornell technical and administrative staff from the downtown office to the new building. The problem was not resolved until January 1962. Despite the frustrations, Fig. 4.15 showing the state of the project in mid-December 1961 highlights the real progress that had been made. Tower T-12 and its anchor block were completed, T-4 was half finished with completion at the end of the month, and the concrete for its anchor block was being poured. The foundation and pedestal for T-8 were finished, the very difficult access road to its anchor site completed and a start made on the anchor block. While far from finished, the excavation and fill work for the bowl was clearly making progress. The excavation work was largely finished and the last two large fill areas were being worked on. Two hundred and seventy tons of the steel for the suspended feed structure had been delivered by Bethlehem Steel and were stored in San Juan awaiting the solution to the Lake Cayugita problem.

Away from the construction site, the receiver and data acquisition systems were making real progress at the downtown office (Fig. 4.16). New developments in receiver technology were reducing their additive noise, improving the signal-to-noise ratio for the radar system and, hence, its sensitivity. Despite a number of problems that were delaying the project, Radiation at Stanford was making reasonable progress on the 430 MHz transmitter. Similarly, TRG was suffering delays in the fabrication of the 430 MHz line feed but at this stage there did not appear to be any insurmountable technical difficulties.

4.5 Schedule and Budget Issues

On 31 August Thomas Walsh, AFCRL's Administrative Contracting Officer, sent a letter to Cornell pointing out that the university had not imposed the required financial penalties on the excavation and tower contractors for not finishing their work in April and May, respectively, and requesting "that you show cause why consideration should not be given at this time to termination of the subject contract

Fig. 4.16 The receiver and data acquisition systems group in the downtown Arecibo office in front of the developing data acquisition system in August 1961. From left to right: Domingo Albino, transmitter engineer, Joseph Pierluissi, engineer; José (Pepe) Lopez, the first technician hired; George Peter; Hilda Pagan, secretary; Luis O. Rodriguez, engineer; Miguel Feyjoo, receiver engineer; and Tom Talpey. Miguel Feyjoo was the first local engineer hired (on 9 January 1961), José Lopez was hired on 2 February 1961, and Domingo Albino on 29 May 1961. All three remained on the staff of the Arecibo Observatory until they retired. Miguel Feyjoo succeeded Merle LaLonde as head of the electronics department in 1965. Credit: Cornell University

(i.e. the one between AFCRL and Cornell)."[4] A reply was demanded by 15 September. The problem from Bill Gordon's viewpoint was that he knew the contractors were working as hard as they could and that any imposition of financial penalties would limit their ability to finish the work due to a lack of funds. On 11 September Hastie wrote Walsh that the delays were at least partly due to the problems with the access road, the completion of which was the responsibility of AFCRL since they had contracted the Corps of Engineers, that penalties would be imposed at the time of final payment and that the government should be pleased that such a unique and complex instrument was being constructed at minimum cost.[5]

[4] Letter from Thomas Walsh to John Hastie dated 31 August 1961. Cornell University Archives.

[5] Letter from John Hastie to Thomas Walsh dated 11 September 1961. Cornell University Archives.

Hastie did not mention that the delays were also a product of the very unrealistic schedules for the work.

Walsh's threat to terminate AFCRL's contract with Cornell prompted Dale Corson to consider hiring someone with experience in contracting and contract administration to help with the interaction with the contractors. After a search, Ralph DeSimone, an experienced construction consultant, was contracted to, in Corson's words, "Make a special study of the operation and performance of our subcontractors, and to assist us with negotiations arising from engineering problems in work with our subcontractors." The total effort was estimated to be about 45 days between 1 February 1962 and 31 October, the then scheduled project completion date. DeSimone's hiring was at least partially to reassure AFCRL that Cornell was responding to their concerns.

On 28 September 1961, General Holtzman, the Commander of AFCRL, telephoned Dale Corson suggesting that he come to Cornell to, in Corson's words in a memo dated 29 September, "get acquainted and to discuss with me means for getting the Arecibo project back on the track of expeditious and cooperative work aimed towards an early completion date." He (Holtzman) said that the "Indians" had become much too excited and that he was quite unhappy about the exchanges of recriminating letters." The visit would be on 10 October and they agreed that Holtzman would come to Corson's office. This visit did not happen. Instead, a much larger meeting starting Monday, 11 October in San Juan was called by General Hooks, Commanding Officer of the Office of Aerospace Research. It was primarily aimed at addressing the significant delays in the telescope construction work. A visit by Sir Robert Watson-Watt, one of the pioneers in the development of radar in Britain, and others on the Wednesday was a complication. The meeting was attended by virtually everyone with an involvement in the project: General Hooks, General Holtzman, Blacksmith, Bouché and Murphy (AFCRL), Dr. Hollingsworth of the Air Force Electronics Systems Division (ESD), plus the contract officers Walsh and Idonasi, who worked for the ESD, Dr. Knox Millsaps, Head of AFOSR, accompanied by Dr. Wood, Mr. Adams and Mr. Dietz, Bolton and Cook of ARPA, and Corson, Hastie, Gordon and Johnson of Cornell.[6] The main issue was the work delays and the need for a realistic schedule. Gordon said that one had been prepared and that it showed completion of the project at the end of May 1962 although there would be an additional period of testing. It was agreed that 3 months should be added to this schedule to allow for delays and this was the basis for the schedule sent by Gordon to Blacksmith on 30 November showing a completion date of 15 October 1962. It also showed a completion date for the Corps of Engineers work of 15 December 1961. This schedule reset the excavation and tower contracts completion dates to 15 February and 15 March 1962, respectively, the feed support structure to 15 July and the reflector to 15 August. Blacksmith must have approved this new schedule as it appeared on his January 1962 report to ARPA. However, given the amount of work still needing to be completed, it remained an aggressive schedule.

[6]Memorandum to File by John Hastie dated 18 October 1961. Cornell University archives.

At the San Juan meeting there was also discussion of the withholding of contractor payments due to accumulated damages caused by their delays. It was agreed that withholding should occur subject to reconsideration at the end of the contracts based on claims by the contractors. However, clearly one of the main reasons for the meeting was to get everyone talking to each other, especially Walsh and the Cornell people, to try to damp down the never-ending, and often petty, issues between the Contract Officer and the University. After the meeting, a smaller one was held between Walsh and Idonasi on one side and the Cornell contingent on the other and a number of festering issues were resolved.

Ever rising budgets: Based on the contracts that had already been let and estimates for the remaining ones plus Cornell's in house receiver work and management costs, Philipp Blacksmith submitted a request to ARPA on 30 September 1960, for approval of an increase in the total cost of the Cornell work by \$656,500 to \$5,176,900, which included a contingency of \$297,600.[7] The cost of the work by the Corps of Engineers was increased by \$140,800 although this included \$150,700 for the Navy Bureau of Yards and Docks under a separate Air Force contract to construct the 13.6 kV power line to the site. The actual work would be done by the Puerto Rican Water Resources Authority, which, slightly confusingly, was the Puerto Rican government's electric power company.

ARPA's then Director, Brigadier General A. W. Betts, approved this request on 1 November 1960[8] and it was reflected in the 1 January 1961, Progress Report to ARPA with the Cornell contract amount increasing to \$5,241,000. It is not clear why there is a discrepancy of \$64,100 but it very likely is related to the theoretical studies, which were funded at a cost of \$62,000. The Corps of Engineers, including the Bureau of Yards and Docks, budget increased to \$912,700 with \$121,500 set aside as contingency. The Cornell amount was based on a budget analysis included in a request for a contract extension from 1 September 1961 to 31 October 1962, sent to AFCCDD (AFCRL) by Cornell dated 20 December 1960, and signed by Gordon as Principal Investigator, and Henry Booker as Associate Director of CRSR.[9] Curiously, the monthly reports that AFCRL sent to ARPA did not reflect this contract extension until the January 1962 report.

This was not the end of the requests to ARPA for additional funding in 1961. Only a few months later, on 10 May 1961, Phil Blacksmith again wrote the new ARPA director, Dr. Jack Ruina, who had been appointed Director on 20 January 1961, requesting an additional \$913,400 to cover increased costs for the Cornell contract.[10] About \$720,000 of this was ascribed to bad estimates for the work still to

[7]Letter to the Director of ARPA from Philipp Blacksmith dated 30 September 1960. Cornell University Archives.

[8]Supplementary funding for Ionospheric Research Facility at Puerto Rico dated 19 October 1960. Cornell University archives.

[9]A proposal for the continuation of contract AF19 (604)-6158 dated 20 December 1960. Cornell University archives.

[10]Letter from Philipp Blacksmith to the Director of ARPA dated 10 May 1961. Cornell University archives.

be contracted after the last budget revision in late 1960. The electrical, mechanical and RF Transmission System contracts came in at $388,000 over the estimates. The reflector $150,000 over the estimate. Transmitter related costs increased by $71,000 related to shipping, spare parts and the duplexer, the device that switched the line feed connection from the transmitter power to the receiving system in microseconds to allow the ionospheric radar echo to be collected. Cornell requested $158,000 in additional funds to cover its costs during the schedule extension but was allocated only $70,000 with the comment that the remainder should really be paid for by the operations budget from the Air Force Office of Scientific Research (AFOSR). This raised the issue of the transition from construction to operations: how is it defined? Cornell's attitude was that once the work under a given contract was completed and accepted then any required repairs or maintenance such as for the site roads, after completion of the excavation contract would be paid for under the operations contract.

On 29 May 1961, ARPA director Ruina signed a letter to the Air Force Systems Command increasing the Cornell related contract amount by the requested $913,500 from $5,544,400 to $6,457,900.[11] This amount included AFCRL's costs and contingency. Cornell's contract amount increased from $5,241,000 to $6,175,000. The total allocated funding for the project including the Corps of Engineers work was now $7,370,600 including $60,016 under a separate contract for theoretical studies at Cornell related to the future ionospheric science program. This exchange highlights the critical role that Phil Blacksmith played in the project, keeping it moving and funded. However, apart from the monthly reports to ARPA and requests for budget increases, there is virtually no correspondence that reflects Blacksmith's role. All formal communications with Cornell were through Thomas Walsh. We do know that Blacksmith visited the Arecibo site about once every month and there are occasional references to these visits.

4.6 Disaster Averted

It is very likely that it was at this time that the whole project almost came to an end. In Jack Ruina's own words during an American Institute of Physics oral interview in 1991[12]; "The one thing that's interesting that relates to sort of basic science is that under the Ballistic Missile Defense Research Program at ARPA we were building a big radar in Puerto Rico, the Arecibo radar that you may know about. The original intent of the radar was to measure the properties of the ionosphere, particularly

[11] Letter from ARPA Director Ruina to Headquarters, Air Force Systems Command dated 1 June 1961. Cornell University archives.

[12] Interview of Jack P. Ruina by Finn Aaserud on 8 August 1991, Niels Bohr Library & Archives, American Institute of Physics, College Park, MD USA, www.aip.org/history-programs/niels-bohr-library/oral-histories/35154

electron density profiles. The staff, the ARPA staff came to me and said they really can't justify that program anymore; it wasn't that important and was getting very expensive. On the other hand, my impression from talking to astronomers and people at Cornell, in particular, was that this was going to be a marvelous instrument for planetary astronomy and that it was going to be unique, the largest telescope of its kind. And so I felt very badly about closing down this facility, yet I couldn't justify it under Ballistic Missile Defense. So I remember going to Harold Brown (Director of Defense Research and Engineering of the Department of Defense as of 3 May, 1961 and Ruina's boss) and saying, 'Harold, I really think we ought to keep it open and it means exaggerating a little bit about how important it is. It's not totally irrelevant; it was built for that purpose.' He said, 'You do what you think is best,' and so we kept that facility open. And so I consider myself a hero for having that radar. Without me there'd be no radar. Because nobody in ARPA wanted it; they all said, you know, 'Close it down, we need the money for other things.' ". Ruina really was the hero, possibly due to his background. Before coming to Washington, initially as Deputy Assistant Secretary of the Air Force for Research and Development, he was at the University of Illinois, where he was Professor of Engineering and Director of the Radar Group. This combination of experience with both basic research and radar meant that he could appreciate the Arecibo telescope's potential for radar studies of the ionosphere and the terrestrial planets independent of any contribution to Project Defender. It probably also helped that Harold Brown was a physicist, who had been Director of the Lawrence Livermore Laboratory, the nuclear weapons laboratory managed by the University of California. He was the first scientist to become Secretary of Defense serving from 1977 to 1981 during the Carter administration. How realistic closing down the project would have been is not clear, the cost of closing out the contracts and refurbishing the telescope site would have been very high, possibly approaching the cost of finishing the project. Harold Brown's appointment as Director of Research and Engineering on 3 May 1961, meant that the issue of continuing the project was not raised within ARPA significantly before this date and may have been prompted by Blacksmith's 10 May request to increase its budget by almost 1 million dollars.

This was still not the end of requests for additional funding in 1961. In October Blacksmith submitted a request for an additional $458,100 to cover underestimates in the final contract prices, to fund some omissions in the statements of work for some contracts, and to cover Cornell's internal costs due to the extension of the completion date from 31 March 1962 to 15 October 1962. The request was recommended internally at ARPA for approval on 14 December 1961 and final approval by the ARPA director was given on 2 January 1962.[13] The silver lining for ARPA in this request is that the delay in the completion date meant that starting telescope operations would also be delayed requiring less funds for the operations

[13] Arecibo Ionospheric Observatory—Additional Funding for Completion, ARPA approval signature document dated 14 December 1961, with last date 2 January 1962 and including details of the Cornell request forwarded to ARPA by AFCRL. Cornell University archives.

contract with Cornell administered, as described in Chap. 3, on behalf of ARPA by the Air Force Office of Scientific Research (AFOSR). $315,000 was removed from the proposed $2,200,000 operations budget, which was under negotiation at this time, to offset the construction increase. The new Cornell contract amount was $6,584,000 and the overall Arecibo Ionospheric Observatory construction budget was $7,884,000.

Chapter 5
Overcoming the Problems

By early in 1962 there was very visible progress with the terraforming of the bowl, the construction of the towers and anchors, the delivery of steel for the construction of the suspended platform, the start of the rim wall needed to support the reflector, and the completion of the operations building and the Visiting Scientist Quarters. The downtown staff were about to move to the new operations building, which must have been a significant morale booster. The transmitter was delivered in August after passing its tests at the manufacturer. A temporary large horn was constructed allowing the detection of radar signals reflected from the moon and, later, the lower ionosphere. Meanwhile, the suspended structure was under construction in the bowl. Its lift in October was a very big day for the project and all involved. However, the schedule delays and cost increases continued leading to some changes in the project management at Cornell.

5.1 Significant Progress at Last

After a very difficult 16 months since the beginning of construction, in early 1962 prospects must have looked much brighter. The access road was finally paved, the problems of Lake Cayugita were on their way to being solved, the excavation work was approaching completion, two towers and their anchor blocks were completed and a start made on tower T-8 and its anchor, hundreds of tons of steel for the suspended feed structure had been delivered to San Juan and were being moved to the site, and a start was being made on the concrete rim wall for the reflector by the contractor Farnsworth (Fig. 5.1). Moving from the downtown office to the completed operations building was imminent and the completion of the Visiting Science Quarters (VSQ) with its six rooms and a cafeteria meant that visitors would no longer need to stay at the Hotel Mir on the plaza in Arecibo (Fig. 5.2). It helped that January and February are the traditional dry months in Puerto Rico, especially in the mountains, so rain was likely not a problem at this time.

D. Campbell, *The Arecibo Observatory: A History of Innovation and Discovery*, Historical & Cultural Astronomy, https://doi.org/10.1007/978-3-031-75225-4_5

Fig. 5.1 9 February 1962 photo of the center of the bowl being covered in several feet of soil to get it above the water level with the moat to the right. Steel for the feed structure is being delivered and work on the concrete reflector rim wall is in progress. Half round corrugated pipes can be seen down the side of the bowl intended to control water drainage so that it does not erode the steep sides of the bowl. Credit: Cornell University/ARPA

A snapshot of the state of the project at the end of March 1962 gives a good idea of the now relatively rapid progress. The excavation and fill work were essentially completed with some relatively minor finishing details being handled by Raymond Construction. The grading in the center of the bowl was completed (Fig. 5.1) and the drainage moat was nearing completion with the concrete base for the pump being poured and the pipe over the ridgeline almost completed. The moat and pumping system were in operation in April. Tower T-8 was on the verge of completion (Fig. 5.3), the three anchors were done and the concrete anchors for the six platform tie-down cables were being poured. Painting of the towers with white paint was progressing as a means of minimizing asymmetrical heating of the concrete and possible resulting distortion of the towers. It also gave them a dramatic look.

380 tons of steel had been positioned in the bottom of the bowl, enough for the feed platform, the azimuth girder and part of the feed arm. Some assembly work on the feed platform had started. Work had also begun in the bottom of the bowl on

Fig. 5.2 A photo from 28 February 1962 of the completed six room Visiting Scientist Quarters (VSQ) and cafeteria. The center tepee tent was probably not for additional accommodations but to provide shade for a lunch table. The whole scene now looks like part of an established installation. Credit: Cornell University/ARPA

pouring the concrete supports for the assembly of the feed platform and the large concrete bases for the lifting equipment.

Pouring of the concrete for the reflector rim wall (Fig. 5.4) and the anchors for the ten 1.25 inch lower cables was making rapid progress and was about two thirds completed. The pouring of the anchor blocks on the sides and bottom of the bowl for the tie downs needed to force the lower cables to conform to the spherical shape of the reflector was also proceeding rapidly. One problem was Farnsworth's inability to find mesh that met the tolerance specifications of the contract. Cornell provided Farnsworth with $3000 to engage a Connecticut based supplier to see if they could adjust their machines to meet the tolerance requirements. This eventually solved the problem.

The RF transmission line waveguide and electrical contractors had largely completed what work they could do from the operations building to the base of the catwalk. In the absence of a catwalk, much less the suspended feed structure itself, there was little more that they could do apart from lights and lightning arresters on the towers.

Fig. 5.3 The transmitter bay of the new operations building on 22 March 1962, showing part of the electric sub-station and the first part of the waveguide run to the feed platform. Tower T-8 in the background is approaching completion. Credit: Thomas Talpey and Cornell University/ARPA

Work on the receiver and data acquisition systems at the downtown Arecibo office came to an end in late March as preparations were made to move the equipment to the new operations building at the telescope site (Fig. 5.3). This was accomplished in the first 2 weeks of April, after which work resumed in the new and more spacious location.

The one disadvantage of the move to the new operations building was that communications with the outside world became even more difficult, there were no telephone lines to the site. A microwave link to Ramey Air Force Base was eventually established which linked to Patrick Airforce Base in Florida and then into the regular phone system. However, even when it worked the system was so poor that it was almost impossible to carry on a two-way conversation. The project resorted initially to occasional use of an amateur radio link to Cornell until a frequency assignment was obtained for a permanent two-way link.

Fig. 5.4 Reflector rim wall under construction on 27 March 1962. The workman at right is standing on the concrete ballast platform, the weight of which counteracts the load of the 3/8-inch upper reflector cables attached near the top of the wall, acting to pull it over. Credit: Cornell University/ARPA

5.2 The Transmitter and the First Test Results

The major objective for 1962 was the progressive completion and lifting into place of the triangular feed platform, azimuth girder and feed arm by the contractor Harada. However, three other key projects were approaching completion at this time: the transmitter, the receiver and data acquisition system, and the 430 MHz line feed. At the end of March, the transmitter was under test at Radiation at Stanford with most of its problems solved. The main residual problem was with the dummy loads, the devices that soak up the power from the transmitter when it is under test. The original ones were proving unsatisfactory and the decision had been made to have the supplier construct new ones of a different design. New dummy loads were in place in April and full power testing started plus preparation for shipping. By May full peak pulse power of 2.5 megawatts (MW) at a full average power of 150 kilowatts (kW) had been achieved for pulse lengths of 60 μs. For a pulse transmitter, there is a large capacitor bank whose energy storage is recharged during the period between pulses, 94% or more of the time for the Arecibo transmitter, and discharged during the 6% or less of the time of the transmitter pulse. In June preliminary acceptance tests were run on the transmitter at Radiation at Stanford's plant and observed by representatives from Cornell and AFCRL. The

Fig. 5.5 The console of the new 430 MHz transmitter installed in the receiver room of the operations building. Credit: Cornell University/ARPA

transmitted was then packed and shipped and arrived at the observatory in August. After installation, the final acceptance tests took place in September (Fig. 5.5).

Getting the output power from the transmitter in the operations building to the 430 MHz line feed hanging from one of the carriage houses moving on the feed arm involved not just the 1400 ft of waveguide but also the complexity of getting the power through the central rotary joint about which the feed arm rotated in azimuth and getting it into the moving carriage house. Clearly, a rectangular waveguide cannot be used at the rotary joint; a circularly symmetric device such as a coaxial transmission line, which has a circular outer wall and a central round conductor, is needed. The solution was to have the waveguide transition to a short section of coaxial line through the rotary joint and then transition back to a waveguide to take the power to the carriage house. Getting the power to the carriage house was a more complex problem. A waveguide with a slot along the bottom (Fig. 5.6) was attached to the 150 ft length of the bottom of one side of the feed arm. A device called a collector, a complex waveguide based device that turned the signal through 90 degrees (Fig. 5.7) into the carriage house and thence to the line feed. Since the collector was attached via the waveguide to the carriage house, it moved on small wheels along the interior of the slotted waveguide as the carriage house moved on the elevation rails. The manufacturer of the collector, RCA, had continuing problems

Fig. 5.6 Merle LaLonde with his arm in the slotted waveguide that is attached along the bottom of the feed arm. This photo was taken just before the feed arm was lifted in March 1963 and attached to the trolleys running on the azimuth girder rails. Credit: Cornell University photo by Thomas Talpey

Fig. 5.7 The collector, which directs the transmitted signal from the slotted waveguide along the bottom of the feed arm to the carriage house and thence to the 430 MHz line feed. Since the collector is physically attached to the carriage house, it has to run along the inside of the slotted guide as the carriage house moves on the elevation rails along the bottom of the feed arm. It is upside down in this photo. Credit: Cornell University/ARPA, photo by Thomas Talpey

Fig. 5.8 The horn antenna used to test the receiver system and, later, the transmitter. It was constructed in the first months of 1962. This photo was taken on 11 April 1963, and shows the waveguide and power, control and signal lines from the operations building at the base of the cliff to the catwalk with an offshoot of the waveguide going to the horn. A waveguide switch could direct the transmitter power in the direction of the catwalk or to the horn. The VSQ can be seen at center right. Credit: Cornell University/ARPA

getting it to work reliably at high powers. It finally met specifications by pressurizing it with sulphur hexaflouride gas.

In order to be able to test the receiver system and the transmitter before the telescope was completed, a test antenna was needed that looked at the sky. To solve this, Merle LaLonde organized the construction of a large horn antenna with a small "receiver shack" to hold the equipment (Fig. 5.8). It was located on the hill above the operations building next to the waveguide run from the operations building to the base of the future catwalk next to tower T-12. With an opening size of 16 ft (5 m) its resolution on the sky at 430 MHz was approximately 8 degrees so that it could view a strip of the sky directly overhead with this resolution as the Earth rotated. Any celestial object in this strip, such as the Moon for a few days per month, could be observed for about 32 min. A waveguide switch was installed in the waveguide run to the base of the catwalk to divert the transmitter power to the horn.

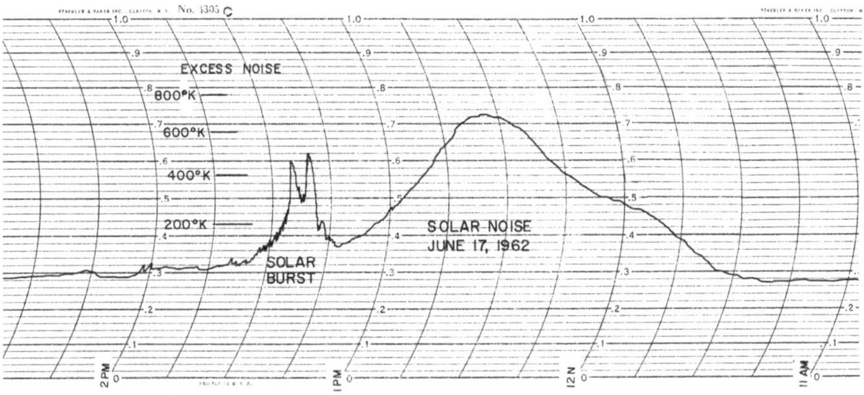

Fig. 5.9 One of the first observations of radio emission from the Sun from Arecibo using the horn antenna at 430 MHz. Time is from right to left with the Sun drifting through the beam of the horn antenna as the Earth rotated. A burst of radio emission can be seen between about 1:15p and 1:30p probably associated with a flare on the Sun. Credit: Cornell University

With the exception of the collector and slotted waveguide, all the equipment needed for the use of the 430 MHz transmitter and line feed for ionospheric measurements was incorporated in the horn test system. In other words, the horn just substituted for the line feed. Included was a five-port turnstile junction, the interface between the waveguide run from the collector into the carriage house and the line feed, that can be adjusted to transmit and receive either circular or linear polarization. Normally a circularly polarized wave is transmitted and the received ionospheric radar echo has the opposite circular polarization (i.e. the opposite helicity) and this emerged from the turnstile on the port opposite that of the transmitted signal, where it is coupled into the receiver system.

In May radio emission at 430 MHz was detected from the Sun using the horn and the new receiver. This allowed the angular response function of the horn to be measured as the Earth rotated. An attempt was also made to detect radar echoes from the Moon using a 50 Watt transmitter that had been built "in house" in March. This was not successful due to instabilities in the receiver; its gain fluctuated as the ambient temperature varied. Clearly, a temperature-controlled cabinet was needed. In June radio emission from the Sun was monitored daily (Fig. 5.9), the Sun reached its maximum northerly excursion so was in view of the horn for the entire month near noon and, at night, radio emission from the Milky Way was observed. By July the temperature-controlled cabinet was finished and installed greatly improving the stability of the receiver system. This led to the detection of the well-known radio "star" in Taurus. A new parametric amplifier was installed in August but this month and September were busy periods installing and testing the new 430 MHz transmitter.

In October the new transmitter was used to obtain radar echoes from the Moon. The echoes were displayed on an oscilloscope but, unfortunately, they were not photographed or the photos have not survived. There were some problems and the

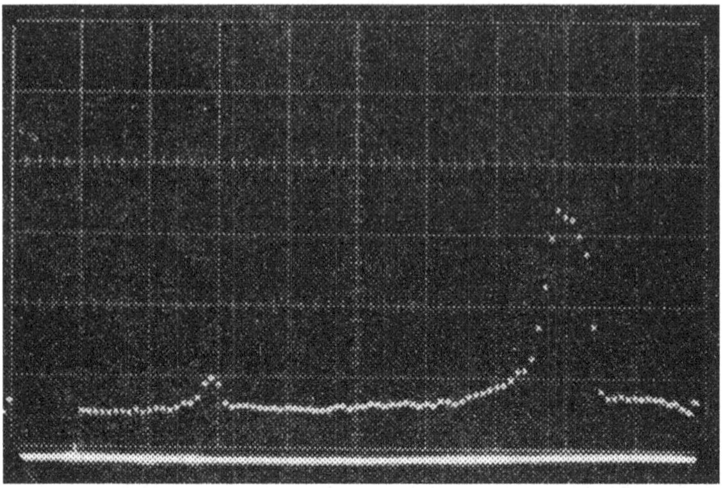

Fig. 5.10 A radar echo from the Moon obtained using the transmitter and the horn antenna on 14 December 1962. It takes close to 2.5 s for the transmitted pulse to travel to the Moon and back. The large peak at the right is the lunar echo and the small one at left is a calibration pulse. The transmitter peak power was 1.8 MW and the pulse length 1.0 ms. Time is from right to left and the spacing between the grid lines is 2 ms. This photo of the echo displayed on an oscilloscope is likely the only surviving record of the first lunar radar echoes (Figure 1 in Research Report RS 45, Summary Technical Report No. 6, Center for Radiophysics and Space Research (CRSR), Cornell University, 31 December 1962. Available in the Cornell University library). Credit: Cornell University

signal strength was not as high as expected. This persisted into November but for measurements on 14 December the signal strength of the lunar echoes was "in good agreement with predictions" (Fig. 5.10). On 19 December radar echoes were obtained from the Echo 1 satellite, NASA's first communications satellite launched on 12 August 1960. It was a 100 ft (30 m) diameter silvered Mylar balloon that could be used to reflect communications signals transmitted from one location to another, distant, location. Since the size of the balloon was known and it was an almost perfect reflector of radio waves, it could be used to calibrate the sensitivity of radar systems such as the horn based one at Arecibo in late 1962. Probably much more excitement was generated by the first detection of radar echoes from the ionosphere on 27 December. In January an improved temperature-controlled cabinet was installed to house the receiving system in the shack by the horn. It could control the temperature to 1 °F (0.6 °C) giving further improvements in the stability of the receiver system. With this system, tests on the Moon and the ionosphere continued. One result of these successes was that AFCRL declared that the transmitter and receiving system were completed and no more construction funds were to be used to improve them; all such work was to be funded out of the operations budget.

5.3 Progress on the Line Feed

The Technology Research Group (TRG) had contracted to provide the 96 ft (29 m) line feed, the turnstile junction, the transition section connecting the turnstile junction to the feed, and the steel mount for the feed in the floor of the Carriage House. By May 1961 they had settled in conjunction with Cornell and AFCRL on the specifications for the feed and were working on both the electrical and mechanical design. By July they had designed the transition section and a short test section to use in high power tests at the nearby Lincoln Laboratory's Millstone Hill high power test facility. Use of this facility was facilitated by AFCRL. However, they were also expressing concern that they could not achieve the specification that the self-supported feed would not deflect at its tip by more than 2 inches (5 cm) when suspended at 20 degrees to the vertical, a requirement related to illuminating the surface of the 1000 ft reflector with the right electrical phase distribution.

By September they had settled on a mechanical design that had a tip deflection of 2.1 inches. It weighed almost 5 tons and was made of aluminum, the most cost-effective way to meet the structural requirements and minimize deflections due to asymmetrical heating by the Sun. By November TRG felt that they had a viable design based on successful high power tests of a test section and electric field tests of a scale model of the feed. They began work on detailed drawings to go out to bid on the fabrication of the feed. Unfortunately, after multiple rounds of bids, the cost continued to significantly exceed the budget estimates so that in May 1961, with the acquiescence of Cornell and AFCRL, TRG decided to fabricate the feed themselves. In June TRG estimated that fabrication and testing would take until the end of February 1962. Actual machining of the feed and other components did not begin until September.

By February TRG had largely completed the turnstile junction and the transition section but they informed Cornell and AFCRL that the feed itself would not be finished until October 1962. Given the scheduled 31 October completion date for the telescope, this was unacceptable to Cornell and AFCRL; they wanted a 15 June date. TRG replied with a mid-July date. Why the insistence on a June date is not clear except to meet a bureaucratic requirement since it must have been obvious in March to both Cornell and AFCRL that the 31 October completion date could not be met given how much work remained. Four of the 12 feed sections were completed by the end of March, nine by the end of May and all of them by the end of June. In July the completed feed was placed in a test fixture to check that it met the stiffness requirement (Fig. 5.11). Electrical testing was also satisfactory with each section being tested for phase delay, the amount radiated from the slots (see Fig. 5.11), and its polarization properties. Starting with the top section, as each subsequent section was finished, it was added to the previous section and these same tests performed. For all 12 sections, close to 99% of the power was radiated leaving only a very small amount emanating from the open end of the feed. Through section 6 of the feed the results looked almost perfect. For the end sections, the ones that illuminated the outer regions of the reflector, the radiated power varied somewhat from the design values

Fig. 5.11 The 96 ft self-supporting line feed in TRG's test facility. The far, larger end is 32 inches (81 cm) in width and the near end is 14 inches (36 cm). Power to illuminate the telescope reflector is radiated from the slots. Alan Francis Kay, the designer and TRG project manager, is at right holding the scale model. Credit: Philipp Blacksmith

(by up to 50%) but, in general, the results looked good. The polarization isolation looked perfect.[1]

5.4 Raising the Feed Platform

With all of the feed platform steel delivered, work on assembling the 260 ton platform in the bottom of the bowl began in March 1962 and was finished in May with detailing work and painting still to be done (Fig. 5.12). After these finishing touches were completed, the next task was to lift the platform 480 ft (145 m) into the air, clearly the most delicate and difficult task of the entire construction effort. Once in place and supported by the main and backstay anchor cables, the feed platform would provide a stable platform for the lifts of the azimuth girder and feed arm.

For the lift, pairs of temporary rigging cables were strung from the top of the towers to a steel block centered over the feed platform. Three of the five backstay anchor cables for each tower would be needed to stabilize the top of the towers under

[1]Construction of the Department of Defense Ionospheric Research Facility –V, Cornell University CRSR Research Report RS 45, 31 December 1962. Available in the Cornell University library.

Fig. 5.12 Assembly of the 260 ton, 216 ft on a side feed platform was completed in May 1962 except for detailing and painting. Pieces of the azimuth girder can be seen to the right. Note the progress on the reflector rim wall at top and bottom. Some of the concrete tie down blocks for the lower main cables of the reflector can be seen at upper left. Credit: Cornell University/ARPA

the load. These 3.25 inch diameter cables had to be progressively tensioned via 150 ton jacks to 100 tons each as the loads of the feed platform and main cables were taken up so that the tower tops would not deflect more than the allowed 2 inches (5 cm). This would be monitored with vertical looking theodolites. Since only three backstay cables per tower were needed, the remaining six were to be used as part of the pairs of temporary cables with the remaining distance being made up of specially fabricated cables. Pulley blocks were positioned on the rigging cables directly over the corners of the triangular platform. Nine-part 1.125 inch diameter lines (i.e. each pulley had four spools) were to be used with the lower pulley blocks connected to the lower corners of the triangular platform (Fig. 5.13). The pull line at each corner would pass through a 60 ton fairlead—basically a pulley—anchored to a 100 ton block of concrete and then to a heavy duty winch also anchored to a large block of concrete (Fig. 5.14). Two of the four main cables to each tower would be needed to support the load of the platform once it was in place. These would be attached to the

Fig. 5.13 The corner trusses of the platform (see Fig. 5.12) were removed to allow the vertical lift rigging to be connected and also to save weight. One of these can be seen on the ground at lower left. These trusses have no structural role except for the attachment points at the top for the pairs of tie-down cables that stabilize the platform against rotation. The bottom pulley and multi-part lifting cable can be seen at top center. Extending from the bottom are the four 20 ft long steel U-bolts (called hairpins because of their shape) that hold the main cables going to the tower. In this photo the central initially unused pair are supported by wooden beams and the gin pole that can be partially seen to keep them roughly horizontal for the future attachment of the second pair of main cables. Credit: Cornell University/ARPA

lower corners of the triangular platform with their load being taken by winch lines from the tower tops as the platform was raised.

By June all of the backstay and main cables had been delivered from Bethlehem steel. The three service cables that supported the platform end of the cable car and catwalk had also been delivered and most of the erection cables were on site. The 60-ton saddles had been lifted to the tops of the three towers (Fig. 5.15). By July, three backstay cables had been connected for all three towers. However, one of the T-12 backstays was damaged and had a broken strand. Fortunately, Bethlehem steel had a spare cable that was almost a perfect match. It was socketed and shipped in late July and replaced the damaged one when the remaining backstay cables were installed in January 1963. During August and September, the erection cable rigging from the tower tops to the center block and the vertical multi-part cables to the corners of the platform (Fig. 5.15) were both completed and work was proceeding on installing the cables that would haul the ends of the main cables to the top of the towers as the platform was raised. These preparations took considerable time but everything was in place for the lift by mid-October (Fig. 5.14). 25 October was the big day with the platform lifting off its concrete support blocks (Fig. 5.16). As liftoff

Fig. 5.14 The feed platform ready to be lifted. Lift cables are attached and the winches in place; one can be seen at lower far right. Two main cables are connected to each corner (see foreground) with their ends attached to winch lines from the top of the towers. One of the winches can be seen at lower right. Painted sections of the azimuth girder are ready to lift and sections of the feed arm are lying on the ground. Credit: Cornell University/ARPA

Fig. 5.15 The 60 ton steel saddle at the top of either Tower T-4 or T-8. The saddles have nine U-shaped slots into which the socketed ends of the four main cables to the platform (left) and the five backstay cables are inserted. The smaller cable at top is one of the service cables that support the platform ends of the catwalk and cable car. Heavy steel bolts buried in the concrete at the top of the towers hold the saddles. Photo from April 1963 after all the cables had been installed and painted. Credit: Cornell University/ARPA

Fig. 5.16 Merle LaLonde making sure that the platform had lifted off and clearly confident that it would not drop down. 25 October 1962. Credit: Cornell University

occurred, the backstay tensions were adjusted to keep the towers within the 2 inch sway limit and checking continued during the lift. After ascertaining that all was well, the platform was lifted 375 ft vertically over about 8 h at a rate of 9 inches/min (Fig. 5.17). Success with the most difficult operation of the entire project was greeted with tremendous relief.

The initial position of the platform was 40 ft below the desired height. At this time the remaining three pairs of main cables were lifted into place with the exception of the fourth cable to tower T-8. During installation it was discovered to be about 9 ft short, a fabrication error by Bethlehem steel. Bethlehem did not have a spare cable and it would have taken several months to fabricate and deliver a new one. No one wanted to delay the work for that long. Bethlehem did have short pieces of cable so one of these was cut to the right length, socketed, shipped and connected to the existing cable. The connection was encased in a block of nickel to protect it from corrosion.[2] Twenty four 150-ton jacks were attached to the hairpins of the 12 main cables at the corners of the platform allowing the cables to be tensioned simultaneously via a central control system. This system was used to transfer the load of the platform to the main cables from the erection cables allowing the vertical erection cables to be removed. The platform corner trusses were then re-installed and the six platform tie-down cables, which connect to the tops of these corner structures, installed. This was immediately followed by the installation of the azimuth girder in six separate lifts while at the same time the azimuth arm was being assembled on

[2]This "fix" allowed work to proceed but it had serious repercussions years later during an upgrading of the telescope in the 1990s and during an earthquake in May of 2014.

Fig. 5.17 The lift in progress showing the erection cables with the multi-part, lines connected to the corners of the platform. The two main cables to each of the bottom corners of the platform were hauled up until their sockets could be inserted into the slots in the tower saddles as shown in Fig. 5.15, once the platform had reached its initial required height. Credit: Cornell University/ARPA

the ground (Fig. 5.18). Once the azimuth girder was in place the pairs of erection cables from the top of each tower were removed freeing up the remaining six tower backstay cables so that they could be installed and tensioned using jacks. At this point the platform jacking system was used to raise the platform to within 5 ft of its final height while adjusting the tensions of the backstays to keep the tower top deflections within the allowed limit. The three service cables coming to a steel block, which held the platform ends of the catwalk and cable car support cables, were then installed (see Figs. 4.11 and 4.13). The support cables were also installed at this time allowing the construction of the catwalk to begin.

While this work was going on, the feed arm assembly was being completed in the bottom of the bowl and preparations made for it to be raised and connected to the two azimuth trolleys that run on the rails of the azimuth girder. However, there had been considerable delays in Harada's, and its sub-contractors', fabrication of the azimuth trolleys and hydraulic drive systems. The trolleys were finally shipped in early

Fig. 5.18 4 December 1962; one of the six lifts of the azimuth girder above the partially assembled feed arm. The lift cage, the only way to get personnel to the platform until the catwalk was constructed, can be seen just to the right of the azimuth girder. The reflector rim wall is completed at least in the section in the photo. Credit: Cornell University/ARPA

February 1963 again by a MATS (Military Air Transport System) flight organized by AFCRL's James Murphy due to a maritime strike that severely affected shipping to Puerto Rico. Once the trolleys arrived it took about 2 weeks to install them and the rigging for the lift. By the end of February, the feed arm had been raised 5 ft off the ground and the slotted waveguide installed. It was then raised to 15 ft allowing the carriage houses to be installed (see Fig. 4.11). On 3 March the 160 ton feed arm was raised and connected to the trolleys. While many details remained such as the installation of the aluminum siding on the carriage houses and the hydraulic drive and electrical systems, the suspended feed structure of the telescope was finally in place, an amazing and unique civil engineering achievement.

5.5 Delays, Cost Increases and a Management Shakeup

After the multiple budget increases and schedule slips during 1961 as discussed in Chap. 4, 1962 saw more of the same. Starting with the November 1961 monthly report, Cornell was required by AFCRL to include a schedule for the completion of the telescope construction. At the end of 1961, the Cornell contract amount was $6,584,000 and the completion date in Cornell's schedule was 15 October 1962. The May 1962 schedule saw a change in the estimated completion date to 15 January

1963, despite AFCRL's report to ARPA still showing 31 October 1962. In July the estimate went to 1 February 1963, in August to 31 March 1963, and in October to 30 June 1963. Delays meant extra costs for Cornell's management activities and for The Joint Venture's supervision of the contractors' work.

On 4 August 1962, Cornell President Deane Malott received what can only be described as a scolding letter from ARPA Director Ruina. The letter was politely worded but it made it clear that ARPA was extremely concerned about the never-ending cost increases and schedule extensions for the Arecibo project and wanted Cornell to give the project "the most serious attention of the University and that firm measures should be taken immediately to improve the project management in order to bring the project to a successful, early and economical completion."[3] Ruina must have been seriously concerned to bypass AFCRL and write directly to the president of Cornell. Curiously, there is no indication that the letter was copied to anyone at AFCRL. The timing of the letter may have been triggered by the cancellation in July of the US Navy's 600 ft (180 m) Sugar Grove telescope project in West Virginia. It is highly likely that this would have raised concern within ARPA about heightened attention to the significant delays and large cost increases of ARPA's Arecibo telescope project.

Apparently AFCRL was not considered responsible for the delays and cost increases as on 31 May Ruina had written to the Commanding Officer AFCRL commending Philipp Blacksmith for "outstanding initiative and competence in dealing with the many and varied technical and administrative problems which have arisen in the past two years."

There was warning that Ruina's letter to Malott was coming. On 31 July, John Hastie, the head of Cornell's Research Office, had travelled to Washington to meet with Bolton at ARPA. Hastie pointed out that Cornell had taken measures to beef up the management of the Arecibo project by contracting Ralph DeSimone, a construction consultant and the hiring of Phil Waldron as the Business Manager of the Arecibo Ionospheric Observatory as of 15 May 1962. The impression from Hastie's memo to Corson, Gold and Gordon recounting the meeting is that Bolton was not fully convinced that Cornell was hiring the "experienced hard-fisted management to keep to schedule".[4] Bolton also pointed out that, while there were good, informal communications with Philipp Blacksmith, there was a lack of formal reporting to Thomas Walsh. It was agreed that Cornell would send a monthly letter to Walsh providing a detailed update on the progress of each contract and reasons for any further slippage.

Hastie recommended that on receiving Ruina's letter, President Malott should respond saying that "we believe that we are now properly staffed for tight control, that we regret that Ruina has felt that there was cause for concern, and that we think

[3]Letter from ARPA Director J. P. Ruina to Cornell President Deane Malott dated 2 August 1962. Cornell University archives under Arecibo Ionospheric Observatory.

[4]Memo to Corson, Gold and Gordon from J. W. Hastie dated 1 August 1962. Cornell University archives under Arecibo Ionospheric Observatory.

the record will show past performance to be, on balance, highly credible to all concerned." Probably fortunately, Malott did not take this advice and responded to Ruina saying that "I wish to assure you that firm and immediate steps are now being undertaken to assure that the project will progress towards the earliest possible completion date as economically as possible."[5]

The Cornell organization for managing the Arecibo project was clearly deficient. The first issue that Malott addressed was the Cornell campus organization of the Arecibo Project. Since its beginning, the business aspects had been handled through the office of the Coordinator for Research, John Hastie and his deputy Bob Robinson, and the contracting by Richard Hufnail, the CRSR Administrator. This was the first large off-campus project that the University had been involved with and there were no precedents. There is no documentation showing that President Malott or the Vice-President for Research was involved much less the Vice-President for Business. At that time there was no facilities department at the university that handled contracting. How the construction of large buildings on the university campus was handled is not clear. Malott promptly assigned responsibility for the administration of the project to John Burton, Vice-President for Business. After a search a civil engineer, Everett Lenderman, who worked for Detroit Edison, was hired as the Project Manager-Construction. He would live primarily in Puerto Rico for the anticipated year before completion of the construction and would be in charge of the construction aspects of the project, so replacing Bill Gordon and Phil Waldron in dealing directly with the contractors on site. Gordon remained Project Director and in charge of the myriad details of interacting with AFCRL and ARPA, overseeing the instrumentation development and planning for research operations. Needless to say, Gordon and Waldron were far from happy with this development especially as it seems that they were not consulted beforehand. It is not clear if the change made any practical difference, but it did serve to reassure ARPA and AFCRL that Cornell was taking their concerns seriously. At a meeting at AFCRL in the week of 26 November Cornell requested concurrence with the formation of the new management team, Burton, Gordon, Lenderman, Wood (a construction specialist) and a lawyer specializing in construction matters. As of 14 November, the estimated cost to complete the Cornell work was $7,087,000 and the completion date was 31 July 1963.

Fortunately, the first meeting of the Arecibo Advisory Committee[6] took place during last week of November 1962. In addition to the committee chaired by Ronald Bracewell of Stanford University, Jack Ruina and ARPA Associate Director Charles Hertzfeld attended as did Knox Millsaps and Lloyd Wood of AFOSR. Corson, Gold, Salpeter and Cohen attended from the Cornell campus, Gordon Pettengill, the future Associate Director of the Observatory, was also there as were Bill Gordon and other

---

[5]9 August 1962, letter from Deane W. Malott to Dr. J. P. Ruina, Director, ARPA. Cornell University archives under Arecibo Ionospheric Observatory.

[6]The committee members attending were Ronald Bracewell (Stanford University, Chair), Colin Hines (University of Chicago), Arthur Waynick (Pennsylvania State University) and Herbert Friedman (Naval Research Laboratory).

Arecibo site personnel. In a 6 December letter to President Mallott, Dale Corson expressed great satisfaction with the meeting and the positive response of the ARPA and AFOSR representatives to both the science discussions and the state of the construction work. By this time, it must have been clear that the end of the construction was in sight even if the 31 July 1963 date slipped a bit more. The platform triangle was in the air, the azimuth girder was being installed and the azimuth arm was being erected in the bowl. The reflector rim wall was completed and the installation of the lower reflector cables had begun. The transmitter had been installed and tested and the 430 MHz line feed had been delivered. With the success of the Advisory Committee meeting and the very positive response of the ARPA management, 1962 ended on a very optimistic note.

Chapter 6
Finishing Up and a Celebration

At the beginning of 1963 it was clear that the project was approaching completion even if there were still numerous tasks to be completed, putting in some doubt the 31 July completion date. The raising of the azimuth arm in March cleared the way for the completion of the installation of the heavy and upper light reflector cables. This, in turn allowed the installation of the wire mesh surface to commence in early April. April also saw the completion of the installation of the waveguide along the catwalk and the raising of the platform to close to its final height, allowing the catwalk and cable car to be used to access the platform. Apart from some minor adjustments, the reflector surface was completed at the end of August and August also saw the raising of the line feed. Work on the telescope was essentially complete by the end of September and AFCRL ended their activities as contract monitors. Five years and 4 months after Bill Gordon's May 1958 seminar at Cornell University, his dream had come to fruition. A celebration was held on 1 November.

6.1 The Final Details

The reflector: At the beginning of 1963, the major construction tasks remaining were the completion and lift of the feed arm to the suspended structure and the completion of the reflector. At the end of August 1962, Farnsworth had completed all of the concrete work, the rim wall and about 600 concrete anchorages for the lower cable tie downs (Fig. 6.1), except in the center of the bowl where the ongoing work on the feed platform prevented access. By the end of September all of the mesh had arrived and preparations begun for the on-site fabrication of the cables that tie the lower reflector cables to the concrete blocks set in the ground. The ten lower cables were received from Bethlehem Steel in November, but there were delays in fabricating the 300 light upper cables. Installation of the cable system was expected to start immediately after the installation of the feed arm, scheduled for January 1963, was completed so that the entire bowl would be available to Farnsworth. However, the

D. Campbell, *The Arecibo Observatory: A History of Innovation and Discovery*, Historical & Cultural Astronomy, https://doi.org/10.1007/978-3-031-75225-4_6

Fig. 6.1 The completed reflector rim wall in September 1962. The attachments for the 3 ft spaced upper cables can be adjusted to set the correct heights of the cable ends. One anchor block for a heavy lower cable can be seen at left center with four of the concrete blocks for its tie downs at lower left. Credit: Cornell University/ARPA, photo by Thomas Talpey

delays in the lift into March and pressures to keep to a schedule meant that Harada and Farnsworth would need to share the space with Cornell acting as mediator. The outer two of the ten lower cables were installed in December and, with the completion of the fabrication of the tie down cables, work stated on pulling these two cables into the right shape to form the spherical reflector. By the end of January, four more lower cables had been installed and a start made on installing the upper cables, most of which had now been delivered by Bethlehem Steel to the site. The crates of reflector wire mesh were distributed around the rim road. During February, 75% of the 300 upper cables were installed and two thirds of them painted. The lift of the feed arm in early March cleared the center of the bowl allowing the installation of the remaining cables to be finished. However, work was hampered by 11 inches of rain falling in a 46 hour period starting on 24 March, leading to flooding of the bowl. The water was cleared in less than 2 days via the pump system and percolation into the ground. In early April most of the U-bolts connecting the upper cables to the lower ones were installed and cable painting largely finished (Fig. 6.2).

Installation of the roughly 18 acres of ½ × ½ inch galvanized wire mesh started on 22 April. The mesh was in sections 12 to 18 ft wide and 5 ft long,[1] spanning four to six of the upper cables that are spaced about 3 ft apart. It was felt important to

[1]A description of the mesh installation can be found in Kavanaugh and Tung (1965).

Fig. 6.2 A view of the telescope on 27 April 1963. The 10 lower cables with their concrete tie down blocks can be clearly seen. The 300 lighter upper cables are at right angles to the lower cables. The first section of reflector mesh can be seen just to the left of the top of Tower T-12 in the foreground. The feed arm was installed in March and most of the (white) waveguide has been installed on the catwalk. Credit: Cornell University/ARPA, photo by Thomas Talpey

electrically connect the mesh sections and this was achieved by clamping them together at close intervals during installation. The installation was done from platforms above and below the reflector supported by cables strung above the reflector and anchored to two weighted steel frames on the rim road just outside of the rim wall. These frames were on wooden rollers that allowed them to be moved along the rim road as the individual strips spanning the reflector were completed. A separate cable was used to deliver the mesh to the installers (Fig. 6.3).

In his 30 April report to ARPA and AFCRL, Gordon estimated that it would take another 10 weeks at a minimum for the reflector contractor, Farnsworth, to complete the installation and adjustment of the reflector mesh. This meant mid-July at the earliest, only 2 weeks before the estimated 31 July completion date for the telescope. Blacksmith was highly skeptical and in his report to ARPA stated that the 31 July date could slip by at least a month. In fact, the 31 May report indicated an installation rate of 154 sheets of mesh per day, which translated into a 31 July completion. The 30 June report was still optimistic that the mesh would be installed by 31 July but

Fig. 6.3 The mesh erection platforms above and below the reflector supported by two cables. It took a significantly-sized crew to do the installation and it was clearly a precarious task. The top cable is supporting the mesh delivery system and the long rods are galvanized ballast weights to bring the 100 ft spans of the light cables to the correct curvature. One of the ballasted steel frames supporting the cables from which the platforms are suspended can be seen above the rim wall. Credit: Cornell University/ARPA, photo by Thomas Talpey

stated that required adjustments would continue into August. However, the 31 July report still had 3% of the mesh to be installed and, with adjustments taken into consideration, completion was expected on 23 August. By the end of August the reflector surface was in place but considerable work remained to adjust and repair some areas and a small work crew remained on site through September to complete these tasks. One major issue was how well the surface conformed to the required spherical shape both on the large and small scale. In many places the mesh was sagging between the cables. A tool was developed that crimped the mesh, shortening it and removing the sag. On the large scale, theodolites were used to check positions for a number of locations where targets had been placed. This roughly validated the overall shape but Cornell was developing a stereo photogrammetric system to do a complete check.

Finishing up: While the feed structure was being assembled and raised and work on the reflector was proceeding, smaller, less visible but critical aspects of the

Fig. 6.4 Waveguide installation on the catwalk on 23 March 1963. The waveguide was supported on the unpainted horizontal steel bars while the ladder like structure above the waveguide carried the cable tray for the electrical cables. Installing the waveguide must have been a scary activity as there was no fencing on the sides of the catwalk and, because the platform had not yet been jacked to close to its final design height, the end of the catwalk was out in space high above the platform and 500 ft above the center of the reflector. Credit: Cornell University/ARPA, photo by Thomas Talpey

telescope construction were proceeding. As of March 1963, the platform was still considerably lower than its final design height, but bringing it to the exact height required that the hydraulic drives, electrical system and waveguide, including the rotary joint, be installed to bring the platform to its final weight. The 5 ton line feed would not be installed but this could be simulated by placing 5 tons of ballast weight in Carriage House 1 and, to counterbalance this load to keep the platform level, 5 tons also in Carriage House 2. Installation of the catwalk and cable car had moved ahead during the work on the feed arm and by mid-March waveguide was being installed on the catwalk (Fig. 6.4). This was completed by the end of April followed by the installation of fencing on the catwalk, making walking along it a less perilous experience. In April, the platform was jacked to within 10 inches of its design height allowing the catwalk and, soon, the cable car to be used to access the platform. This was just in time as the contractor's vertical lift cage had been declared a safety hazard by the Puerto Rican Department of Labor. Apart from a few small sections,

Fig. 6.5 Installation of the waveguide rotary joint. The input waveguide at the top, which is connected to the waveguide run in the foreground, transitions to a circular coaxial line in two parts that can rotate about each other. There is then a transition back to waveguide at the base for the run to the slotted guide along the bottom of the feed arm. The hollow center conductor of the 430 MHz coaxial line is the outer wall of another coaxial line for the future 40 MHz transmitter power. Its input is the black circular area at the very top. One of the azimuth trolleys can be seen at upper right. Credit: Cornell University/ARPA, photo by Thomas Talpey

all that was now missing from the waveguide run was the rotary joint, which was in final tests at the manufacturer, and the collector, which, as discussed earlier, was breaking down at high transmitter powers. The biggest concern were the hydraulic drive systems, which were still under test although tests of the azimuth drive had shown that they could achieve the required 10 arc second positional accuracy.

By the end of May, with the exception of the final lift to its design height, the suspended structure, platform and cables were essentially completed with just minor adjustments and clean up still to do. This did not include the installation of the drive motors, which arrived on site on 21 May. They would require several weeks to install and test. Considerable work remained for the electrical systems, primarily the installation of all of the electrical cables from the operations building to their end locations on the platform and a concerted effort was started to have this completed by 13 July. The rotary joint was finally installed on 17 June (Fig. 6.5), the last significant weight to be added to the structure apart from the 5 ton line feed. At that time, Bill Gordon was still hopeful of meeting the scheduled completion date of 31 July, but it was clearly going to be very difficult.

Fig. 6.6 Tom Talpey inspecting part of the damage from the 29 May fire in the electrical sub-station next to the operations building. Credit: Cornell University

A small disaster struck at 8:30 p.m. on 29 May; there was a fire in the electrical sub-station next to the operations building. The main switch, distribution switches and some cables were destroyed (Fig. 6.6). Despite the extensive damage, the fire did not cause significant delays in the work, but it must have prevented operation of the transmitter. It took about 3 months for all the repairs to be completed.

The final adjustment of the platform to bring it to its design height was accomplished on 15 July. The line feed had not yet been installed but the 5 tons of lead ballast added to each of the carriage houses compensated for this. Carriage House 2 was a counterweight to Carriage House 1 so they had to have the same weight and have identical movement on their respective elevation rails to prevent tipping of the platform and, hence, pointing errors during observations. On 17 July Cornell accepted the results of the final positioning and leveling of the platform, a major milestone. The jacking system was removed by early August. All that remained to complete Harada's steel structural contract was the shimming of the elevation rails along the bottom of the feed arm to bring them into the correct curvature. However, Harada's mechanical and electrical work was still in progress at the end of July. Work was almost completed on the hydraulic azimuth and elevation drive systems, but telephone cables and 30 R.F. coaxial cables still had to be installed from the operations building to the feed arm. The waveguide system was complete except for the collector, which was still under test.

Raising of the line feed and connecting it to its carriage house took place on 14 August (Fig. 6.7), completing the major components of the telescope (Fig. 6.8). Initial observations of a radio source indicated an angular resolution of 9.5 arc minutes, close to the expected value of 10 arc minutes, but the associated gain

Fig. 6.7 Raising the line feed on 14 August 1963. This was a Cornell task. The person on the left is Gordon Pettengill and the one on the right is Merle LaLonde. A specially designed truck provided a swivel connection for the feed as it moved from a horizontal position to a vertical one during the lift. The concrete bridge-like structure and an identical one 90 ft to the right took the load of the light upper cables, giving a large gap in the reflector to allow winch lifts of equipment to the suspended structure. Credit: Cornell University, photo by Thomas Talpey

measurement was lower than expected. This was only a small concern as it was just an initial measurement.

By the end of September, the only major piece of work remaining was the checking of the elevation (i.e carriage house) drives and an overall test of the system. Apart from this, minor adjustment and repair work continued on the reflector. The waveguide system was complete and repair of a transformer was the last task for the electrical system. Cornell's contract with AFCRL ended on 30 September and Blacksmith saw no reason to extend it. Gordon did request that AFCRL continue to provide technical support for the following 6 months on "further checkouts and other improvements of the antenna system." Any further work, such as the pointing system that involved costs, would be done under the operating contract with the Air Force Office of Scientific Research (AFOSR). Five years and four months after Gordon's 29 May 1958, seminar his dream of building the world's largest radio/radar telescope had turned into reality (Figs. 6.8 and 6.9).

Fig. 6.8 In this 15 August 1963 photo with the line feed in place the telescope looks essentially complete. Work still remained primarily to get the carriage house and azimuth drive systems operating. Credit: Cornell University/ARPA

Fig. 6.9 Bill Gordon in August 1963 looking very happy. Credit: Cornell University

6.2 Analog Versus Digital

Telescope pointing: The early 1960s was a time of rapid development of digital technologies and computers but the capabilities were still limited so decisions had to be made as to whether to go analog or digital for the telescope's tracking and data acquisition systems. In the case of the data acquisition systems, the decision was where in the receiver and data processing chain to convert from analog signals to digital ones. The issue for the telescope tracking system was similar. Almost all optical and small radio telescopes used what is known as an equatorial mount, where one axis of the telescope is parallel to the Earth's axis about which the telescope rotates in the opposite direction to the rotation of the Earth compensating for the Earth's rotation. A second axis allows the declination, the angle of the location of the point in the sky that the telescope is tracking relative to the plane of the Earth's equator to be set. With the exception of the Moon and other solar system objects, all of which move on the sky as they orbit the Sun, the declination is constant so no movement of the declination axis is required as the telescope tracks a point in the sky. This means that the telescope only moves about one axis at a constant rate as it is tracking and this can be achieved with a simple mechanical clock drive.

An alternative to an equatorial mount is an elevation-azimuth mount, where the azimuth rotates in a plane parallel to the local horizontal plane and the elevation axis is parallel to this plane. For large telescopes this type of mount is structurally much easier to implement, but it means that when tracking a location in the sky, the telescope has to move about both axes at non-constant rates. To achieve this requires rapid continuous conversion from sky coordinates, Right Ascension and Declination, the equivalent to longitude and latitude on the Earth, to azimuth and elevation coordinates. For the Arecibo telescope an equatorial mount was not an option; by the nature of the construction it had to be an elevation-azimuth system with the feed arm rotating in azimuth and the carriage houses moving in elevation.

In the mid-1950s when the 250 ft Jodrell Bank telescope in England was being designed and built, an equatorial mount was not an option due to the telescope's size. Digital computers were not capable of doing the coordinate conversion from right ascension and declination to elevation and azimuth fast enough and an electro-mechanical analogue computer was the only option. This computer consisted of 10 large racks of equipment, an ingenious but clearly costly solution.

In the late 1950s when the National Radio Astronomy Observatory (NRAO) was initiating construction of a 140 ft diameter radio telescope in Green Bank, West Virginia, the Australian Commonwealth Scientific and Industrial Research Organization's (CSIRO's) Parkes 64 m (210 ft) telescope was being designed and the design of the Arecibo 1000 ft telescope was initiated, three different solutions to the mount and coordinate conversion problem were used. NRAO chose to use an equatorial mount for the 140 ft telescope, a decision that led to escalating cost and schedule problems due to the massive size of the required tilted polar axis (Fig. 6.10). The CSIRO chose an elevation-azimuth mount for the Parkes telescope (Fig. 6.10) and solved the coordinate conversion problem very elegantly by

Fig. 6.10 The NRAO 140 ft telescope (left) and the CSIRO Parkes 64 m telescope (right). Note the massive equatorial mount of the 140 ft telescope compared with the much lighter and structurally simpler elevation-azimuth system for the much larger Parkes telescope. Credit: NRAO and CSIRO

installing a small "master equatorial" telescope mounted on a structurally isolated pillar aligned with the vertical azimuth axis and servoed the 64 m telescope to this. The idea came from Barnes Wallis of World War II Dambusters fame.[2] For the Arecibo telescope, the initial approach to the tracking issue seemed to be to just ignore it; there is virtually no discussion of the issue in the early documentation. This seems strange as the initial completion date for the telescope was in mid-1961 and tracking the planets and cosmic radio sources was needed to meet the scientific goals. However, decisions on this difficult issue may have been deferred partly because the prime mission of the telescope was for incoherent scatter radar measurements of the ionosphere, and a tracking system was not required, and partly because there was no obvious cost effective solution. What was required was knowledge of the position of the feed arm in azimuth and the carriage houses in elevation. The azimuth was measured using a gear wheel attached to the feed arm rotating on finely spaced teeth on a rack gear attached to the azimuth girder. A similar system was used to measure the elevation of the carriage houses with a gear wheel running on the large rack gear used to move the carriage houses in elevation.

The first recorded idea for a tracking system was a suggestion by Gordon Pettengill to use an analog system similar to the one Lincoln Laboratories developed for their Millstone Hill antenna just west of Boston. $50,000 was included in the

[2] A short but interesting account of the building of the Parkes Telescope and Wallis' role can be found at: https://www.atnf.csiro.au/outreach/about/history/building-parkes/index.html

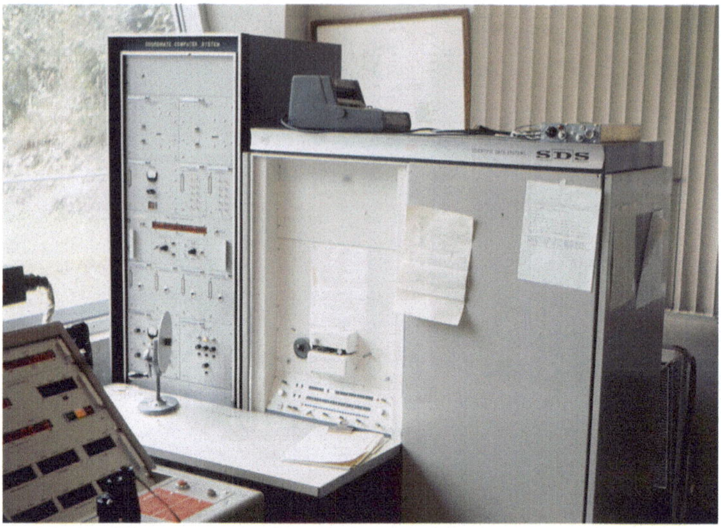

Fig. 6.11 The SDS computer with its paper tape input at right and the MILGO system at its left used to read the digi-switches and Nixie tubes on the control panel at lower left. The hand-cranked device on the shelf was for winding up the paper tape. Credit: Gordon Pettengill

1961 proposal for operational funding to the Air Force Office of Scientific Research (AFOSR). However, the Millstone Hill system did not have very precise tracking accuracy and the idea was discarded. By mid-1962 it was clear that digital computers would be capable of doing the coordinate conversion at the required rate. A contract was let to the MILGO Corporation for $107,517 to design and implement the system. A Scientific Data Systems (SDS) 910 computer was selected. This computer had 2048 24-bit words (i.e. 6 kilo bytes) of magnetic core memory, a punched paper tape input reader and the ability to read digi-switches, that is decimal settable analog indicators that were convertible to binary form for input to a computer, and write to Nixie tubes, decimal output indicators (Fig. 6.11). This meant that the SDS 910 could read a desired position of the telescope in either (zenith angle, azimuth angle) or (right ascension, declination) and give you a visual indication of its current location via the use of Nixie tubes in either of these coordinate systems. It even had a parallax correction in which you could dial in a distance, say to the Moon, and it would compute the apparent right ascension and declination. Since telescopes are on the surface of a rotating Earth, 6378 km from its center, except when the Moon is at zenith, there is a small difference, the parallax correction, in the direction of the Moon relative to its direction from the center of the Earth, which changes as the Earth rotates. Having this ability, especially in 1963, was almost certainly unique to the Arecibo telescope. In fact, the entire system was likely the first time that the required coordinate conversion for tracking a source was done digitally. The SDS based pointing system worked well and was very reliable, remaining in operation until replaced by a Datacraft computer during the first telescope upgrading in the early 1970s.

Data acquisition systems: Selecting an approach to the data acquisition systems faced the same quandary as the pointing system, analog or digital. The solution was both. The requirements were driven by the radar studies of the ionosphere and of the Moon and planets. For a height resolution of 15 km in the ionosphere, Bill Gordon's original goal, the transmitter pulse must be no longer than 100 μs (the time for a radio wave to travel 15 km and return) and the power in the reflected echo must be sampled every 100 μs, the range gates, over the time interval that contains the ionospheric echo. A system that could do this would provide the electron density as a function of height, but Gordon and his colleagues wanted to also measure the temperature and determine the composition of the ions. To do this they needed to form a frequency spectrum of the received echo from each 15 km height interval in the ionosphere. As discussed in Chap. 1, the motion of the ions, with higher velocities corresponding to higher temperatures, results in a small shift of the frequency of the reflected radar echo to higher and lower values, the Doppler shift, and the measure of the distribution of the echo power with frequency provides information about the temperature and ion composition at each height. Forming the frequency spectrum for each height interval was a very challenging task at the time and it was only partially solved.

On arriving in Arecibo and setting up an office in the summer of 1960, the Cornell group had the immediate task of designing and constructing, or procuring, the data acquisition and analysis system. The telescope was scheduled for completion before the end of 1961. Three approaches were adopted, two primarily analog systems and one with the digitization occurring at a much earlier stage. To measure the electron density as a function of height, an analog system was designed and built that measured the radar echo power in 100 separate range gates averaged over a few seconds. A multiplexer took 2 s to convert these powers one at a time to a digital number which was stored on punched paper tape for future processing in a computer. The powers could also be displayed on an oscilloscope screen to give a real time display of the received power as a function of height. This system was also used for early radar observations of the Moon.

Generating a spectrum of the received echo for each range box was a much more difficult task. The Raytheon Corporation had developed a device called a Rayspan, **Ray**theon **Sp**ectrum **An**alyzer, which could produce a spectrum of the radar echo in one range gate at a time. A variation of this device was selected by Ken Bowles for the incoherent scatter radar that the then National Bureau of Standards was building in Peru, and the Cornell project people made the decision to do something similar. The Rayspan as specified by Cornell had one hundred 100 Hz bandwidth filters which could be placed adjacent to each other to cover a 10 kHz band or distributed over 20, 50 or 100 kHz bands but only partially filling them. A contract was let to Raytheon in November 1961 for $15,000 to provide a Rayspan to Cornell's specifications. Delivery was expected in March 1962. Unfortunately, despite the extensive lead time, it was mid-1964 before the device worked correctly allowing it to be used for observations of the ionosphere. Since it could only form a spectrum for one range gate (i.e. one height in the ionosphere) at a time, the Rayspan was a very inefficient device. One solution was to record the input signal to the Rayspan on an analog tape recorder and play it back multiple times through the Rayspan choosing a different range gate each time. The Rayspan was an inventive solution to generating spectra, but the digital revolution caught up with it relatively quickly. By the end of 1965 spectra for multiple range gates were being obtained entirely digitally and, despite all the work that went into its production, the Rayspan was obsolete.

The third data acquisition system was an analog-to-digital converter made by the Adcom Corporation that converted the radar echo signal into 8-bit digital numbers at a rate no faster than once every 100 μs. The digital data stream would be fed to a computer for future processing. Once completed and in use, this system rapidly replaced the 100 range gate system for obtaining ionospheric density profiles and it was also critical to the success of the planetary radar program.

The first data acquisition and analysis computer: It was clear very early that at least a small, modern computer would be needed for initial data analysis and also for tasks such as photometric surveys of the reflector surface of the telescope. Given the Adcom data acquisition system, any computer would also need to be capable of accepting a digital data stream from the Adcom analog-to-digital converter at rates up to 10,000 samples per second. The initial idea was to purchase the computer, and this was included in the initial operating budget proposal to AFOSR. However, leasing a machine made more sense as it would more easily allow upgrades as needed. A Control Data Corporation (CDC) 160-A computer was selected and a lease agreement received from CDC in early April 1962. The lease included:

Model 160-A computer with	Cost per month
1. Paper tape input/output	
2. 8192 12-bit words of core memory	
3. Normal and automatic buffered input/output channels	
4. Program interrupt	2250.00
Model 161 input/output	
Typewriter	262.00
Model 163-2 Magnetic tape system with:	
1. Two 30k characters per second tape handlers	
2. 12-bit Assembly/disassembly logic	1482.00
Model 168 Multiply/divide module	390.00
Total	4384.00
Maintenance contract per month (remote location)	1500.00
Total cost per month including a full time maintenance person	5884.00

It had been anticipated that the CDC 160-A would be installed at the observatory in October 1962. However, for unknown reasons, it did not arrive until late May 1963 (Fig. 6.12). Over the next year it was used extensively especially for analysis of the radar echo data from the Moon and planets and for analyzing the photogrammetric data used to determine the accuracy of the telescope's reflector surface. In general, the ionospheric observations made use of the 100 range gate and Rayspan (after June 1964) systems and the radio astronomers primarily used multi-channel

Fig. 6.12 The CDC 160-A computer in, probably, 1964. The tape drives on the left are the 163-2 units in the table of costs. These were relatively slow at 30 characters per second and were soon replaced with the much faster vacuum column (the two vertical slots) units on the right. The 163-2 units were not there when the author arrived at the observatory in early February 1965. Credit: Cornell University/ARPA

chart recorders. This was just as well as the 160-A could do only one task at a time, either data taking or processing.

Computers, especially the early ones, needed programmers to convert the scientists' desires in terms of telescope pointing, data acquisition, and data processing into reality. Arecibo was fortunate early on to have recruited two excellent programmers, Ray Brose and Hiram Cabassa. Cabassa was a graduate of the University of Puerto Rico's engineering campus in Mayaguez on the Island's west coast. He was a real whiz at developing data acquisition programs especially for the planetary radar group.

Wrapping up: Costs continued to increase in 1963. In March the Cornell construction contract with AFCRL was increased to $7,087,000 and in August was increased again to $7,369,000. In its September report, Cornell estimated that its final construction costs would be $7,102,000 plus $33,917 for the theoretical studies. It is possible that the difference was a provision against claims by the various contractors for compensation for their increased costs due to delays that they claimed were outside of their control. Negotiations took place after the completion of the construction resulting in payments to all the main contractors and a small amount to Cornell ($53,000). The final cost of Cornell's work was $8,410,150. The final cost of the project, the work that AFCRL was responsible for that included its costs and that of Cornell and the Corps of Engineers, was $9,426,000.

6.3 Celebrating the Completion

A dedication ceremony was planned for 1 November 1963, organized by the Air Force with assistance from Cornell. Invitations to attend were issued by the new ARPA Director, Robert Sproull, a former faculty member and administrator at Cornell University. The original plan was to invite representatives of the contractors but this was dropped with the exception of the telescope's main designer, Thomas Kavanagh. Most of the mainland invitees were from the Congress, the administration, air force units, AFOAR, AFCRL and AFOSR, the NSF, Harvard University and, of course, Cornell University. Curiously, Bill McGuire, Ward Low, who had moved from the IDA to Sylvania Electric Products in Massachusetts, and Phil Blacksmith did not attend even though it appears that they were invited. The mainland invitees met at Andrews Air Force base outside of Washington on 31 October and flown to Ramey Air Force Base. From there they flew to San Juan for a reception at La Concha hotel sponsored by the Department of Defense and Cornell University. Island government representatives, prominent locals, and observatory senior staff would have also attended the reception. After the reception the mainland visitors were flown back to Ramey for the night and next morning bused to the observatory for the dedication ceremony starting at 10:30 am. After a lunch and technical discussions with observatory personnel, the visitors returned to Ramey where they boarded a flight to Andrews Air Force Base for a late evening arrival; clearly an exhausting trip.

The governor of Puerto Rico, Luis Munoz Marin, was not able to attend the early part of the celebration and speeches at the Observatory (Fig. 6.13) and the Resident Commissioner for Puerto Rico in the U.S. House of Representatives, Dr. Antonio

Fig. 6.13 From left to right; Dr. Thomas Gold (Cornell University), Major General Ostrander (AFOAR), Dr. Robert Sproull (ARPA), Congressman George Miller, Dr. Antonio Fernos Isern (Resident Commissioner for Puerto Rico in the U.S. Congress), Eugene Fubini (Assistant Secretary of Defense), Dr. Dale Corson (Provost, Cornell University), Brigadier General B. G. Holmes (AFCRL), Dr. William E. Gordon (Director, AIO), Lt. Richard Bonicek (U.S. Navy Chaplin). Credit: Cornell University

Fig. 6.14 Major General Ostrander, Governor Luis Munoz Marin and William Gordon in front of the control room. Credit: Cornell University

Fernos Isern, gave the welcoming speech in his place. As mentioned in Chap. 1, Bill Gordon's speech did not discuss his role in building the telescope and it was Dale Corson who made Gordon's pivotal role clear. The Governor did arrive and spent time with the guests (Fig. 6.14).

Reference

Kavanaugh, Thomas C., and David H.H. Tung. 1965. Arecibo Radar-Radio Telescope – Design and Construction. *Journal of the Construction Division: Proceedings of the American Society of Civil Engineers* 91 (CO1).

Chapter 7
First Observations and Some Disappointment

The first ionospheric measurements gave a very impressive measurement of the electron density as a function of height. However, test observations of radio sources indicated that the sensitivity using the 430 MHz line feed was only about 40% of the expected value and that the main beam had very high sidelobes. The good news was that the polarization properties of the main beam were excellent so the only effect on incoherent scatter measurements of the ionosphere and on radar studies of the Moon and planets was the decreased sensitivity. During 1964 there were very productive measurements of ionospheric properties, Venus was detected by the radar in February, Mercury in April, Mars in November and the Moon whenever it was within the declination coverage of the telescope. Occultation and scintillation programs were started to measure the small scale structure of radio astronomical sources. One serious issue was the initial very small scientific staff in all three research areas. During 1964 and 1965 there was a significant increase in the scientific staff helped by an influx of Australian trained young scientists, who came under the newly formed Cornell Sydney University Astronomy Center. Growing tensions between William Gordon and Thomas Gold were an issue and may have contributed to the reasons for Gordon's departure from the observatory in September 1965.

7.1 Measuring the Telescope's Performance

Ionospheric observations began in mid-October 1963, just an hour here or there depending on work on the telescope. Organizing the 1 November 1963, dedication of the telescope was a major distraction from both checking out the telescope and initiating the research program. After the dedication, work started on testing the performance of the telescope by observing strong cosmic radio sources and using the 430 MHz transmitter to do measurements of the distribution of the electron density in the ionosphere with altitude. All the work in developing the receiver and data

D. Campbell, *The Arecibo Observatory: A History of Innovation and Discovery*, Historical & Cultural Astronomy, https://doi.org/10.1007/978-3-031-75225-4_7

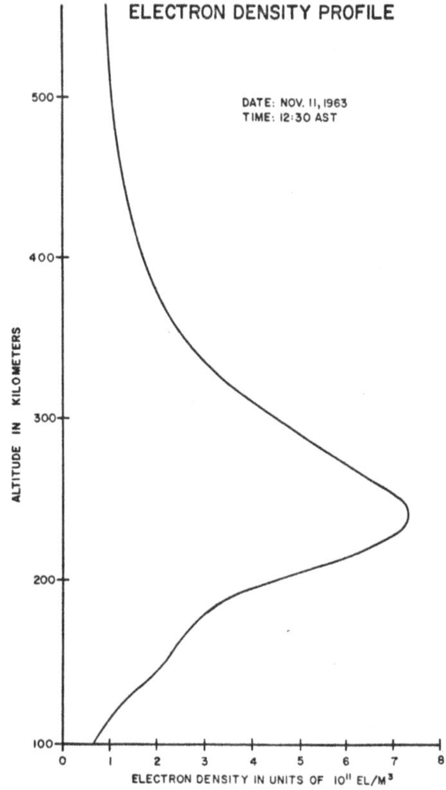

Fig. 7.1 The electron density as a function of height measured on 11 November 1963. Plot is Fig. 13 from the Footnote 1 reference

acquisition systems and their testing with the transmitter using the horn antenna paid off. One of the early ionospheric electron density versus altitude profiles was obtained on November 11 (Fig. 7.1).[1] It is a noise free profile up to an altitude of 550 km, clearly showing that Bill Gordon's dream of studying the properties of the ionosphere via incoherent scatter radar had become a reality.

The most urgent task was to check out the telescope's performance. Was the sensitivity when pointing towards the zenith and its behavior as a function of the zenith angle out to the maximum 20° as expected, and did the beamwidth and sidelobe levels agree with the predictions based on the specifications for the telescope and, especially, the line feed (Fig. 7.2)?

The specifications in the request for bids for the design and construction of the 430 MHz line feed included a requirement that the efficiency with which the feed illuminated the reflector surface should be no less than 53%. In other words, at least 53% of the power at 430 MHz falling on the reflector surface from a radio source is collected and available to the telescope's receiving system. For radio telescopes it is

[1] Construction of the Arecibo Ionospheric Observatory, Arecibo Puerto Rico – Final, CRSR Report RS55, 30 November 1963. Cornell University library.

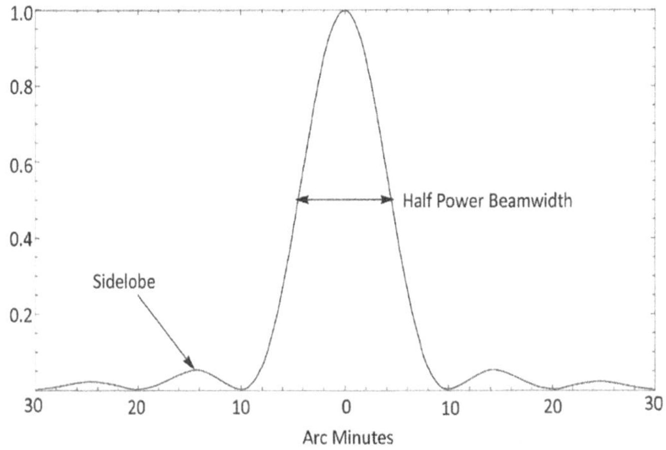

Fig. 7.2 The approximate expected angular response function (beam shape) of the 1000 ft (305 m) diameter antenna at a frequency of 430 MHz. The angle between the 50% sensitivity points, the beamwidth, for this antenna should be about 10 arc minutes or 1/6 of a degree. The sidelobes mean that the telescope is sensitive to radio waves arriving from directions other than the one the antenna is pointing to but with very much lower sensitivity. Keeping the level of these sidelobes low is desirable but is a balance since lowering them also lowers the sensitivity of the telescope in the direction it is pointing and increases the beamwidth. Credit: Lynn Baker

not possible to effectively use the entire area of the reflector surface. The inner part is efficiently used but the efficiency tapers off towards the edge so that the overall effective area of the telescope, the aperture efficiency, is typically about 55% of its physical area. This percentage can be increased by the use of specially designed feeds and one advantage of the line feeds required for spherical reflectors is that, in theory, they can be designed to very efficiently illuminate the reflector with aperture efficiencies of over 70%.

Early observations of radio sources gave mixed results. A scan of the radio source 3C33 mapping out the angular response function, the beam shape, (Fig. 7.3) showed that it was far from the ideal pattern (see Fig. 7.2) that had been specified. The beamwidth is close to the expected 10 arc minutes but the first sidelobes are clearly much larger and less "clean" than those of Fig. 7.2. More disturbing is that the sensitivity, denoted by the antenna temperature, corresponds to an aperture efficiency of about 22%, much less than the minimum of 53% specified in the TRG contract.

The only possible explanations for the poor performance had to be related to: (1) The line feed design and/or its construction; (2) The reflector surface setting, was it within the specified 0.10 ft (30 mm) mean tolerance, was the radius of curvature correct and were the suspended structure and line feed correctly positioned with respect to the spherical reflector. With the reflector surface tolerance and the overall geometry of the telescope possible explanations for the telescope's poor performance, all eyes turned to Robert (Bob) Forrest, who had been hired by Bill Gordon

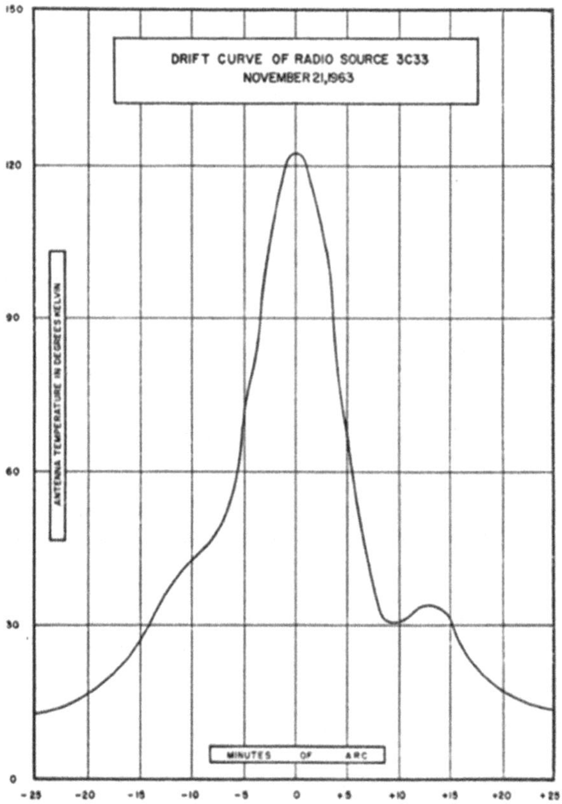

Fig. 7.3 The response of the telescope at 430 MHz on 21 November 1963, as the radio source, 3C33 (3C33 is the 33rd source in the 1962 revised 3C Cambridge University Catalog of radio sources derived from a survey of the northern sky at a frequency of 178 MHz) is scanned by the telescope mapping out the angular response function. It should have looked very similar to the response function of Fig. 7.2. Plot is Fig. 10 from the Footnote 1 reference

to check on the work of the contractors to make sure that these would not be a problem.

Bob, his wife Janet and their son John arrived in Arecibo in June 1963 staying, initially, at the Mir Hotel, then the VSQ at the Observatory before finally settling into a house in Radioville, two doors from the LaLonde's. Radioville was an urbanization named after the local radio station tower not the observatory despite many of the observatory's expatriate staff living there. Bob had been a graduate student in the Geodetic Sciences Department at Ohio State University applying digital techniques to photogrammetry, the precise measurement of large-scale structures in three dimensions. His boss at Ohio State had been contacted, likely by Bill Gordon on the advice of Don Belcher or Art McNair of Cornell's Civil Engineering Department, looking for someone to survey the telescope independently of the contractors. Bob, a finishing graduate student, was recommended due to his experience with newly developed digital techniques. Bob and Janet visited Arecibo, liked what they saw of the Island and the work and accepted the job. They never regretted it. As Janet

Fig. 7.4 Bob Forrest at the console of the CDC 160A computer on 8 August 1963, preparing to analyze the results of the photogrammetry measurements. Credit: Cornell University

said in a video interview, "They were the two most wonderful years".[2] When he arrived in Arecibo, Bob had still not finished his PhD thesis on ice movement in Antarctic and this occupied his spare time until the summer of 1964.

Bob's task was to survey the shape of the reflector both on large and small scales. This involved setting up a coordinate system centered on a surveying monument mounted on a concrete pillar in the center of the reflector. The coordinate system covered the entire observatory as well as the telescope's reflector. In setting up the system, standard surveying techniques were employed using Wild T2 theodolites. Analysis of the measurements was done using the CDC 160-A computer (Fig. 7.4). For the detailed measurements of the reflector, photographic techniques were used. 600 4-inch black and white aluminum targets were distributed over the reflector, some above the 10 lower main cables and some in between these cables where the light cables supporting the mesh tended to sag over the 100 ft spacing between the main cables. Targets were also placed at thirty-three ground control points and their positions carefully surveyed using theodolites. On 28 October 1963, an 80 lb KC1-B aerial photogrammetric camera was mounted on a platform below Carriage House 2, and the carriage house moved as far as was then possible out on the feed arm, 130 ft versus the 150 ft length of the arm from the center. Pairs of images were taken for 6 different azimuth positions 60° apart. Each image covered 1/3 of the reflector surface and images to the left and right were taken for each azimuth position. The

[2]Video interviews with Bob Forrest about his work and with Bob and Janet about living in Arecibo are available on e-Commons at the Cornell University library https://ecommons.cornell.edu/handle/1813/33201

9-inch square photos were developed at the Air Force Ramey base and glass plates prepared. The X and Y positions of all the targets were then read off the plates using a precision comparator at Syracuse University in upstate New York. The final analysis,[3] done on the CDC 160-A, indicated that some areas of the reflector, primarily to the south and west, had significant deviations from the desired spherical shape. The overall mean error from the required radius of 870 ft was ±30 mm (0.12 ft). This error is only 4% of the 70 cm wavelength of the 430 MHz radar and would result in only a very small reduction in the sensitivity of the telescope.

This first set of measurements was made before the discovery of the poor performance of the telescope. As Bob recounted, the discovery of the poor sensitivity created quite a stir with the "big shots" descending on the observatory and doubt being cast on his measurements. As he said, a reasonable concern since he was a newly arrived outsider and probably thought of as inexperienced. However, he convinced them that his measurements were fine, the reflector surface was within the specifications and the overall geometry of the telescope was correct. Adjustments were made to the tensions of the light upper cables and, in places, the mesh was crimped to shorten it and bring it closer to the design shape. Bob did a second set of measurements and additional surface adjustments were made to bring the surface tolerance down from its value after the first measurements and adjustments of 21 mm to 18 mm, although no additional measurements were made immediately to confirm this.

The Forrest's left Arecibo in May 1965 after selling their car, a Peugeot 403, to the author. Living in Arecibo and the involvement with a large telescope construction project had been one of the great experiences of their lives. Bob did return on several occasions to consult during future settings of the telescope's reflector.

With the telescope's reflector surface tolerance and the telescope's overall geometry not to blame for the sensitivity issue, this left the line feed as the only possibility. Multiple observations confirmed that the aperture efficiency was only about 22%. However, this corresponded to a telescope with a diameter of about 660 ft and an aperture efficiency of 53%, still by far the largest radio antenna and most powerful radar for both ionospheric incoherent scatter measurements and studies of the Moon and nearby planets. Some radio astronomical observations would be hurt by the large sidelobe levels, but at 10 arc minutes the beamwidth was not affected by the poor aperture efficiency. The nature of the problem with the feed also meant that the polarization properties of the main beam were good. That is, the transmitted signal was almost purely circularly polarized and two independent circular polarizations could be received. All of this meant that the planned scientific program could go ahead while the problems with the line feed were studied and, hopefully, a solution found. As will be discussed in Chap. 8, a very elegant solution was finally found but

[3] A more detailed description of the photogrammetric measurements and their analysis can be found in Section F, Reflector Survey by R. Forrest in CRSR Research Report RS 56, 31 December 1965 and in CRSR Report 204, Calibration of the 1000-ft Arecibo Radio Telescope, Robert B. Forrest, September 1965. Cornell University library.

it took almost four years and then, due to bureaucratic and funding issues, another four years before it was fully implemented.

7.2 Recruiting Scientific Staff

Not surprisingly, 1964 was a momentous year for the new observatory with many notable scientific achievements in all three fields of research, ionospheric measurements, planetary radar studies and radio astronomy. This drove a need to recruit new scientific staff, the beginning of an influx of an eclectic mix of new arrivals from the United States and other countries. While the research opportunities of the observatory were a major incentive to join its staff, coming to Arecibo to live, especially if you had a family, was a somewhat daunting prospect that required a sense of adventure. There were many positives about living in Arecibo, the wonderful weather, great beaches, easy access to other places in the Caribbean and very friendly people, but this had to be balanced with the difficulty in finding employment for spouses, the lack of schooling in English unless you were prepared to send the children on the long trip to Ramey Air Force base each day, routine loss of electricity and water, the difficulty in getting phone service and the lack of local cultural events unless you were prepared to make the long journey to San Juan. Nonetheless, people came and the shared experience of working at the Observatory and living in Arecibo was a bond that lasted over many people's lifetimes.

Marshall Cohen, Professor of Astronomy at Cornell, spent a year at the AIO starting in mid-1963 as head of the radio astronomy group (Fig. 7.5). He arrived about the same time as Gordon Pettengill greatly augmenting the then miniscule scientific staff. In January 1964 Rolf Dyce joined the planetary radar group under Gordon Pettengill beginning a long association with the observatory during which he served as Associate Director for many years and Interim Director on several occasions. Rolf, his wife Sherry and children were fixtures of the observatory community until 1979.[4] The British radio astronomer Cyril Hazard joined the staff in July 1964 coming from a faculty position at the University of Sydney, Australia (Fig. 7.6). Cyril was famous for using the lunar occultation technique with the Parkes 64 m telescope in Australia to pinpoint the position and delineate the structure of the radio source 3C273. This position was used by Maarten Schmidt and Bev Oke of the California Institute of Technology (Caltech) to identify the source of the radio emission as an object that looked like a star, but subsequent spectroscopic measurements showed that it was over two billion light years away making it the first observation of a new class of astronomical objects dubbed quasi-stellar objects or quasars. This was the beginning of studies of very distant objects in the Universe.

[4] See the video interviews with Marshall Cohen and Rolf Dyce on the Cornell Library e-Commons site https://ecommons.cornell.edu/handle/1813/33201

Fig. 7.5 Marshall Cohen (left) and William Comer in the observatory's receiver room on 8 August 1963, a few days before the 430 MHz line feed was installed. The analog tape recorder in the right background is likely the recorder used for radio astronomy observations. Credit: Cornell University

Fig. 7.6 Cyril Hazard, likely in the early 1960s, giving a seminar on the theory of lunar occultations. The sketch at upper left on the board shows the resulting diffraction pattern as a radio source emerges from behind the limb of the Moon. Since the position of the limb of the Moon is known accurately, the precise position of the radio source and some information about its structure can be determined. Credit: Thomas Gold

Kjell-Olaf and Kerstin Yngvesson arrived from Sweden in 1964 with Kjell joining the ionospheric group. Ilya Kazès arrived from France with Ilya working with the radio astronomy group, and Mukul Kundu, associate professor of astronomy at Cornell, came for a year to further his studies of solar phenomena. Yervant Terzian, who hailed from Egypt and was a newly minted PhD from the University of Indiana, arrived in Arecibo in early January 1965, the beginning of a long association with Arecibo and the Cornell astronomy department. Don Campbell (the author) arrived from Australia in late January, also the beginning of a long association with the observatory and Cornell. Later in 1965, Dan Harris joined the staff after sailing across the Atlantic with fellow astronomers Venkataraman Radhakrishnan, a well known Indian astronomer, and David Morris.

The scientific staff was augmented by, primarily, Cornell graduate students pursuing their thesis research. The initial five students were Herb Carlson (ionospheric studies), Harry Hardebeck (radio astronomy), George Thome (ionosphere), Thomas (Tommy) Thompson (planetary radar) and Walt Zandi (Electrical Engineering). After finishing their PhD degrees, Carlson, Thompson and Zandi joined the Observatory staff. There were at least eight other graduate students at Cornell who were involved with Arecibo research including Francis Perkins, who would play an important role in the ionospheric program and who went on to make substantial contributions to fusion research at Princeton University. Slowly, researchers from institutions other than Cornell started to make use of the telescope. Tom Carr at the University of Florida sent the student, C. F. Tiberi to study radio emission from Jupiter and, in late 1964, he left being replaced by another student, Sam Gulkis, who joined the Observatory's staff after he completed his PhD. John Nesbit of the University of Pennsylvania became involved in the ionospheric program and Jules Aarons and D. A. Guidice of AFCRL started a program of low frequency observations of radio sources.

Solving the office problem: This increase in the scientific staff through 1964 and 1965, while not enough, seriously strained the available office space. The operations building housed the observatory management, the scientific staff, the electronics staff and the small library in the offices and laboratory behind the control, computer and transmitter rooms. There were multiple scientific staff to a room. To alleviate this situation, a three story office building to be built on the slope between the operations building and the VSQ was designed with construction starting in late 1965. Its completion, which included space for a large library cum meeting room, solved many problems. A fourth floor was added in the 1970s.

7.3 Early Research Results

Ionosphere: Cornell's plan for staffing the research program was to have a small number of scientists and a large number of graduate students, primarily from Cornell. However, with three research areas, ionospheric measurements, planetary radar studies and radio astronomy, a significant research staff was required. In late

1963 and early 1964 the ionospheric program was staffed by Bill Gordon plus Cornell graduate students George Thome and Herb Carlson. The ever-versatile Merle LaLonde also turned his hand to this area of research, and supporting theoretical studies were done by Francis Perkins, Ian Axford, Henry Booker, Ed Salpeter and others in Ithaca. Most of the early work involved studies of electron density fluctuations and their behavior with time in the height range 150 km to 600 km in the ionosphere using the range gate system (Chap. 6). The strongest fluctuations, 10% to 20% of the electron density, were observed in the 150 km to 350 km region. These traveling wave disturbances, as they were known, had been observed at altitudes below the peak electron density in the F region at about 300 km but not previously above it. The Arecibo measurements were done by George Thome and showed that the disturbances travelled from the north or northeast at a speed of about 100 m/s (Thome 1964).

Frustratingly, the Rayspan's continuing problems, which were not resolved until mid-1964, prevented obtaining spectra of the reflected signal from different heights in the ionosphere to determine the ion composition and electron and ion temperatures. Attempts were made using a single narrow passband filter and sweeping the reflected signal across this filter, but this was not very satisfactory. Once the Rayspan was working, Herb Carlson began a series of measurements of the spectra of the incoherent scatter radar echo as a function of height and time of day to determine the ionic composition and the electron and ion temperatures given by their mean thermal speeds. For a given radar frequency, the frequency width of the spectrum is determined by the inverse square root of the mass of the dominant ion species assuming that there is just one, and the mean thermal speed of the ions (Fig. 7.7). If there is more than one species, then the spectra are a combination of the expected spectrum for each species weighted by their relative abundances. The ion temperature can be determined by the width of the spectrum and the ratio of the electron to ion temperatures by the depth of the central dip in the spectrum. At altitudes of a few hundred kilometers, oxygen ions are the dominant ion species with molecular oxygen, o_2, being disassociated into atomic oxygen, o, by solar extreme UV photons and then ionized by these same photons to form o^+. The residual energy of the free electrons propels them to high speeds corresponding to much higher temperatures than the oxygen ions.

Sample spectra from about 5:00 pm on 1 December 1964 can be seen in Fig. 7.8 showing the transition from ionized atomic oxygen at F-region heights to ionized atomic hydrogen at heights above 700 km with a previously unknown and significant layer of ionized helium between 400 km and 600 km. By the end of 1964, it was clear from these and other measurements that the Arecibo telescope and radar system was living up to Gordon's predictions and producing a great deal of new information about the Earth's ionosphere. However, one thing was missing, any work related to Project Defender, the detection of wakes in the electron density distribution of the ionosphere related to the detection of intercontinental ballistic missiles, the nominal reason for ARPA's funding the construction of the telescope and radar system.

Project Defender: There is not a single reference in any of the early reports to measurements directly related to Project Defender. One possible explanation is that

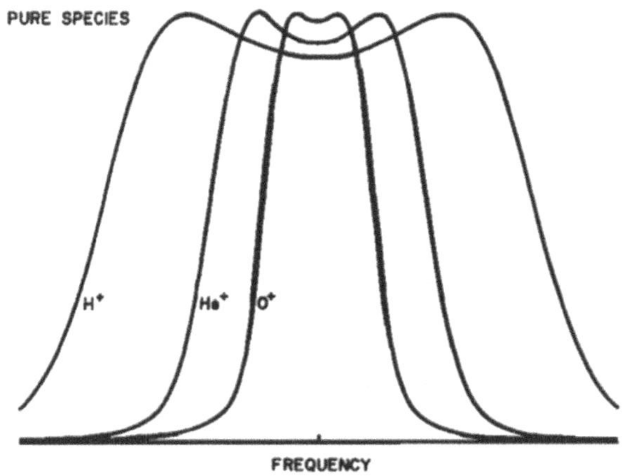

Fig. 7.7 Model incoherent scatter radar spectra for different ion species. The width of the spectrum is proportional to the inverse square root of the mass of the ion. For the frequency of 430 MHz for the Arecibo radar and the expected temperature (i.e the mean speed of the particles), the width of the spectrum if all of the ions are singularly ionized atomic oxygen, O^+ (i.e. one electron has been stripped off by solar UV radiation) is about 10 kHz. Since a helium ion, He^+, has a mass only ¼ that of an oxygen ion, its spectra will be twice as wide, about 20 kHz. Similarly, an atomic hydrogen ion, H^+, has ¼ times the mass of a helium ion so its spectrum will be twice as wide and four times that of the o^+ spectrum. In the absence of bulk motions in the ionosphere due to winds, the spectra are centered on 430 MHz. If there is bulk motion towards or away from the radar the entire spectrum will be shifted slightly in frequency, allowing wind speeds to be measured. The detailed shape of the spectrum, especially the dip at the center, gives information about the ratio of the ion to electron temperatures (i.e. the ratio of their mean thermal speeds). Figure 1 from Carlson and Gordon (1966)

the issue had already been settled utilizing probes on orbiting spacecraft such as Arial I and Explorer 31 plus measurements from the Canadian Aloulette II satellite, which was equipped with a downward looking ionosond. A review of these measurements can be found in Willmore (1970). It is also possible that it was realized that the very coarse height resolution, 75 km, of the early measurements would mask a relatively small change in the electron density distributed over a very small region. However, it is strange that there is no mention anywhere in the early reports of observations directly related to Project Defender. Herb Carlson, who was an important member of the ionospheric research group starting as a graduate student in 1964, had no recollection of any such measurements.

Planetary Radar: For planetary radar studies, the staff was Gordon Pettengill (Fig. 7.9); the newly-arrived in January 1964 Rolf Dyce; a lecturer at the University of Puerto Rico, Andy Sanchez; and Cornell graduate student Thomas (Tommy) Thompson. This group had a very exciting time. Observations of the Moon showed strong radar reflections from a number of impact craters including the youngest large crater on the Moon, Tycho. Venus came within the declination coverage of the telescope in February and was immediately detected. Subsequent observations at 4-day intervals determined the distance to the planet from the time it took for the radar pulse to travel to the closest point on the surface of Venus and return. These

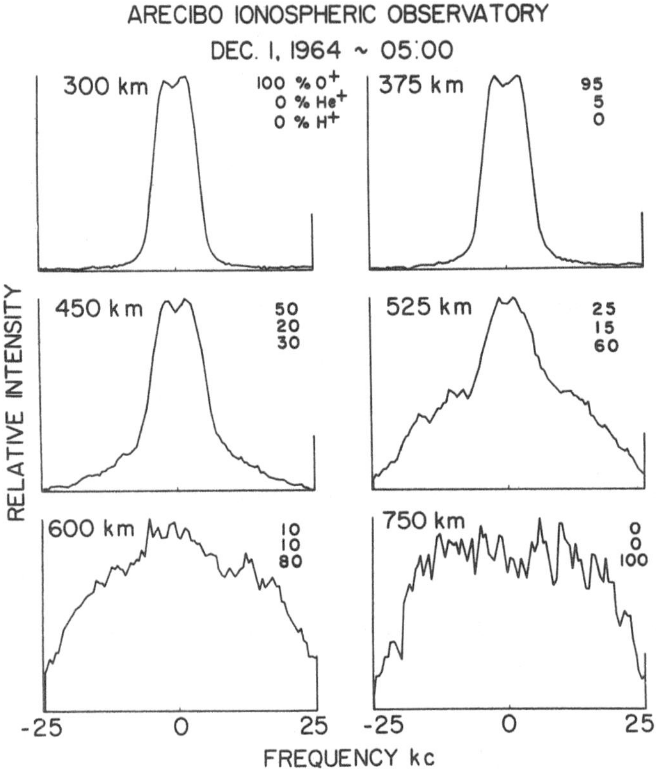

Fig. 7.8 Examples of spectra from six different altitudes at about 5:00 am local time showing the transition in ion species as a function of altitude. Atomic oxygen ions (o^+) dominate in the F-region, helium ions (He^+) are present at about 500 km and atomic hydrogen (H^+) is the sole species above about 700 km at this time of day. Figure 4 from Carlson and Gordon (1966)

measurements were the beginning of a significant refinement of our knowledge of the scale size of the solar system represented by the Astronomical Unit, the mean distance between the Earth and the Sun, crucial for navigating spacecraft to the planets.

Mercury was detected for the first time on April 7 with subsequent observations showing significant discrepancies from its predicted orbit. Mars was a longer wait as for much of the year it was below the plane of the Earth's equator and so beyond the 20° zenith angle coverage of the telescope. On 19 November the planet was detected at a distance of 1.5 times the distance to the Sun, close to Mars maximum distance from Earth. Attempts were made to detect the Sun starting in July 1964 by Vahe Petrosian, a Cornell graduate student working with Edwin Salpeter, but these were not successful.

Venus was both the most interesting planet—what lay below its dense cloud cover?—and the easiest one to observe with radar from Earth. It is the planet closest to Earth, about 30% of the distance to the Sun at each close approach to Earth. The

Fig. 7.9 Gordon Pettengill in the control room, likely talking to someone on the platform. The absence of any feeds on Carriage House 2 indicates that this photo was from late 1963 or early in 1964, perhaps during the first radar observations of Venus. Credit: Cornell University

strength of the radar echo decreases as the fourth power of the distance so the closer the planet is to Earth the better the imagery of its surface. These close approaches occur every nineteen months and it was good luck that one of them was due in June 1964. Venus is also the easiest planet to image with Earth based radar partly due to its closeness to Earth but also because it is rotating backwards compared with all the other planets except Uranus. This retrograde rotation was first detected by radar measurements done in 1962 during the planets previous close approach to Earth with NASA's Goldstone Deep Space tracking facility in the Mojave Desert east of Los Angeles. They measured a very slow retrograde rotation rate with a period of 250 ± 40 days. Measurements with the new Arecibo radar from May through September 1964 significantly refined this, giving a rotation period of 247 ± 5 days (Fig. 7.10) close to the currently accepted value of 243.02 days.

Radio Astronomy: Despite its very high sensitivity, the use of the 430 MHz line feed for radio astronomical observations was hampered by its high side lobe levels and its single frequency. This prompted the construction of multiple feeds at other frequencies ranging from 20 MHz to 1420 MHz, which were mounted on Carriage House 2. These were point feeds so their use provided much lower sensitivity than a line feed. Cyril Hazard wanted to do lunar occultation studies especially of 3C

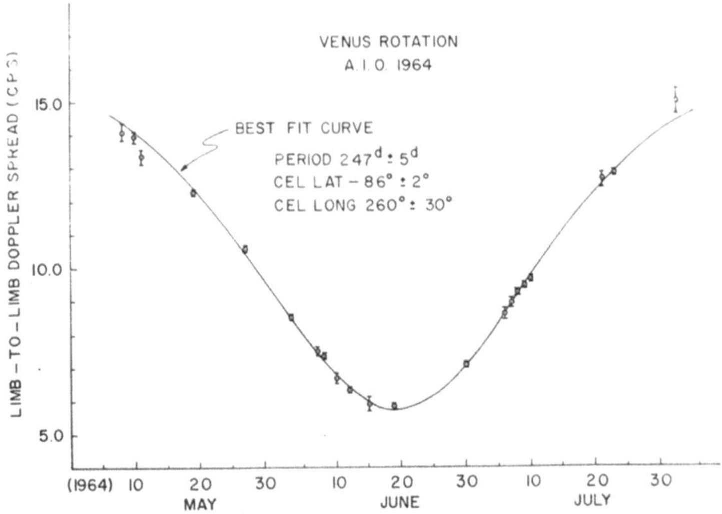

Fig. 7.10 The reflected radar echo from the surface of Venus is spread in frequency due to the Doppler shift from different areas moving towards or away from Earth caused by the apparent rotation rate of the planet. The total Doppler broadening in frequency, the limb-to-limb Doppler spread, is determined by the velocities of locations furthest from the rotation axis. The apparent rotation rate is made up of the sum of the intrinsic rotation rate and an apparent one due to the relative orbital motions of Earth and Venus. For a prograde (i.e. in the same direction as the other planets) rotation, these two effects add and the limb-to-limb Doppler spread would be a maximum at the date of close approach to Earth, June in 1964. For retrograde rotation they subtract so there would have been a minimum in June, which is clearly the case. Minimizing the Doppler spread also maximizes the sensitivity of the radar imaging so Venus was ideal for the imaging of its surface using a radar system on Earth. Fig. 2 from Dyce et al. (1967)

273 on 7 September 1964, and 7 June 1965. He nested simple dipole feeds constructed at 41 MHz, 195 MHz and 430 MHz so that they could all be used simultaneously. The effective diameters of the illuminated areas on the reflector were 1000 ft, 500 ft and 350 ft at the three frequencies, respectively, larger than any other existing single dish telescope. The AFCRL team of Aarons and Guidice constructed a feed that covered the range 20 to 40 MHz for observations of radio emission from Jupiter, and a log periodic feed tunable over the frequency range 100 MHz to 1500 MHz was constructed. Ilya Kazès used a horn feed at 1420 MHz to test the performance of the telescope at the frequency at which atomic hydrogen, the major constituent of the interstellar medium of our galaxy, emits. The sensitivity of this system was about 40 times less than that of the 430 MHz line feed due to a combination of the inability of a point feed, a horn, at this frequency to illuminate a spherical reflector with the correct phase distribution and the large amplitude of the reflector surface irregularities relative to the 21 cm wavelength.

The primary objectives of the early radio astronomy program were the study of radio emission from the planets, primarily Jupiter and Venus; a survey at 430 MHz of radio emission from a large number of galaxies, accurate positions and structures

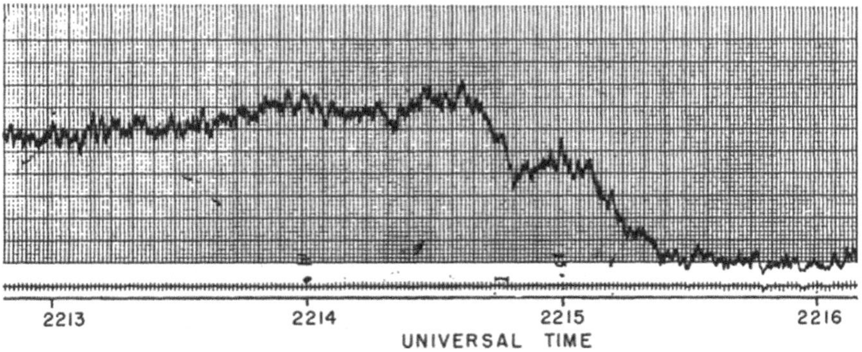

Fig. 7.11 The chart recording of the diffraction pattern as the limb of the Moon occulted the radio source 1101+11 on 15 June 1964. The two broad peaks indicated the presence of two components to the source separated on the plane of the sky by approximately 9 arcseconds. The effective angular resolution of the scan was about 2 arcseconds. Figure from Cohen and Kundu (1966)

of radio sources via lunar occultation measurements and the detection of radio sources with very small angular sizes via the study of scintillations caused by the interplanetary medium close to the Sun. These scintillations studies also gave information about the structure of the interplanetary medium in the vicinity of the Sun.

The success of Cyril Hazard's first lunar occultation for 3C273 with the Parkes telescope and the discovery that the source was over two billion light years from Earth, the first identified quasar, was a tremendous impetus to use the new highly sensitive Arecibo telescope to do as many lunar occultation observations of radio sources as possible. The positional accuracy that this technique could provide, 1 to 2 arc seconds or about 1/3600th of a degree, could not be achieved by any other radio astronomical technique at that time and was good enough to allow optical counterparts such as a distant galaxy or quasar to be identified. In early 1964, Marshall Cohen initiated a survey of the parts of the predicted path of the Moon on the sky that would be observable with the Arecibo telescope searching for currently unknown weak radio sources that would be occulted in the near future. Initially very few of these were found but one, initially labeled S1 and later designated 1101+11 (right ascension and declination, respectively), was predicted to be occulted by the Moon on 15 June 1964. Figure 7.11 shows the radio flux from 1101+11 as it passes behind the Moon. Marshall Cohen's return to Cornell in August 1964 coincided with Cyril Hazard's arrival in Arecibo continuing the emphasis on lunar occultation observations. The search for radio sources in the track of the Moon continued now with considerable success and 1965 saw about 80 occultation observations including one of 3C273 on 7 June that provided additional information on the structure of this quasar. Lunar occultations would remain an important technique for high angular resolution determinations of the positions and structure of radio sources until it was superseded by the development of Very Long Baseline Interferometry (VLBI) in the late 1960s and early 1970s. Marshall Cohen and David Jauncey, who joined the

Ithaca CRSR staff in 1965, would play an important role in the development of this new technique.

7.4 Cornell Sydney University Astronomy Center

Seven resident scientific staff plus five graduate students spread over the three fields of research were not enough to fully utilize the new telescope and radar system. Part of the problem was the lack of experienced researchers in the three research areas, two of which, radar astronomy and incoherent scatter studies of the ionosphere, were very new disciplines. As mentioned earlier, this was also partly true of radio astronomy, which was pursued in Australia, Great Britain and Holland immediately after World War II but not in the United States until sometime later except for some early work at Cornell University and the Naval Research Laboratories. In the late 1950s there were few radio astronomers in the United States with the stature of Joseph Pawsey, John Bolton, W. N. (Chris) Christiansen, Paul Wild, Frank Kerr and Bernard Mills in Australia, Bernard Lovell, Martin Ryle, Antony Hewish, and Francis Graham-Smith in Great Britain and Jan Oort in Holland. With some exceptions, most U.S radio astronomy was conducted in electrical engineering departments such as at Cornell and Ohio State University, from which Marshall Cohen received his PhD in 1952. One exception was Harvard University where Bart Bok, a well-known optical astronomer, realized the importance of radio astronomy and, in the 1950s, was instrumental in the construction of a 60 ft radio telescope at Harvard's Agassiz Station west of Boston and in fostering the training of graduate students in radio astronomy. Among others, this program graduated David Heeschen, who went on to be director of NRAO for many years, and Frank Drake, who, after working at NRAO and the Jet Propulsion Laboratory (JPL), joined the astronomy faculty at Cornell in 1964. Drake was already well known for Project Osma, the first serious attempt to detect signals from extra-terrestrial civilizations (SETI). (See Kellermann et al. 2020 for a more detailed discussion of early radio astronomy in the US and around the world). Shortly after arriving at Cornell, he was made associate director of CRSR, the beginning of a long association with the Arecibo Observatory.

The difficulty in recruiting staff, especially radio astronomers, for Arecibo prompted Thomas Gold to look abroad. He had been invited by Harry Messel, the very entrepreneurial head of the School of Physics at Sydney University, Australia to give lectures at the Physics Summer School, which Messel had started some years earlier. It was a fortuitous meeting, Gold needed radio astronomers and ionospheric physicists and Messel's School of Physics was producing them in the form of newly minted PhDs. This led to Messel suggesting the creation of an association between Cornell University and Sydney University that would facilitate the interchange of personnel between the two institutions (Gold 2012). The result was the creation of the Cornell Sydney University Astronomy Center (CSUAC) in 1964 funded by the U.S. National Science Foundation, under which a large number of, mainly, young Australian scientists moved to either Arecibo or Ithaca. Cyril Hazard was the

Fig. 7.12 CUSAC related visitors from Sydney University at the front door of the Arecibo operations building in late 1964 or very early 1965. Left to Right; Harry Messel and R. Hanbury Brown (Sydney University), William E. Gordon, Bernard Mills (Sydney University) and Gordon Pettengill. Credit: Thomas Gold

forerunner followed to Arecibo in early 1965 by Tony Bray and Don Campbell and, in the latter part of 1965, by Ron and Judy Wand, with Ron joining the ionospheric group. John and Ann Sutton arrived in the summer of 1966 and Graeme Nelson in 1969. A number of young scientists joined CRSR in Ithaca, Ian Johnston, David Jauncey, Les Sharp, David Crawford, Michael Yerbury, Uris Ulrichs and Malcolm Campbell, who had been working at NASA's Goddard Space Center in Maryland. Ulrichs and Malcolm Campbell worked on Apollo related lunar research and not directly on Arecibo observations. Senior Australian scientists also took the opportunity to visit Arecibo and Cornell (Fig. 7.12). Most of the Australians stayed for one to two years with some staying longer. Cyril Hazard and Ron Wand transferred to Ithaca with Hazard then moving to Cambridge University and Wand spending several years with the ionospheric group at the Massachusetts Institute of Technology's (MIT's) Lincoln Laboratories Millstone Hill facility west of Boston before returning to Sydney. Dave Jauncey remained at Cornell involved with radio source counts and the very new field of Very Long Baseline Interferometry (VLBI) before returning to Australia in 1974. Mike Yerbury became the Director of the Danbury Laboratory at Cornell, which provided instrumentation for Arecibo, before returning to Sydney.

On 1 October 1969, the Arecibo observatory was transferred from ARPA to the National Science Foundation (NSF) and the NSF, no longer seeing the point of CSUAC, cut off any funding. CSUAC faded away over the following few years.

While Wand, Sharp and Nelson did work on ionospheric studies, CSUAC did not solve the problem of adequately staffing this group, a cause of considerable concern for Gordon, Gold and AFOSR.

7.5 Tensions at the Top and a Change

1964 also saw rising disagreement between Bill Gordon and Thomas Gold. Who was really in charge of the scientific direction of the observatory? Gordon had built Arecibo and was the AIO Director, but he reported to the CRSR Director, Thomas Gold, and CRSR was responsible for the management and operation of AIO under Cornell's contract with the Air Force Office of Scientific Research (AFOSR). However, the quarterly reports that were submitted to AFOSR were written by Bill Gordon and listed him as Project Scientist. In his 1979 self-interview Gordon commented that he and Gold had "some pretty good arguments" but what is not clear is how they interacted over appointments to the scientific staff since such appointments had budget implications. Gold seems to have appointed the early radio astronomy staff while Gordon appointed the ionospheric staff with the likely exception of Wand, Nelson and Sharp, who came under the auspices of CSUAC. Gordon Pettengill would have had considerable say over the appointment of staff for the planetary radar program, especially that of Rolf Dyce. Gordon was concerned about the creation of CSUAC in mid-1964 as he probably saw it as potentially undermining his authority over the Arecibo scientific staff. Certainly, the letter dated 23 November 1964, which the author received offering him a one year position at the Arecibo Observatory was written by Gold. However, Gordon seems to have realized the benefits of the Australian influx under CSUAC and he was very welcoming and supportive after I arrived in Arecibo.

The fundamental problem between Gordon and Gold, one that was never settled, was control of the observatory and its scientific program. Should it be run from the Cornell campus or, as Gordon put it in his 1979 self interview, "There should be a scientific leader, who resided in Arecibo and presided over the observatory and all of its functions, scientific, engineering, and budgetary." Gold was also primarily interested in the Arecibo telescope for its potential to investigate the Moon and planets and contribute to the still relatively new field of radio astronomy and he had little interest in ionospheric studies,[5] which were the primary motivation for Gordon having worked so hard to build the telescope. That ARPA through AFOSR was funding the entire costs of operating the AIO primarily for ionospheric research nominally in support of Project Defender was an inconvenient truth. Gold was also frustrated by the 430 MHz line feed situation and felt that Gordon was not taking it

[5] In 1959 Thomas Gold suggested the name "magnetosphere" for the magnetic bubble surrounding Earth due to its magnetic field. Gold clearly had an interest in this field of research; he just was not that interested in Arecibo's potential for ionospheric studies.

seriously enough. This was understandable. Having worked for over five years to build the Arecibo telescope, Gordon wanted to use it to study the ionosphere, the major objective of its construction, and he undoubtedly appreciated its unique capabilities to study the Moon and planets. That its use for radio astronomical observations was compromised by the high sidelobes and generally poor performance was likely of secondary interest. As he discusses in his autobiography, Gold (2012) felt strongly that he, as Director of CRSR, was being blamed by ARPA for the problems with the 430 MHz line feed although it is not clear that this was true since he was not involved in the selection of TRG and Alan Kay to design and build the line feed. Nonetheless it was Gold who vigorously pursued the feed issue over the following eight years until a new, very efficient dual polarized 430 MHz line feed was finally installed on the telescope in early 1972.

Another feed issue came between Gordon and Gold, which Gordon described in an interview with Andrew Butrica[6] and which clearly upset him at the time. Plans were developed at Arecibo and at AFCRL—it is not clear who initiated the idea—to construct an approximately 40 ft diameter Gregorian reflector to be mounted below Carriage House 2, which would allow coverage of frequencies between 430 MHz and 1420 MHz, the frequency of emission from atomic hydrogen, for radio astronomy observations. Philipp Blacksmith was involved and, according to Gordon, was willing to find the needed funding although final approval by AFOSR was needed since it required the use of ARPA funds under the operating contract with AFOSR. The feed was designed by AFCRL likely based on a design by Joseph Pierluissi,[7] who had moved from the observatory's engineering staff (see Fig. 4.16 of Chap. 4) to Cornell as a graduate student, and permission was requested in July 1964 from AFOSR to go out for bids on its construction at an estimated cost of $30,000. This was approved with bids being received in October and a contractor selected at a cost of $35,743. It wasn't until a request for permission to proceed with a contract reached Cornell that Gold was aware of the project. He had the recently-hired Associate Professor Frank Drake take a look at the plan since he distrusted AFCRL after the 430 MHz feed debacle. Drake's analysis of the feed design was quite critical leading Gold and, according to a hand written note, the AIO Technical Committee to say no to the plan. This is the only clearly recorded occasion on which Gold vetoed an initiative at the observatory supported by Gordon.

In the first half of 1964 there were suggestions, which many people believed were initiated by Thomas Gold, that it was time to have Gordon recalled to his academic position on campus as Professor of Electrical Engineering. The argument was that no faculty member should be allowed to be absent from campus for more than two years and Gordon had been gone for almost four years. By this time Dale Corson was

[6]Interview of William Gordon by Andrew Butrica on 18 November 1994, Niels Bohr Library & Archives, American Institute of Physics, College Park, MD USA, www.aip.org/history-programs/niels-bohr-library/oral-histories/22789

[7]J. Pierluissi, A Theoretical Study of Gregorian Radio Telescopes with Applications to the Arecibo Ionospheric Observatory, Research Report RS 57, CRSR, 1 April 1964.

provost of the university, the chief academic officer, so the issue landed on his desk. As a good friend and colleague of Bill Gordon, he was in a difficult position. In the absence of anyone raising the issue the rule had been ignored for Gordon, but it was difficult to do so once it was pointed out. Corson finally decided that Bill Gordon should return to campus. He and his wife Nellie made a trip to Arecibo, where they stayed with the Gordon family, to tell Bill that he had to return. This was not something that Gordon wanted to hear, as he recounted to Andrew Butrica, "I was mad at the time." However, he did accede to the request setting a departure date for one year in the future, September 1965. Gordon was "sure that Gold had talked him (Corson) into it" but Corson insisted to the author that it was his decision alone.

It is not clear that there was any public search for a new AIO Director but clearly a search was begun as a request was made to AFOSR on 27 January 1965 for approval of a one year appointment of Dr. John W. Findlay as Director of the AIO beginning 1 September 1965. Approval was received on 29 January and an announcement was made during the meeting of the Arecibo Advisory Committee on 2 February. At that time, John Findley was Deputy Director of the National Radio Astronomy Observatory (NRAO) and a radio astronomer experienced in telescope construction; he had organized the construction of the 300 ft transit radio telescope at NRAO's Green Bank, West Virginia site. There were reactions in both Arecibo and Ithaca to the announcement that Gordon would be leaving. In Arecibo, where Gordon was held in tremendous esteem, Gold was blamed. In Ithaca, Gordon's mentor, Henry Booker, was very critical of Cornell's action. In 1965 Booker left Cornell for the recently established University of California at San Diego, where he had been offered the opportunity to head a new department of Applied Electrophysics. It is not clear whether Cornell's treatment of Gordon was related to Booker's move.

References

Carlson, H.C., and W.E. Gordon. 1966. Radar Spectroscopic Estimates of Ionic Composition from 225 to 1,400 Kilometers for Solar Minimum Winter and Summer Conditions. *Journal of Geophysics Research* 71: 5573–5578.

Cohen, M.H., and M.R. Kundu. 1966. Lunar Occultation of the Radio Sources 0938 + 18 and 1101 + 11. *Astrophysical Journal* 143: 621–625.

Dyce, R.B., G.H. Pettengill, and I.I. Shapiro. 1967. Radar Determination of the Rotations of Venus and Mercury. *Astronomical Journal* 72: 351.

Gold, T (ed. Simon Mitton). 2012. *Taking the Back off the Watch*, 123. Springer.

Kellermann, K.I., et al. 2020. *Open Skies; The National Radio Astronomy Observatory and Its Impact on U.S. Radio Astronomy*, Springer Historical & Cultural Astronomy Book Series. Springer.

Thome, G.D. 1964. Incoherent Scatter Observations of Traveling Ionospheric Disturbances. *Journal of Geophysics Research* 69: 4047.

Willmore, A.P. 1970. Electron and Ion Temperatures in the Ionosphere. *Space Science Reviews* 11: 607.

Chapter 8
Feeds, Planning the First Upgrade and the Transfer to the NSF

Over the second half of the 1960s there were three issues which would determine the future of the AIO: developing efficient line feeds, including a dual polarization 430 MHz feed, the development of plans to upgrade the telescope's reflector to allow operation at frequencies up to 3 GHz, and the future organization of the observatory's management and funding source(s). These were all intertwined. No institution was going to fund the upgrading of the telescope's reflector until Cornell demonstrated that efficient feeds could be produced. ARPA's interest in funding the non-ionospheric programs at the AIO was waning and another source of funding was needed, either the National Science Foundation or NASA. However, NSF's participation in the upgrading of the reflector and/or taking over the management and funding of Arecibo was dependent on a demonstration that a successful radio astronomy feed could be constructed.

8.1 Looking for a Solution to the Line Feed Problem

Immediately after the 1 November 1963, dedication of the observatory, work started on characterizing the 430 MHz feed.[1] Multiple observations of radio sources, primarily by graduate student Harry Hardebeck, clearly showed that the gain was at least a factor of two smaller than anticipated, corresponding to an aperture efficiency of about 22%, but the beamwidth at 10 arc minutes was close to the expected value. The first side lobes levels were large and asymmetric being factors of only about 3 and 5 below the on-axis peak instead of a factor of about 30. By mid-December it was clear that detailed measurements of the amplitude and phase of the field produced over the reflector when transmitting from the feed were needed to

[1] See Appendix V by L. M. LaLonde of the Feed Sub-Panel of the Arecibo Evaluation Panel report of 5–6 October 1967. Cornell University archives.

D. Campbell, *The Arecibo Observatory: A History of Innovation and Discovery*, Historical & Cultural Astronomy, https://doi.org/10.1007/978-3-031-75225-4_8

understand what was going on. Marshall Cohen, Merle LaLonde and Joseph Pierluissi organized these measurements by equipping a sled with a receiving dipole that was pulled radially from the center to the edge of the 1000 ft reflector. The phase measurements were inconclusive, but the amplitude ones showed distinct local differences in the amplitude of the transverse and orthogonal linear polarizations, which should be equal to form circular polarization. Ron Bracewell, the chair of the Arecibo Advisory Board, was spending a year at Cornell on leave from Stanford University. He did an analysis of the measurement results and concluded that the slots on the four faces of the feed were radiating efficiently in the plane of the slots, the transverse direction, but not in the orthogonal one, resulting in a mismatch of the powers in the two linear polarizations. For the main beam, these polarization differences averaged out over the 360° of azimuth but they did contribute to the loss of sensitivity.

With the feed pointing to the zenith, measurements were made in mid-1964 of the amplitude of the field at a fixed location on the reflector as the feed was rotated in azimuth. This was an eye opener; the amplitude of the field in the transverse direction fluctuated by a factor of up to 30 and had the shape in azimuth of a four-leaf clover. It was determined that this was related to having just 4 radiating slots, one on each face of the square waveguide feed, combined with the slots being displaced from the central axis of the feed, unavoidable given the required dimensions of the square waveguide. Similar tests were done on the 5 GHz model feed that Alan Kay and TRG had used in their original tests (Fig. 5.12). This scalloping of the field meant that the reflector was not being fully illuminated, significantly reducing the aperture efficiency. In his autobiography Kay claimed that the Arecibo 430 MHz line feed that TRG designed and built worked "almost perfectly" (Kay 2008). It is possible that he failed to rotate either the production feed or the scale model to measure the field as a function of azimuth angle, a measurement that would have told him that the design would not give satisfactory performance. However, despite its low efficiency and high sidelobes, Kay's line feed did perform well enough to allow most of the telescope's planned research programs to proceed. Given the learning curve that took place before a line feed that gave the desired aperture efficiency was produced in the early 1970s, it is not clear that any of the line feed designs proposed in early 1960 would have worked any better than TRG's.

At its meeting on 11 June 1964, the Arecibo Advisory Committee concluded, "As a result of the defect in the feed, the radar work on the ionosphere and planets has not so far been seriously handicapped otherwise than by a reduced signal-to-noise ratio. Radio astronomy has, however, been hindered by the existence of bad sidelobes, and in due course the potentialities for radar will be seriously affected. It is therefore important to pursue the feed problem vigorously and to allow for the possible cost of solving it. The problem is of sufficient importance and complexity that action should be taken to ensure that additional outstanding antenna experts are consulted."

At this time, it was clear that little could be done to correct the problem by modifications to the current feed. No modification would provide a roughly uniform azimuthal power pattern, the clover leaf pattern was intrinsic to having only four circumferential transverse slots on the four faces of the feed and, at the time, this was considered to be the fundamental problem with the feed. The good news was that within the main beam (i.e., in the direction the telescope was pointed) by a stroke of electromagnetic good fortune, the transmitted power was almost purely circularly polarized even if it only contained about 40% of the power put out by the transmitter. Much of the remaining 60% went into those large sidelobes, which radiated the power into unwanted directions. This meant that the telescope would essentially operate as designed for ionospheric and planetary radar observations, just with reduced sensitivity. For radio astronomy, observations of the polarization properties of "point" sources (i.e. ones with angular diameters smaller than the 10 arc minutes beamwidth of the telescope at 70 cm wavelength) could be done successfully, but observations of extended sources (i.e. ones with angular sizes much bigger than the beamwidth) would be problematical due to the high side lobes and their uncertain polarization.

Tom Gold thought that the "radar work is, in fact, seriously handicapped by the loss of more than a factor of 4 in power due to the inadequacy of the present feed." In fact, a fully efficient feed would increase the sensitivity of the ionospheric measurements by a factor of between 2.5 and 6 and for radar observations of the planets by a factor of about 6.[2] In a 14 December memo to Jack Lowe, the Project Coordinator in Cornell's Research Office, Gold pointed out that the understanding of the problem with the current 430 MHz feed was such that "no successful solution can be found using the current feed. It is therefore necessary to proceed to the design, the model experimental studies, the mechanical design, and the fabrication of a new feed." Gold indicated that it was planned to do this in close consultation with Professor Vic Rumsey of the University of California at San Diego, and Professor Ron Bracewell of Stanford University. The expected cost of the initial design work was $60,000. Lowe wrote the AIO contracting officer at AFOSR, Captain A. R. Kardoskee, requesting that $60,000 be added to the AIO budget to initiate the feed work. On 4 March 1965 Lowe again wrote Kardoskee requesting permission to use some of the $60,000 for a small sub-contract with the California office of the National Engineering Science Company, for the services of Allan Love to also perform a study of the 430 MHz feed problem. As recounted in Chap. 2, Love had worked with Carl Wiley of Wiley Electronics in Phoenix on feeding spherical reflectors and, with Wiley, had proposed for the contract to build the first line feed that was awarded to TRG.

In early 1965, a contract was let to the School of Electrical Engineering at Sydney University at a cost of $60,000 for feed development work. The contract was with the School of Electrical Engineering under the supervision of its Director, W. H. (Chris) Christiansen, a noted radio astronomical antenna builder, with the work being conducted by Richard Wielebinski with Vic Rumsey, who was spending a leave in Sydney from the University of California at Berkeley, acting as a consultant. Rumsey's proposed design was based on a cylindrical waveguide with a series of

[2]For planetary observations, the low antenna efficiency lowers the amount of power that falls on a planet and it also lowers the sensitivity to the received echo so the total loss in sensitivity was very significant. Since the ionosphere's angular extent is vastly larger than the telescope's beamwidth, the situation was a bit more complicated.

Fig. 8.1 A section of waveguide with eight flat surfaces to simulate a cylindrical waveguide with eight circumferential transverse slots. The large number of slots gave a close to uniform azimuthal field distribution, but it was not understood until later testing at Autonetics that there was a second major problem related to the radiation patterns of the transverse slots. University of Sydney, Arecibo Line Feed Group Annual Report, Thomas Gold files in the Cornell University archives

8 transverse slots around its circumference. Eight slots would reduce the azimuthal variations in the radiated power to a negligible amount. The design could be implemented via a tapered cylindrical waveguide (Fig. 8.1) or via a simulated waveguide using 8 metal rods forming an octagonal cage. To give stiffness the rods would be threaded through circumferential rings spaced one half to three quarters of a wavelength apart. Model sections of both designs operating at 3 cm wavelength were built and tested and shown to give satisfactory results. With this, the Sydney group's work was done and plans were made to go out for bids based on the results of the Sydney measurements to design, build and test a full length, scale model operating at a wavelength of 10 cm.

In early 1966 bids were solicited from three companies with a cost plus contract being given to the Autonetics Division of North American Aviation Inc. in Anaheim, California at an estimated cost of $105,500. Gold had visited Autonetics in December 1965 as Allan Love had joined the Division again teaming up with Carl Wiley, who was the relevant sub-division director. A contract was signed with Autonetics effective 15 October 1966, with the work to be completed in 9 months. The decision was made to construct an octagonal tapered waveguide feed, not the rod and ring option, operating at a frequency of 2900 MHz. It would be a manageable 4.27 m (14 ft) long and based on the University of Sydney results with 8 transverse slots around its circumference. Love would construct a system to measure the phase and amplitude of the radiated field close to the feed, the so called near-field, and these would be translated into projections of the amplitude and phase of the field at the theoretical scaled location of the reflector using a technique developed by Vic Rumsey.

When Merle LaLonde and perhaps others, visited Autonetics in December 1966, the project was on schedule with another 8–12 weeks expected for the field pattern measurements and any necessary adjustments to optimize it. However, in a 31 January 1967 report, LaLonde pointed out "the aperture efficiency attainable with this feed should be approximately 50%". He then went on to say that "even higher aperture efficiency should be attainable with a really 'optimum' feed. This would require the use of both axial and transverse slots" rather than just the

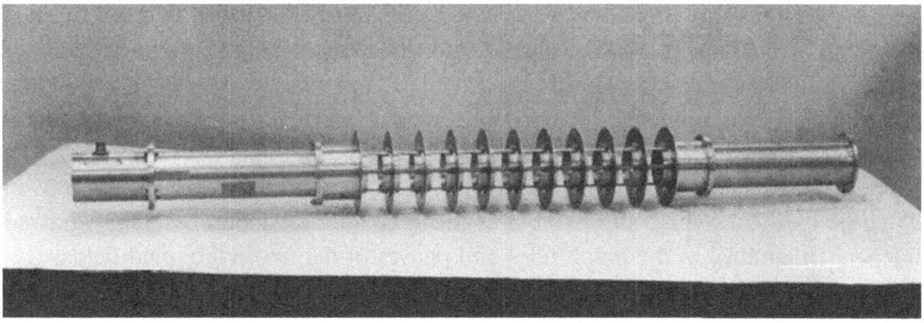

Fig. 8.2 A section of circular waveguide with six slots around its circumference and the radial disks that change the phase of one of the linear polarizations by 90°. This is not Love's 1967 scaled test section but rather the lower end one from the final design effort in 1969/1970 to show the general concept. From Arecibo 430 MHz Feed Development Study, Final Engineering Report, Autonetics, April 1971. Cornell University Archives

transverse slots. A short while later Love reported that the feed design that he was testing would have a maximum aperture efficiency of 40%, clearly an unacceptable number even though it was a significant improvement on the 22% of the square feed. What Love realized, and what was hinted at by LaLonde's January report, was that the loss of aperture efficiency was due to two roughly equal problems, the non-uniform azimuth illumination pattern of the square feed and a second effect related to the transverse slots not radiating equally in the two planes parallel and perpendicular to the axis of the feed. Having eight transverse slots solved the azimuth illumination problem but the continued use of transverse slots did nothing for the second one. Given the measurements of the fields of the full sized and model square line feed and the measurements by the Sydney group on their model feed sections, there must have been a fundamental misunderstanding of how line feeds work. It is not clear given Bracewell's analysis of the early feed measurements why this was not recognized sooner.

Alan Love's realization of the problems with the Sydney design led him to request a suspension of the contract with Cornell for the feed work so that he could "go back to the drawing board" to see how to resolve the issues. When Tom Gold informed Major Dowe of ARPA of this, Dowe insisted that Cornell cancel the contract and that a competition be initiated to procure a new line feed. A stop work order was issued to Autonetics on 21 April 1967. However, Love had some ideas as to how to solve the problem and wanted to try them out so continued work supported by Autonetics. The problem with a rectangular slot that would radiate roughly equally in the two planes is that the relative phases of the fields are not correct for producing circular polarization. Love came up with an ingenious idea to solve this problem, place circumferential disks between each ring of six rectangular slots. The effect of these disks is to change the phase of the field in the plane perpendicular to the feed axis by 90 degrees without affecting the orthogonal plane, the requirement that the two linearly polarized fields combine to form a circular polarized electric field (Fig. 8.2). Having come up with what he considered a very elegant solution to the feed design problem, Love was very disappointed that there was no funding immediately available to pursue model feed measurements. He wrote to Gold to say that it was not clear that he would be available if and when there would be funding under the new procurement plan as Autonetics may require him to move on to other tasks.

Vic Rumsey had also come up with a novel idea for building a workable feed in which he used transverse and axial slots in separate flat waveguides. This allowed independent control of the amplitudes and phases of the two orthogonal fields.

Rumsey's idea was to use flat waveguides, some with transverse slots in the center of one side of the wide face and some with axial (parallel to the axis of the feed) at the edge of one of the wide faces. This would be an octagonal feed system with eight circumferential transverse slots interspersed with eight axial slots. However, Major Dowe's decision to do a competitive procurement for a new feed also meant that there were no funds to pursue Rumsey's design and it, too, was put on hold until there was a general request for proposals.

The issue was how to proceed. Major Dowe wanted to start over and go out for bids to procure a new feed. However, the first action was to set up a Feed Advisory Committee or, as it turned out, two advisory committees. Tom Gold wished to set up a committee to "evaluate present and future suggestions and proposals for a new high power line feed." The committee, which would report to Cornell, consisted of John Ruze of MIT Lincoln Laboratory, Hanbury Brown and Bernard Mills of Sydney University and CSUAC, Richard Honey of Stanford Research Institute, Vic Rumsey and Frank Drake. ARPA wanted its own committee and a sub-panel of the Arecibo Advisory Panel chaired by John Evans of Lincoln Laboratory was formed to advise ARPA on the feed procurement. Its membership in addition to Evans was Phil Blacksmith (now at Rome Air Development Center (RADC)), S. Holt (AFCRL), C.L. Pankiewicz (RADC), Vic Rumsey, John Ruze, Captain J. Torres (ARPA) and W. White (AIL Division of Cutler-Hammer). This committee met on 5 and 6 October, 1967, with a large number of attendees including ones from ARPA (Dowe), AFOSR (Harrington), NSF (Hurlburt) and Cornell (Gold, Drake, Dyce, LaLonde). Presentations were given by Gold, Rumsey, Love, LaLonde, Drake, Ruze, and Dowe amongst others. At the end Dowe reviewed ARPA's proposed feed procedure, which had RADC involved with the "technical and administrative monitoring of any new feed development effort."[3] Gold objected—he wanted the procurement process under Cornell's control. Dowe stated that "funds for the new feed could be secured by following his proposed procedure." In other words, he would only fund the procedure that ARPA preferred. Gold responded by saying that the procurement procedure "was not acceptable to Cornell." From Gold's point of view, Dowe's suggested procedure was just a return to the original late 1959 one that had ended up with a feed that did not work. There was now a good understanding of the issues with the original feed and, after several years of work,

[3]Minutes of the Meeting of Dr. Evans' Sub-panel, M.I.T. Lincoln Laboratory, 5–6 October, 1967. Cornell University archives.

two believable plans for building one that worked correctly. Why start all over again?

There was general agreement within the sub-panel that a new feed would give an improvement in sensitivity of a factor between 2 and 2.5. The panel also recommended that two concurrent but separate efforts be funded. These would examine Rumsey's and Love's schemes and/or others that seemed attractive. Despite the last phrase, this seemed to indicate that the sub-panel members basically supported Gold's contention that the sensible path to a new feed was to proceed with testing Love's and Rumsey's approaches by building and testing scale models. There was a possibility that both Autonetics' and Rumsey's designs were proprietary and would not be available to other proposers, so it is likely that any procurement process would result in one or both of them being selected to do the design and testing of a model feed. However, the impasse between Gold and Dowe as to which institution would control the procurement process meant that no progress would be made on developing a new feed for almost 2 years.

8.2 The Reflector Upgrading: Plans and Funding

The feed issue was inextricably linked to plans to upgrade the telescope's reflector to allow observations up to a frequency of 3 GHz (10 cm wavelength). Testing had been done at 1.4 GHz, but the approximately 3 cm mean error in the setting of the reflector's wire mesh surface meant that its aperture efficiency at this frequency was less than 10% of that for a perfect surface. Considerable effort had been put into trying to reduce the mean error of the wire mesh surface, but it was a lost cause; the changing sag with temperature of the upper light cables over the 100 ft span between the heavy lower cables, 3 mm per degree Celsius, could not be corrected for. There was also the issue of the 12 mm (0.5 inch) mesh size; a maximum size of 1/20th of a wavelength is required without a serious loss in sensitivity and this is only 5 mm at 10 cm wavelength. Clearly, any plans to move to higher frequencies required a new surface for the reflector.

Before planning could commence, a second issue, the stability of the suspended platform under wind loading had to be determined. In August 1966, John Findley had completed a year as the Director of the AIO while on leave from NRAO and decided to return to NRAO's Green Bank site. Frank Drake was appointed to take his place arriving in Arecibo with his family in early September. One of the issues on Drake's mind was the need to upgrade the telescope's reflector surface to allow it to observe at higher frequencies, especially at 1400 MHz, the frequency at which atomic hydrogen radiates. However, there was no point upgrading the surface if the position of the suspended structure was not stable to a fraction of the wavelength of the highest frequency planned, about 1 cm for the 10 cm wavelength corresponding to a frequency of 3000 MHz. Wind speeds are normally benign in the tropics, 16 to 24 km/h being the norm, so hurricane Inez, a very powerful storm which passed south of Puerto Rico on 28 September 1966, shortly after Drake

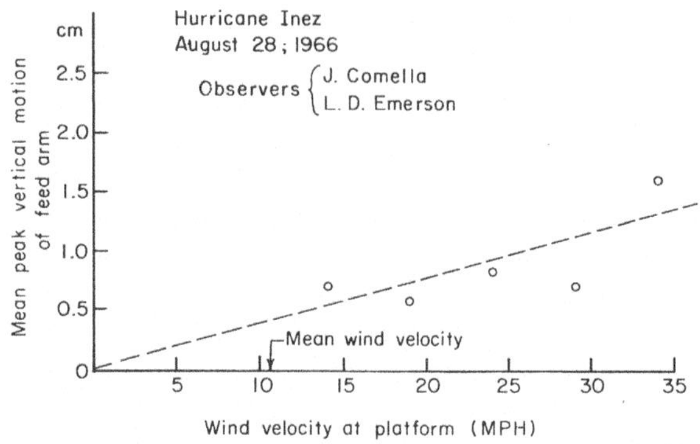

Fig. 8.3 The measured vertical motion of the platform as a function of wind velocity during Hurricane Inez, which passed just south of Puerto Rico on 28 September 1966. The August date in the figure is a mistake, it was September. From 31 January 1967 Cornell University Center for Radiophysics and Space Research report RS70. Cornell University Library

arrived in Arecibo, offered an opportunity to measure how the suspended structure behaved in high winds. In Frank Drake's own words "we taped a scale on the platform and set up a theodolite in the control room looking at the scale."[4] Drake and two Cornell graduate students, Linda Emerson and John Comella, stayed in the control room monitoring the movement of the platform during the hurricane. Peak wind speeds at the height of the suspended structure were about 80 km/h (50 miles/h) and the horizontal movement of the suspended structure was less than 0.6 cm. Vertical motion at 55 km/h (35 miles/h) was about 1.5 cm and about 4 mm at 16 km/h (10 miles/h) (Fig. 8.3).

The suspended structure was amazingly stable and any concerns about observations at frequencies up to 3 GHz were gone. Similarly, the reflector was also very stable under wind forces. The next requirement was a detailed engineering plan and cost estimate for replacing the wire mesh surface with a solid aluminum surface that would have a mean surface error of less than 3 mm.

A proposal was submitted to the National Science Foundation for $43,000 to fund a design study for resurfacing the reflector to allow operation up to 3000 MHz. It was anticipated that the cost of the resurfacing would be approximately 3 million dollars. On 15 September 1967, Cornell issued a request for proposals for an "Engineering, Design and Cost Study involving the replacement of the reflector surface at the Arecibo Ionospheric Observatory located at Arecibo, Puerto Rico." The basic requirement for the study was that the surface consist of aluminum panels which must conform to a sphere of the required radius with a mean error of less than 3 mm and for which the corners of each panel could be set to an accuracy of 3 mm. An additional requirement was the need to replace the current surface piecemeal so as to allow operation of the 430 MHz transmitter while construction was taking place. The

[4]A video interview with Frank Drake can be seen on the Cornell library's e-Commons site https://ecommons.cornell.edu/handle/1813/40748

proposal request was sent to nine firms including the Rohr Corporation, Westinghouse Electric Company, Philco Corporation and North American Aviation. A bidders conference was held at the Observatory on 3 October 1967. The Rohr Corporation was the successful bidder at a contract price of $41,800 and delivery of the study by 1 July 1968. Rohr's design involved replacing the wire mesh and light upper cables of the reflector with 100 ft trusses at 14 ft intervals spanning the distance between the heavy lower cables. The trusses would support 14 ft square panels made up of a light aluminum surface supported by an aluminum framework. Estimated cost was $3.5M.

Cornell's plans for upgrading the Arecibo 1000 ft telescope were initiated at a time of considerable discussion within the United States astronomical community, especially the radio astronomical one, as to what new large radio telescope or telescopes should be constructed. Arecibo, the world's largest single dish telescope, had been completed in 1963 but was limited to frequencies below 600 MHz and by the difficulty in building efficient line feeds; NRAO had completed a 300 ft (91 m) transit telescope at Green Bank, West Virginia in 1962 and the fully steerable 140 ft telescope was finally completed in 1965. MIT Lincoln laboratory's 120 ft (37 m) antenna in a radome was completed in 1964. Interferometry, the linking of two or more antennas to give angular resolutions equivalent to a telescope with a size equivalent to the largest spacing between the antennas, was making rapid progress with two linked 90 ft (27 m) telescopes at Caltech's Owens Valley Radio Observatory (OVRO) and three linked 85 ft (26 m) antennas at NRAO's Green Bank site. Caltech and NRAO had developed plans for much larger interferometric antenna arrays, a group of Northeastern institutions, the North Eastern Radio Observatory Consortium (NEROC) was proposing to build a 440 ft (134 m) diameter telescope in a radome at a cost of $27.8M that would be used for both radio and radar astronomy, and Cornell was proposing to resurface the Arecibo telescope at a cost of about $3M. In order to set priorities, in July 1967 the NSF convened an "Ad Hoc Panel on Large Radio Astronomical Facilities" chaired by Princeton Physicist Robert Dicke, hence dubbed the "Dicke panel".

The Dicke panel issued its report on 14 August 1967. It was good news for Cornell as the panel in its own words "urges that the proposal by Cornell University for upgrading the 1000-foot spherical dish in Arecibo, Puerto Rico, to permit observations at 10 cm wavelength or shorter, be accepted in its entirety and funded as soon as possible, with an adequate operating budget, and with the proviso that at least 50% of the observing time for astronomy be made nationally available." It went on to say, "It is imperative that there be definitive studies directed towards assessing the potential of large, fixed, spherical dishes with multiple feeds (the Arecibo type) since this approach may lead to instruments of the largest collecting area." The panel commended NEROC for its proposal for a light-weight, fully steerable dish in a radome but stated that the "NEROC proposal should be deferred until more is known of the capabilities of an Arecibo-type spherical dish as a large precision instrument, operating at short wavelengths." However, the Panel did recommend that Caltech's proposal for an 8-antenna interferometric array at the OVRO be accepted. Despite the clear recommendations, neither the Arecibo project nor the OVRO array were included in the FY 1969 NSF budget sent to Congress by the Johnson administration. The Nixon administration did include funding for them in the FY1970 budget, but the House of Representatives Committee on Science and Astronautics deleted it.[5]

One indication of the competitive nature of proposals for future large instruments was a report submitted to the NSF by NEROC "Comparison of Large Spherical and Parabolic Radio Telescopes" in March 1968. The report was nominally in response to the Dicke Panel's statements regarding the NEROC and Cornell proposals: "Specifically, the Panel recommends that ample funds be made available to qualified groups of engineers and scientists to conduct studies of the feasibility of the Arecibo concept for larger sizes and operation at shorter wavelengths (5 cm). Such studies should include comparisons between this and other antennas, especially those of the NEROC type." It went on to say, "The panel hopes that the NEROC group will assist in making comparisons between the potentialities of these two types of antennas" The March 1968 study report was NEROC's response to this last statement. Needless to say, it came down very strongly on the side of building lightweight radome-enclosed steerable parabolic antennas versus Arecibo type spherical antennas. A copy was sent to Frank Drake in Arecibo, who did a detailed and critical analysis pointing out what were, in Drake's view, numerous serious flaws in the study's reasoning. On 2 July Drake wrote to Jerome Weisner, former Chairman of President John F. Kennedy's Science Advisory Committee and now Chairman of NEROC and Professor of Electrical Engineering at MIT, pointing out the flaws in the study and attaching his analysis. On 12 July Robert Fleisher, Head of the NSF Astronomy Division, sent the NEROC study to Tom Gold and Drake requesting that they provide comments. In a 1 August 1968 letter to Bart Bok, his former mentor at Harvard and a member of the Dicke Panel, Drake expressed some reluctance to distribute Cornell's criticisms of the NEROC study in order not to embarrass them; what Cornell wanted was for it to be withdrawn and revised.

Given the lack of progress for 2 years in obtaining funding for one or more large radio astronomical telescopes, the NSF reconvened the Dicke panel on 9 June 1969. This time, its report, issued on 15 August 1969, recommended that a start be made on all four of the leading proposals from the first report; the resurfacing of the Arecibo reflector, the eight antenna array at the OVRO, the construction of the NEROC 440 ft antenna in a radome, and the 36 antenna array, the Very Large Array, proposed by the NRAO. With regard to the Arecibo project the report stated, "The panel is especially disappointed that the new surface of the 1,000-foot spherical dish in Arecibo, Puerto Rico has not been constructed, and feels that the need for it is even more pressing that it was earlier. It wishes to emphasize that the facility should be improved in its entirety and be made nationally available." The Arecibo

[5]For a brief but comprehensive discussion of the issues in the mid-1960s with respect to radio astronomical telescopes and the Dicke Panel see "A Case Study of the Decision to Construct a Large Radio Telescope, the VLA" by George W. Swenson, January 1977. It is available on the WEB.

upgrading project was included in the NSF's FY1971 budget request and this time it was approved.

On 3 January 1968, Everett Hurlburt, the Radio Astronomy program director within the NSF Astronomy Section, wrote to Tom Gold pointing out a resolution of the NSF Advisory Panel meeting of 17 October 1967, "The NSF Advisory Panel will be hesitant to favor the improvements of the surface of the Arecibo dish and/or the undertaking of substantial operating expenses for Arecibo until a successful radio astronomy feed has been constructed and made operational at frequencies low enough that the surface is not critical." However, the second Dicke Panel report of 15 August 1969, stated, "The panel notes, too, that feeds for the Arecibo spherical reflector have been successfully designed, built, and operated with illumination patterns superior in some respects to those currently in use with parabolic reflectors. Continuing studies of spherical reflectors and feed systems for them are highly desirable." What had changed?

The 430 MHz circularly polarized feed program was still stalled due to the differences between Tom Gold and Mike Dowe of ARPA. However, there had been a breakthrough in the thinking about feeds suitable for radio astronomical observations. The initial insight into the problems with the 430 MHz circularly polarized feed related to the distance of the transverse slots from the central axis of the feed; it was this that was responsible for the clover leaf pattern with azimuth angle. This problem could be solved by using a waveguide that was very narrow in one dimension and with broad faces in the other dimension containing the transverse slots, a flat waveguide that has the slots very close to the feed axis. It is possible that this idea originated with Vic Rumsey based on his multiple flat waveguide feed design for the 430 MHz dual polarization feed. The only disadvantage is that such a feed can only transmit or receive a linearly polarized signal, not good for the radar systems, which needed circular polarization, but perfectly fine for most radio astronomical observations.

One of the recommendations of the first Dickie Panel report was for the NSF to fund continuing studies of the Arecibo, NEROC, Caltech and NRAO telescope proposals. Cornell used this NSF funding to initiate the resurfacing study by Rohr and also to initiate a contract with Autonetics for Allan Love to design a flat waveguide feed to operate at 611 MHz. It would be 45 ft (14 m) long and illuminate a 750 ft (229 m) diameter area on the Arecibo reflector. The advantages of using a short line feed were the ease of construction and, most importantly, that the 750 ft illuminated area would not spill off the edge of the reflector until the telescope was observing at an angle larger than 10.5 degrees from the zenith. After that, the sensitivity would progressively decrease as the zenith angle moved to its maximum value of 20 degrees and an increasingly larger portion of the illuminated area spilled onto the surrounding terrain.

While a contract had been let to Autonetics for the 611 MHz feed design, Merle LaLonde decided to build a similar feed at 327 MHz, the frequency of a spectral line from deuterium, atomic hydrogen but with a neutron as well as a proton in its nucleus, thought to be present in the interstellar medium at a much lower concentration than normal atomic hydrogen. LaLonde's feed would be 40 ft (12 m) long and

Fig. 8.4 The 318 MHz flat feed (center left) and the 611 MHz feed (center right) mounted on Carriage House 2 in 1970. The feeds on outriggers on either side of the Carriage House are "point" feeds at a variety of frequencies that illuminate much smaller areas on the reflector and are much less sensitive than the line feeds. This image demonstrates one advantage of spherical reflectors. As long as the feeds lie along a radius of the sphere, they will focus the incoming radio wave. This allows multiple feeds looking in different directions to be utilized simultaneously. Credit: Cornell University/ARPA

illuminate an area with a diameter of 700 ft (214 m) that would spill over the edge of the reflector at a zenith angle of 11.3 degrees. A 1/6th scale model was designed, built and tested. After some adjustments of the spacing between the large flat sides of the feed to obtain the desired phase distribution over the reflector, the design was finalized and a full-scale feed constructed. It was too big and heavy to test in Ithaca so it was shipped to the observatory and installed on Carriage House II in December 1968 (Fig. 8.4). Observations of radio sources showed that it worked almost exactly as predicted although its optimum operating frequency was found to be 318 MHz, not 327 MHz. The aperture efficiency was close to 70% with most of the loss being due to the inaccuracies of the reflector wire surface. For a perfect surface, the planned new surface would be "perfect" at this frequency, the aperture efficiency was expected to be between 80% and 90%. After 5 years of effort an efficient line feed for a spherical reflector had finally been demonstrated. It was only applicable to radio astronomical observations, but it convinced the NSF and the second Dicke Panel that resurfacing the Arecibo reflector would be a very worthwhile endeavor.

Allan Love had similar success with his design and testing of a 1/5th scale model of a 611 MHz feed. He delivered the design details of the scale model feed to Merle LaLonde, who constructed the full sized version. This feed was installed on the Arecibo telescope on 10 October 1969 (Fig. 8.4). In order to minimize the sidelobe levels, Love designed the feed to not efficiently illuminate the outer region of the

750 ft illuminated area. This reduced the aperture efficiency, but the big loss in efficiency was due to the irregularities in the wire mesh surface, much more important at 611 MHz than at 318 MHz due to the 50% shorter wavelength. The predicted efficiency was 26% to 44%, and the measured range after the feed was installed was 22% to 35%. The range was due to the temperature changes from day to night and the resulting sag during the daytime over the 100 ft span of the upper light reflector cables between the large lower cables. Once the reflector was resurfaced, it was expected that the aperture efficiency would be close to 70%, a very satisfactory value. The radio astronomers now had two feeds that had sensitivities far larger than those of any other existing telescope, and funding for the resurfacing project that would allow observations at much higher frequencies was imminent.

As a demonstration that flat feeds could be easily built for the higher frequencies that the Arecibo telescope could operate at after the reflector resurfacing, in 1969 Merle LaLonde built a 16 ft long feed designed to work at 1667 MHz (one of the frequencies emitted by the OH molecule) and illuminate 450 ft of the reflector. The feed weighed only 40 lb (18 kg) and was designed and constructed in only 4 weeks. Its small size meant that no test model was needed; the field of the full size feed could be measured directly (Fig. 8.5).[6]

Planetary Radar: A reflector surface that provided good aperture efficiency up to 3 GHz opened up the possibility of a new high-powered radar for planetary radar observations. High powered transmitters rely on amplifying tubes, klystrons, to produce the power. There are two types of transmitters and klystrons, pulsed systems, where a very high power pulse is transmitted at regular intervals, typically 15 ms for the 430 MHz transmitter on the Arecibo telescope, and CW (continuous wave) systems where a continuous signal is transmitted. For a pulsed system the radar echo from the target, the ionosphere or the Moon in the case of Arecibo, is received in the interval between the pulses. The peak power of the 430 MHz transmitter was 2.5 MW but the maximum average transmitted power was only 150 kW. CW transmitters are typically simpler than pulsed ones, but they can have much higher average powers. For radar observations of the planets, what is important for obtaining maximum sensitivity is the average power, so CW transmitters are favored. However, the target must be far enough away that the time it takes the radar signal to go there and back, the round-trip time, is large. In the case of Venus, our nearest planetary neighbor, the round-trip time when the planet is closest is approaching 5 min. For Venus or more distant targets, the transmitter is turned on and left on until the first echo is about to arrive back at the telescope. It is then turned off and the radar echo received for the equivalent length of time. This is repeated for as long as the telescope can observe the target object.

[6]A brief history of the initial development of flat feeds for the Arecibo telescope can be found in "Three Highly Efficient Line Feeds for the Arecibo Spherical Reflector Telescope," L. M. LaLonde, CRSR report #385, 1970. Available in the Cornell University Library. A detailed description of the 318 MHz feed development and testing is given in Lalonde and Harris (1970).

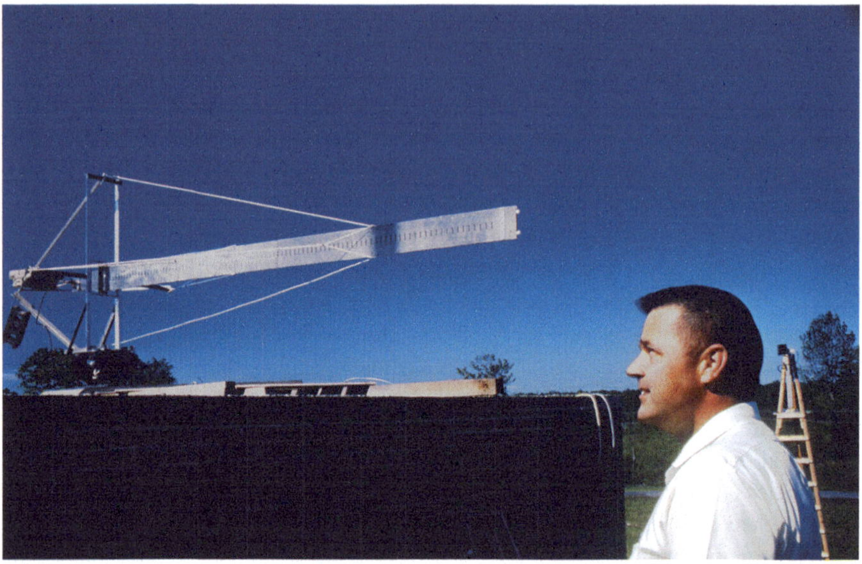

Fig. 8.5 Merle LaLonde and the 16 ft 1667 MHz feed under test in 1969 at Cornell's Zeeman laboratory about 5 miles (8 km) from the Cornell campus. Credit: Cornell University

For a transmitter with the initially proposed average power of 900 kW operating at a frequency of 2.4 GHz, the improvement in sensitivity over the 430 MHz radar system for radar studies of the planets would be about a factor of 50 assuming an efficient line feed at 430 MHz, a factor of about 300 with the then existing line feed. This sensitivity would allow imaging of the surface of Venus under its cloud cover with a resolution of about 2 km, good enough to identify many features on the planet's surface such as impact craters and the role of volcanism and tectonics in forming the surface. How similar was Venus to the Earth despite the large differences in the atmospheric pressure and surface temperature? This capability was potentially of great interest to NASA and discussions were initiated with NASA about funding an enhanced planetary radar capability.

Despite the lack of progress in obtaining funding for the upgrading project, work continued on estimating the costs of resurfacing the reflector, procuring a 900 kW transmitter operating at a frequency close to 3 GHz, building a dual polarized feed at this frequency, and making needed modifications to the suspended structure and Carriage House II to accommodate the transmitter and multiple additional line feeds. Estimates of the costs were: reflector surface \$3.3M, high powered feed at the transmitter frequency \$0.25M, improvements to the platform \$0.4M, transmitter \$1.0M, a new carriage house \$0.04M and two radio astronomy line feeds \$0.1M for a total estimated cost of \$5.1M in 1968 dollars.[7] As we will see, by this date the NSF had agreed with ARPA to progressively take over the funding for the AIO

[7] "A Five Year Plan Report for the Arecibo Ionospheric Observatory", Cornell University archives.

starting in October 1968, and the assumption was that the NSF would fund the reflector resurfacing and some of the modifications to the suspended structure. If NASA wanted the upgrades to the planetary radar system, primarily the transmitter, then it would have to pay for them.

8.3 Transition to the NSF and Becoming a National Center

As early as the Spring 1964 meeting of the Arecibo Evaluation Panel, ARPA expressed a desire to eventually reduce its funding for Arecibo to just the ionospheric program and that Cornell and ARPA should solicit long term funding for the radio and radar astronomy programs from other agencies. On 10 September 1964, and based on previous discussions with the National Science Foundation, Charles Herzfeld, Deputy Director of ARPA, wrote Randal Robertson, NSF Associate Director for Research, proposing a funding profile for the AIO in which the NSF would progressively assume the costs of the astronomy programs in yearly increments starting with \$0.3M in FY1966 and increasing by \$0.3M each year until the NSF's contribution was \$1.2M and ARPA's contribution \$0.5M. On 5 October Robertson replied that the NSF could likely provide \$0.3M in 1966 and enter into a formal arrangement starting with the 1967 budget. He went on to say, "We have formed a high opinion of the observatory so are prepared to explore means of sharing in its support." On 8 January 1965 Bill Gordon had a telephone discussion with Marshall Harrington at AFOSR during which Harrington mentioned that ARPA was budgeting the same amount for 1966 as was in the 1965 Arecibo operations contract and that a meeting was scheduled between ARPA and the NSF on 15 January to discuss "NSF partial support of AIO at an early date, and more complete support eventually."[8] A meeting was held at the NSF on July 6 attended by representatives of the NSF, ARPA, AFOSR, OAR (Office of Aerospace Research of the DoD), NASA and Cornell to discuss future funding for the AIO. ARPA insisted that they wished to continue to maintain responsibility for the AIO and to fund the ionospheric program for the indefinite future but that they would welcome NSF and/or NASA funding for the programs that were not within ARPA's defined mission, the radio and radar astronomy programs. The NSF agreed to consider a proposal for FY 1966 funding via a direct grant or an interagency transfer of funds to AFOSR so that Cornell would not have to separately account for expenditures under the NSF support. Likewise, NASA indicated that they were interested in involving themselves with the AIO via either a direct grant to Cornell or an interagency transfer. It was agreed that a single proposal from Cornell would be best and that a joint technical and administrative steering committee should be formed with a representative from each agency. Each agency would examine Cornell's draft 1966 funding proposal, that was distributed

[8]Letter from William Gordon to Franklin Long, Vice President for Research at Cornell, dated 11 January 1965. Cornell University archives.

during the meeting, and convey the needs of their agency to Marshall Harrington, who would convey them to Cornell for preparation of the final proposal.[9]

Cornell's July proposal for AIO funding from 1 February 1966 to 30 January 1967 was for a total of $2,425,842. Each category—salary related expenses, maintenance, supplies and capital equipment—had estimated costs assigned to the ionospheric, radio astronomy and radar astronomy programs. A 20 September letter from Lloyd Wood, Director of Physical Science at AFOSR, stated that after review by ARPA and the newly formed Arecibo Steering Committee, the proposal would be approved but with reduced funding and that it was expected that the NSF would provide some of the funding. ARPA wanted all capital items larger than $25k removed and applied for separately. A revised proposal dated 11 October was submitted for a total of $1,867,562.

The NSF did not provide funding in 1966. The National Science Board (NSB), the oversight Board for the NSF, had not been asked for their approval and expressed displeasure. The funding plan was supposed to be brought before the Board during their November 1965 meeting, but the NSF in light of the NSB's concerns pulled the item from the agenda. It had been planned to discuss the issue of support for Department of Defense originated projects during the January 1966 NSB meeting. However, the entire meeting was taken up with the report of the review of the NSF by the House of Representatives Sub-Committee on Science, Research and Development of the House Committee on Science and Astronautics—the Daddario sub-committee after it Chair, Representative Emilio Daddario. Leland Haworth, the NSF Director, and Robertson decided that the NSF could not contribute to the AIO budget in FY 1966 out of concern for the attitude of the NSB. However, the NSF had agreed in principle to become involved with the management and funding of the AIO. NASA provided $211k to support the lunar radar work at Arecibo via a direct grant of $116k to Cornell for relevant staff salaries and computer costs and also transferred $95k to AFOSR to cover general running expenses under the guise of indirect costs. However, NASA made it clear that they would not support the general operational expenses on a long-term basis. The now ARPA Director, Charles Herzfeld, agreed that ARPA would continue to support the full operational costs.[10]

In July 1966, Cornell submitted a proposal for the operation and maintenance of the AIO for the period 1 February 1967 to 31 January 1968. The funding request for this period was $2.43M. Due to financial problems of its own, ARPA requested a revised, lower budget. In September Cornell submitted one for $2.166M. This was still too much for ARPA, and they requested that the budget be reduced to $1.8M. Cornell submitted an amended budget of $1,799,932 on 6 October. Another revision was submitted on 6 December again staying within the $1.8M limit but, in his submittal letter to Marshall Harrington at AFOSR, Rod Dennis stated that the budget assumed that ARPA would add approximately $144,000 to cover an expected

[9]J. W. Lowe, Assistant Coordinator for Research at Cornell, memo to file dated 6 July 1965. Cornell University archives.

[10]Gold Memorandum for the Record dated 21 February 1966. Cornell University Archives.

overrun in the present year's budget, which ran until 31 January 1967. On 7 December Gold wrote a four-page letter to Herzfeld detailing Arecibo's achievements and pointing how hard it would be to operate AIO on $1.8M for 1967 when the expenses for 1966 were close to $2.0M. Because of their appointment terms, the earliest that scientific staff could be dismissed was mid-summer 1967 by which time $1.06M of the budgeted $1.8M would be expended leaving only $0.74M for the remaining 6 months; it was an untenable situation. Franklin Long, the Cornell Vice-President for Research and Advanced Studies, and Tom Gold went to Washington on 16 December to discuss the situation with Charles Herzfeld and David Mann (Chief of ARPA's Applied Research Branch and Mike Dowe's supervisor) at ARPA. Gold pointed out the serious problems that the budget would cause the AIO. Herzfeld was sympathetic but said that they just had no money; it was an exceedingly tough year for ARPA. It was agreed that Cornell would look for all possible economies in Arecibo's operations and will approach NSF and NASA for funding including for engineering studies for the reflector resurfacing project. For its part, ARPA promised that it would see what additional funds it could come up with during a budget review in mid-January. In early 1967 Cornell submitted a supplementary budget for $310,000 and Gold wrote Herzfeld on 8 March to point out that they needed to know soon if these funds would be available as Cornell needed to make a decision on renewing appointments for some of the scientific staff. He definitely put "the ball in ARPA's court" as to Arecibo's future. On 28 April 1967 Herzfeld approved the supplement but in his Order to AFOSR emphasized that the supplement was a onetime increase and the underlying budget number was to remain at $1.8M.

For the 1968 budget, ARPA requested that Cornell submit a budget covering 1 February to 30 September, an eight-month period, of $1.2M instead of a full year $1.8M. ARPA also requested a five year Program Plan laying out the future scientific program and the capital improvements needed to accomplish the research goals, such as the upgrading of the reflector and the new transmitter for planetary radar astronomy. The plan was submitted in late June or early July 1968, presumably to both ARPA and the NSF.

On 23 January 1968, the NSF Director, Leland Haworth, wrote to Franklin Long saying that the NSF would provide 50% of Arecibo's budget in 1969 and would support radio and radar astronomy costs in subsequent years at levels to be discussed with ARPA. He also raised the future of the administrative arrangements for the AIO, pointing out that there was sentiment within the NSF for making the AIO a National Center managed by a multi-university consortium similar to the Associated University's (AUI's) management of the NRAO. This was an issue that would continue to be raised over the following several years as Arecibo transitioned to being a National Center under NSF management.

Arecibo may have been a significant research establishment but as far as the Department of Defense (DoD) was concerned it was excess property, land plus capital improvements (i.e. the telescope, buildings, etc), which would be transferred from the DoD to the NSF. Curiously, because of the Ramey Air Force base connection, it was carried on the books of the Strategic Air Command. A DoD audit of the AIO was initiated and the 28 February 1968 draft audit report showed no problems, but it did point out that there was a DoD directive that said that excess DoD property should be offered to other DoD departments first before being offered to other governmental institutions. The audit also pointed out that in 1962 AFCRL had indicated interest in taking over the AIO if ARPA should ever decide that it wished to relinquish control. AFCRL had an interest in both radio and radar astronomy and was a recognized center for DoD research in geophysics, atmospheric sciences and astronomy. It turned out that no DoD agency showed any interest[11] and the transfer to the NSF proceeded.

On 28 June Robert Fleisher, the head of the NSF Astronomy Section, and Fred White, head of the Atmospheric Sciences Section, visited Cornell meeting with Rod Dennis, John Payne, Administrative Director of CRSR, and Yervant Terzian, Assistant Director of CRSR, standing in for Tom Gold, and Frank Drake since Drake was now the AIO Site Director. They had many questions related to AIO and the role of Cornell and CRSR. Of particular interest were the role of the advisory committees, the Cornell-Sydney University connection, the relative roles of Gold and the AIO Director and the payment of salaries in Ithaca. The impressions left by the visit were that the NSF would require that the AIO be structured much more like a National Laboratory such as NRAO, that the AIO Director have much more of a role, that no more funds be expended on the Sydney connection, that Ithaca expenses be substantially reduced, and that outside users of the telescope be encouraged and looked after much better.[12] After talking with Terzian, Drake wrote to Fleischer from Arecibo the next day making a strong case that outside users at Arecibo are treated very well and that anyone with complaints should put them in writing.

On 22 July 1968, ARPA Director Rechtin wrote AFOAR confirming an agreement reached between ARPA and AFOAR that AFOSR would no longer provide technical oversight of Cornell's management of the AIO but would be confined solely to contracting services, primarily the transfer of funds from ARPA to Cornell and financial support of the Bracewell committee. Technical cognizance would be retained by the ARPA program manager. Future funds transfers to Cornell or the NSF would be done through AFOAR. This effectively ended 6 years of AFOSR involvement with the management and operation of the Arecibo Observatory. Lloyd Wood, Director for Physical Sciences at AFOSR, wrote Tom Gold on 2 August expressing gratification at AFOSR having been involved with AIO over the past several years and sorrow at the abrupt termination of the association. Gold responded on 9 August thanking Wood for AFOSR support.

ARPA again set the AIO's budget for its 1969 budget year (now 1 October 1968 to 30 September 1969) at \$1.8M, unchanged from the previous two budgets and, in fact, about \$300k less than the 1967 budget after the \$310k increment. These

[11]F. A. Long Memorandum for the Record, Discussions at ARPA, 16 August 1968. Cornell University archives.

[12]Memo to File dated 26 June 1968 by Rod Dennis, "NSF-Cornell meeting". Cornell University archives.

were difficult times as the United States was involved in the war in Vietnam, which was requiring about 3% of Gross National Product (GNP) or roughly 15% of the Federal budget. The NSF and ARPA went through several drafts of a Memorandum of Understanding for the transfer of the AIO to the NSF. The plan was to set up an Interagency Committee for Arecibo (the ICA) with ARPA in the management role for the first year, 1969, and the two agencies splitting the budget putting in $900k each. Starting in the 1970, the NSF would be the controlling agency and would contribute $1.2M with ARPA putting in $600k. This would hold for FY71 and, perhaps, beyond but the agreement would be for only 3 years initially. The amount that ARPA would contribute beyond the first year fluctuated with the drafts but otherwise there were only small changes. Budgetary details included additional funds from ARPA of $49k to account for the increase in the Puerto Rico minimum wage, funds for painting of the telescope structure, cables and towers and for the new 430 MHz line feed. This last item of about $100k was to be shared with the NSF. One curious and unexplained detail was the setting up of a repository in Puerto Rico that could receive and store classified documents.

In October ARPA sent its latest revisions to the draft document to the NSF. To ARPA's surprise in late November or very early December, the NSF submitted a significantly revised plan that was a major departure from the previous drafts. ARPA would be responsible for the management of AIO in 1969 with NSF contributing $900k through AFOAR. In 1970 (i.e. starting on 1 October 1969) control would shift to the NSF with ARPA contributing $600k. For 1971 ARPA would again contribute $600k and after 30 September 1971 the AIO would be the sole responsibility of the NSF. There would be no Coordinating Committee. NSF indicated that it planned to make AIO a National Research Center similar to its other Centers such as NRAO.

In a 4 December 1968 memorandum to Rechtin, Beard and Mann,[13] ARPA Deputy Director Lukasik wrote "As a foundation for an appropriate joint-management of Arecibo, I feel that the new version is deficient." He requested comments on the NSF draft in preparation for a counter proposal. He then made a telling comment, "A result of calling AIO a National Research Center is that the Director of AIO should report to a higher level in the Cornell management than Gold. NSF has been reluctant to try to push Gold out of the picture in the past, but this may be one of the benefits of going along with this new NSF direction." Gold certainly had had a very poor relationship with Colonel Dowe, AIO's Program Manager until September 1968, especially over the 430 MHz feed issue and clearly Gold also had difficulties with ARPA's upper management.

After some negotiations ARPA accepted the NSF plan. The final agreement had ARPA as the contracting agency for 1969 with both ARPA and the NSF contributing $900k to the $1.8M AIO operating costs. After this the NSF would become the contracting agency with ARPA contributing $600k to the operating costs in 1970 and only $300k in 1971. Capital funds could be provided by either agency in support of their approved programs. If DoD had any continuing interest beyond the initial

[13] Available in the Cornell University archives under Arecibo Ionospheric Observatory.

3 years, then they could negotiate an agreement with the NSF during 1970. No classified work would be done and there would be no classified repository. The agreement was signed on 12 May 1969, by Leland Haworth for the NSF and John Foster for the Office of Defense Research and Engineering. Since the Bracewell Committee, the Arecibo Evaluation Panel, reported to ARPA, the agreement also spelled the end of the Committee. Its last meeting was on 30 January 1970, a little over 7 years since its first meeting in Puerto Rico on 27 and 28 November 1962.

The agreement was signed just in time. Six weeks later, on 27 June 1969, there was article in the Washington Post stating that the House of Representatives had cut the Administration's budget request for the NSF by $80M, from $480M to $400M, $95M less than the previous year. This lower number was the Foundation's budget for FY1970 and it did not recover to $513M until FY1971.[14] Had this massive cut occurred before the AIO agreement was signed, it is not clear that the NSF would have agreed to take over Arecibo.

The remaining issues were between Cornell and the NSF; would Cornell continue to be the sole manager of Arecibo now reporting to the NSF and what would any new management structure look like? While the NSF clearly disliked the then current structure where the AIO Director reported to the CRSR Director, Tom Gold, with little involvement by the University, the fundamental issue was whether it was appropriate for a single university to manage a large national research center.

In the summer of 1968 with the prospect of the NSF taking over the responsibility for the AIO looking very likely to happen, the Cornell administration was focusing on two issues related to Arecibo: the projected flat budget of $1.8M per year and the need to give the AIO more of the appearance of a National Center. This did not mean changing the basic structure of the Observatory director reporting to the Director of CRSR but rather the adding of several advisory groups to CRSR and the observatory. In one plan suggested by Franklin Long, an AIO National Policy Board would provide input to the CRSR Executive Committee chaired by the Vice President for Research. This Board would be established "to develop the policies under which AIO operates and to work with Cornell and the supporting agencies to insure that AIO policies and programs are optimal for a national laboratory." Two committees would provide input to the observatory director—a Users Committee and a Committee on Programs of Applied Sciences. This last committee's role would be to "search on a continuing basis for programs wherein AIO could make a contribution to the applied research and development programs of the US." The National Policy Board would meet three times a year, including once in Arecibo. The CRSR Executive Committee would meet four times a year, at least once in Arecibo. The Users Committee would meet 3 or 4 times a year with half its meetings in Arecibo. It is not clear how often the Committee on Programs of Applied Science would meet but the total number of committee meetings per year would be over a dozen, clearly a costly plan and one that Gordon Pettengill, the newly appointed Director of AIO,

[14] At that time the Federal Government's fiscal year was 1 July to 30 June. The shift to the current fiscal year of 1 October to 30 September did not take place until 1976.

was concerned about. However, it did make clear that at that time Cornell was not planning on changing the basic management structure; the Arecibo Director would still report to the CRSR Director, a far cry from a National Center.

On 9 August 1968, Franklin Long, Cornell's Vice-President for Research, met with Steve Lukasik (ARPA) and Jerry Fregeau (NSF) at Cornell to discuss the arrangement for the management of AIO between ARPA, NSF and Cornell. At that time the plan for the transition still included an interagency committee (the ICA), in practice Lukasik for ARPA and Randy Roberson for the NSF. Fregeau and Lukasik were generally favorable to the plan to continue with the Arecibo Director reporting to the Director of CRSR and the various committees that Cornell proposed. However, Fregeau pointed out that this was a plan for the transition to the NSF and the Foundation had still not decided what institution would manage AIO after 1 October 1971. Long, in his memorandum written after the meeting,[15] had the feeling that the NSF would continue with Cornell. One difficult issue was the budget for 1969. Cornell had submitted one for \$2.4M and then a revised one for about \$2M. Lukasik made it clear that the funds available for operations would be \$1.8M and urged Cornell to submit a budget to this number as soon as possible. Needed capital items such as the 430 MHz feed could be requested separately and either agency could fund them as they wished. Lukasik also made it clear that the budget would likely be flat funded, or close to it, for the following year and, possibly, the next.

Long met with ARPA Director Rechtin and Lukasik in Washington on 16 August bringing the new budget written to \$1.8M with him. Some relief was given in that Lukasik had agreed to add additional funds of about \$50k to cover the costs associated with the painting of the suspended structure and its support cables. Capital items specific to certain projects would be funded by the agency most interested in the project.

On 3 February 1969, Long met with Fregeau (NSF), who stated that NSF "had just about concluded that it wishes to treat AIO as a National Center".[16] Fregeau also said that it is not incompatible with this concept for there to be a single university management and that they are, therefore, willing to give very serious consideration to the operation of Arecibo as a National Center by Cornell. If this happened then Fregeau said that the NSF would definitely expect the Arecibo director to report to Cornell at a high level, most probably something like the Vice-President for Research. Long told Fregeau that "we had not been terribly happy with this somewhat elaborate organization which our initial discussions had led to" but told Freguau that they would like to move ahead with appointing the Users Committee. Fregeau had no problems with this. He went on to say that Cornell should develop a specific financial request to make the handling of visitors feasible.

[15]F. A. Long, Memorandum for the Record, Discussions on AIO with Steve Lukasik from ARPA and Jerry Fregeau from NSF, 9 August 1968. Cornell University archives.

[16]Long Memo for the Record, 6 February 1969. Cornell University archives.

Long's interactions with the NSF in February 1969 must have been interesting as this was the month that he was nominated to be the Director of the NSF. However, he had been a strong critic of the Nixon administration's anti-ballistic missile (ABM) Safeguard program and President Nixon vetoed the appointment explaining at a press conference on 18 February that Long's opposition to Safeguard made it impossible for him to make the appointment. The resulting outcry caused Nixon to reconsider his decision and in April Long was asked through intermediaries if he would resubmit his name for the NSF Directorship. He declined.[17] In June 1969 Long stepped down as Vice President for Research and Advanced Studies at Cornell. He was succeeded by W. Donald Cooke, who, like Long, was a long time Professor of Chemistry at Cornell. William McElroy took the oath of office as director of the NSF on 14 July 1969.

On 1 October 1969, the date the NSF became the contracting agency, Cornell did an end-run around the issue of the Arecibo director reporting to a high level person by appointing Gold as Assistant Vice-President for Research, Arecibo Affairs. It is not clear what the NSF thought about this but they clearly agreed to it. Frank Drake would coordinate both the AIO related Ithaca personnel and the modifications to the suspended structure of the telescope related to the reflector resurfacing project. He had returned to his campus positions as Professor of Astronomy and Associate Director of CRSR in August 1968. Gordon Pettengill, the Associate Director of the Observatory under Bill Gordon, returned as the new Observatory Director on 1 November 1968 remaining in the position until December 1970. Pettengill entered into discussions with Randal Robertson, NSF Assistant Director for Research, about changing AIO's name. The initial suggestion was "Arecibo National Observatory" but, as an acronym, this had distinct problems in Spanish. The next one was "Arecibo National Radio Observatory" but Robertson did not like this as it may be viewed by the ionospheric community as downgrading the importance of ionospheric research at Arecibo. He suggested going back to the original name from before Arecibo was constructed, "Arecibo Observatory (AO)". Pettengill agreed and the name change took place on 1 October 1969.

As of 1 October 1969, the AO was clearly not a National Center in practice, it was still largely a university based observatory with a limited visitor program. As Frank Drake observed, "You can't make a laboratory a National Observatory by just calling it so." The problem facing Cornell was how to create an organization similar to NRAO but managed by a single university rather than a university consortium similar to AUI. The major issue was the old one: was the AO to be run by the director in Arecibo reporting to Cornell management or by Gold or his equivalent in Ithaca? If by the Arecibo director, then would the Arecibo related faculty and technical staff in Ithaca report to the Arecibo director? Other major issues were the need to bring visitor usage from about 20% to greater than 50% of the observing time, keep the ionospheric program at a substantial level, set up a committee structure with national

[17] Political Criteria for Non-Political Jobs, Eugene Rabinowitch, Bulletin of the Atomic Scientists, Science and Public Affairs, Page 2, June 1969.

representation to provide input and direction on management and the needs of the research programs, and the need for a management organization for the upgrading project. All these had to be resolved in the face of budgets that had barely changed since the $1.8M 1967 budget and which were projected to stay flat for several years. Drake attended the 1969 International URSI Congress and reported to Corson and Gold on 3 September that NRAO had described the extensive support they provide to visiting scientists and that there was no way that AO could deliver the equivalent given the funding level.

At an 18 September meeting at Cornell, Gold said that NSF Director McElroy had told him privately that the NSF planned to set up a National Centers office separate from the astronomy and atmospheric programs and Gold recommended delaying any agreements with the NSF until this happened. In early November, the NSF announced a reorganization, which included appointing four new Assistant Directors including an Assistant Director for National and International Programs.[18] Included in this division would be a new Office of National Centers and Facilities Operations to be headed by Daniel Hunt, formerly Special Assistant to the NSF director. Each Assistant Director would have a deputy and on 3 November Thomas Jones, formerly Special Assistant to the Director for Antarctic Affairs, was appointed Acting Deputy Assistant Director for National and International Programs. The Assistant Directors were presidential appointees requiring Senate confirmation and Thomas Owen, a rear-admiral in the Navy and former Chief of Naval Research, took up the position of Assistant Director for National and International Programs in January 1970. Cornell dealt primarily with Dan Hunt and Gold's interactions with him were largely cordial. This was not the case with Owen and Jones.

On 16 December 1969 Gold wrote to all AO staff members and related faculty laying out the new Arecibo Project management plan.[19] The AO would no longer be part of CRSR, and a separate Arecibo Project Office would be set up in Ithaca headed by the Assistant Vice President for Research (Arecibo Affairs) and Arecibo Project Director, both positions to be held by Gold. Assisting Gold would be an Administrative Director (Frank Bergen) and a Chief Engineer (Merle LaLonde). Arecibo related Cornell faculty, research associates and graduate students would form the Ithaca Scientific Group under Frank Drake, who would report to Gold and work with the AO Director. The AO would be under the Observatory Director, Gordon Pettengill, and report to Cornell through the Assistant Vice President for Research (Gold). The technical staff in Ithaca at the Danby Laboratory south of the city and on the Cornell campus would be under the Chief Engineer, who would coordinate the groups engineering activities in support of the AO with the AO Director. Two new committees would be formed, The Arecibo Policy Committee consisting of five

[18] National Science Foundation press release dated 5 November 1969. NSF Director McElroy wrote to Cornell President Dale Corson on 28 November describing the reorganization and enclosing the Press Release. Corson replied on 4 December. Cornell University archives.

[19] Memo to Arecibo Project Staff from T. Gold dated 16 December 1969. Cornell University archives.

members of the scientific community plus the Cornell Provost and the Vice President for Research as ex officio members, and the Scientific Advisory Committee. The Policy Committee would be appointed by the Provost in consultation with the NSF and would report to the Provost. The Scientific Advisory Committee would be appointed by the AO Director in consultation with the NSF and would report to the AO Director. The management organization at least temporarily settled the question of where the management of Arecibo would reside, definitely in Ithaca. However, it did not address the issues related to increasing the number of visitors, maintaining the scientific programs in the face of flat budgets, and the management of the telescope upgrading project.

Tom Gold was focused on the scientific output of Arecibo and was willing to sacrifice other aspects of the operation of the observatory to maintain its scientific productivity. He outlined this philosophy in a letter to Tom Jones on 30 January 1970,[20] in which he pointed out, referring to the budget situation, that "an operation which keeps the door open, keeps maintenance, administrative, technical support and guard personnel, but can hardly afford any scientific work, is obviously most uneconomical, if the scientific work is the sole purpose." However, he went on to say that in the hope that the budget situation would soon improve and because of the difficulty in obtaining trained personnel if staff is let go, he would maintain the current operation.

This all came to a head with the submission of the $2.2M FY 1972 Arecibo operations budget proposal in September. To the great surprise of the NSF, the proposal included the termination of the entire Arecibo based ionospheric group. On 2 October Hunt wrote Gold rejecting this plan and insisting that a viable ionospheric program should be included in an updated budget proposal by adjusting the costs of the astronomy programs, deferring equipment purchases, etc. He did promise to try to provide additional funding but pointed out that all the national centers were suffering due to the constrained NSF budget. Gold replied saying that Cornell should be the judge of what research areas can or should be supported. Invoking his desire for scientific excellence and the budgetary constraints, Gold made the argument that "it is better to concentrate on the areas where we have strength and where the upgraded instrument will be pre-eminent, and abandon the in-house operation where our position is not so impressive." Gold went on to state that the best ionospheric research was being done by visiting teams, especially the one under Bill Gordon, and that Arecibo would definitely continue to support these groups. Basically, Gold felt that the radio and radar astronomy programs were producing excellent results and that the results of the routine measurements of the properties of the ionosphere being done by the in-house group were not productive or very useful.

The threat to terminate the in-house ionospheric group produced a "fire-storm" within that community. On 30 October, Bill Gordon wrote to a number of senior researchers in ionospheric physics providing a draft of a letter to Tom Owen that he wanted them to sign. The fundamental argument in the letter was that the in-house

[20]Gold to Jones, 30 January 1970. Cornell University Archives.

staff are needed to support the many visiting scientists, especially given the new endeavor of studying non-linear effects in the ionosphere due to "heating" by radiating high frequency (3–12 MHz) power that is absorbed by the ionosphere at the height where the plasma frequency is very close to the frequency of the "heating" signal. Included was a table showing the large number of visiting ionospheric researchers and graduate students. Gold saw the letter and responded to Gordon reiterating that the decision was based on the very severe budgetary situation and that the in-house ionospheric program was producing rather routine results compared with the radio and radar astronomy programs. This was compounded by the difficulty in attracting high quality staff in ionospheric physics.

Gordon replied to Gold on 31 December with the first sentence being "It has been my opinion for the past several years that Cornell should be relieved of the management of Arecibo." He went on to admonish Gold for not involving the faculty members in ionospheric physics, Don Farley and Neil Brice, in the operation of Arecibo and stating that Arecibo was held in high regard by the aeronomy community and that they would solve the staffing problem if ionospheric research had the support of management. Gold, of course, responded rebutting Gordon's arguments and, again, pointing out the severity of the budget situation. He also stated that neither Farley nor Brice had shown any interest in using or getting involved with Arecibo. All of these letters were copied to Owen, Jones and Hunt at the NSF and several other members of the aeronomy community including Brice.

These acrimonious letters likely convinced the Cornell upper management that a radical change was required; Tom Gold would never have a good working relationship with the National Science Foundation. On 22 March 1971, Cornell President Corson wrote NSF Director McElroy telling him that Gold would relinquish any role in the Arecibo Project as of 30 June and suggesting that Frank Drake replace him. After seriously considering the NSF's suggestion that Arecibo be directed from Puerto Rico, Corson stated that Cornell felt that it would not be possible to attract and retain a director of high scientific stature to such a position and that the Arecibo Observatory Director should be based in Ithaca and would report to the Vice President for Research. This would also keep close ties between Cornell and its faculty with Arecibo. The University did not want to just manage the operation; it wanted intellectual involvement. The director in Arecibo would be designated as the Arecibo Director of Operations. An organization plan was included with the letter, which also had a detailed list of the responsibilities of the two directors, and showed two committees, the Arecibo Advisory Board (later called the Visiting Committee or VC) reporting to the Cornell President and the Scientific Advisory Committee (later the Arecibo Users and Scientific Advisory Committee, AUSAC) reporting to the Director of Operations. In line with the names of the other national centers, Corson suggested that the name be changed to National Astronomy and Ionosphere Center at Arecibo (NAICA). (This was later thought to be unwieldy and the "at Arecibo" was dropped leaving just NAIC.)

To respond to some concerns at the Foundation about Cornell's ability to manage the upgrading of the telescope, Corson stated that Bob Matyas, the Director of the

Fig. 8.6 Tor Hagfors about the time that he was appointed Director of the Arecibo Observatory. Credit, Cornell University

university's construction office, who had managed the construction of the Cornell synchrotron, would devote up to 100% of his time managing the upgrade.

On 30 March McElroy replied to Corson saying, "I strongly endorse the concepts of organization and management for the National Astronomy and Ionosphere Center at Arecibo that you have put forth." He also promised that the NSF would increase the FY1971 and 1972 budgets to provide support for the ionospheric program.[21] In May 1971, $200k was added to the budget specifically designated to pay for the retention of the in-house ionospheric group. Gold had won but at a high cost personally in that he lost control of Arecibo. The observatory would remain under Cornell management for the next 40 years. Gordon Pettengill resigned as the AO Director in December of 1970 to take up a faculty appointment at MIT and was replaced on 1 August 1971 by Tor Hagfors (Fig. 8.6), a highly respected scientist from Lincoln Laboratories, who worked in the areas of ionospheric physics and radar astronomy. This appointment must have helped to calm the concerns of the ionospheric community. Rolf Dyce was the acting director between Gordon Pettengill's departure and Tor Hagfors arrival.

Two major technical projects continued during the transition from ARPA to the NSF and the creation of NAIC: the development of an efficient dual polarization 430 MHz line feed and the upgrading of the telescope with the replacement of the reflector surface and modifications to the suspended structure, including the installation of a high powered transmitter operating at 2.4 GHz.

[21] Dale Corson to William McElroy, 22 March 1971 and William McElroy to Dale Corson, 30 March 1971. Cornell University archives.

8.4 Success at Last, a Circularly Polarized 430 MHz Line Feed

With the transition in September 1968 from Mike Dowe to Alvin Van Avery as ARPA Program Manager for Arecibo, the impasse over the way forward for the 430 MHz line feed project was overcome. In a discussion with Rodney Dennis on 26 February 1969, Van Avery said that Autonetics should make a proposal to continue the feed work they have started. On 22 March Autonetics submitted a proposal to ARPA to complete the construction and testing of a scaled model feed at 2810 MHz, and to provide a "cook book" from which to build a full sized 430 MHz version. A contract was awarded to Autonetics by ARPA through AFOSR for $93,370 on 22 September 1969, with the final report due 26 January 1971. The design was based on Love's 1967 work, a tapered circular guide with six circumferential openings and disks between the rectangular opening to change the phase of the field of one the polarizations by 90°. Five short sections at various locations along the feed were fabricated based on theoretical calculations of the waveguide diameter, which determines the phases of the radiated fields, and the vertical and horizontal sizes of the openings, which change along the length of the feed depending on the desired aperture illumination. Based on satisfactory results for the measurements of the phase and amplitude of the radiated fields of the test sections, a complete model was constructed and successfully tested (Fig. 8.7; Love 1973).

Fig. 8.7 Love's "ray gun" 15 ft (4.5 m) long scaled 430 MHz model feed undergoing far field amplitude and phase tests. From Arecibo 430 MHz Feed Development Study, Final Engineering Report, Autonetics, April 1971. Cornell University Archives

Fig. 8.8 The new 96 ft long 430 MHz feed installed on the telescope. Credit: Cornell University/ NSF

Autonetics delivered the Final Report on the feed study in April, 1971. It listed Alan Love as Principal Investigator and was approved by his old mentor, Carl Wiley. The report predicted that the feed would provide 70.5% aperture efficiency if the primary reflector was perfect. Given the irregularities in the wire mesh reflector, the aperture efficiency was predicted to be 51%, a factor of 2.3 improvement over the old square feed. Included with the report were the detailed dimensions of every aspect of the approximately 90 slotted waveguide sections and rings needed for the 430 MHz feed. Using this information, Merle LaLonde organized the construction of the final feed by a local Ithaca machining firm, Therm Inc. Unlike the self-supporting square waveguide feed, the new feed needed to be supported by a system of guys in a manner very similar to a sailboat mast. Because of the high RF powers radiated by the feed, the guys could not be made of a conducting material, stainless steel was then the normal guy material for sail boat masts. However, a new material, Kevlar, had just been developed by DuPont and DuPont made some of this available for the feed. LaLonde had the Kevlar fibers encased in a plastic sheath to form the guys, possibly the first use of Kevlar for this purpose and one now common on sail boats.

The feed was completed in late 1971 and mounted on the telescope in early February 1972 (Fig. 8.8). It worked slightly better than predicted giving 53% aperture efficiency, a factor of 2.4 over the square waveguide feed. The sensitivity for radar studies of the planets increased by a factor of about six, 2.4 on transmitting and 2.4 on reception of the echo. While Alan Love, along with Vic Rumsey, Merle LaLonde and others deserve great credit for the development of a successful design for a dual polarized feed for a spherical reflector, Tom Gold also deserves considerable credit for his persistence over 8 years in pushing for the development of the feed. The successful development of the 430 MHz dual polarization feed was critical to the future utilization of the resurfaced reflector as it allowed similar feeds to be built at 2380 MHz for the new planetary radar system, at 1420 MHz for studies of neutral hydrogen in our galaxy and distant galaxies, and at a number of other frequencies. Even though the radio astronomy observations were not as dependent as the radar observations on having dual circular polarizations, the ability to receive two polarizations simultaneously rather than a single linear polarization with flat feeds increased the sensitivity for most observations by a factor of 1.4 and also allowed the polarization properties of the radio wavelength emission from the source to be determined.

8.5 A Welcome Addition to the Observatory

During his tenure as the Arecibo director, Frank Drake persuaded Cornell University to fund the installation of a swimming pool. It was built in the valley across the parking lot from the VSQ (Fig. 8.9). A flat area was created next to the pool for a volley ball court. This may seem like a relatively small event but it had a tremendous impact on the social life of the observatory. All staff and their families plus visitors were welcome to use the pool and the volley ball court was used during lunch hours.

Fig. 8.9 A view of the swimming pool and volley ball/basketball court. Credit: Cornell University

Many events, large and small, were held at the pool with a big tent erected over the volley ball court for large events for the staff and visiting groups.

References

LaLonde, L.M., and D.E. Harris. 1970. A High-Performance Line Source Feed for the AIO Spherical Reflector. *Institute of Electrical and Electronics Engineers* AP-18: 41–48.

Love, A.W. 1973. Scale Model Development of a High Efficiency Dual Polarized Line Feed for the Arecibo Spherical Reflector. *IEEE Transactions on Antennas and Propagation* AP-21 (5): 628–639.

Kay, A.F. 2008. *Militarist Millionaire Peacenik: Memoir of a Serial Entrepreneur*. Cosimo.

Chapter 9
Science and the First Upgrade

The early research results described in Chap. 7 were just the beginning of a flood of results in the lead up to the first upgrading of the telescope, the replacement of the reflector and installation of a high-powered transmitter for planetary radar observations, and a deluge of them post-upgrade. Despite the reduced sensitivity due to the feed problems, Arecibo was one of the most powerful radar systems in the world and it was used to make significant advances in our knowledge of the Moon and terrestrial planets as well as of the Earth's ionosphere. With the discovery of pulsars in 1967 (approximately 20 km in diameter balls of mainly neutrons, neutron stars, with masses about 1.4 times that of the Sun left over from the supernova explosions of massive stars), Arecibo came into its own as a radio astronomical telescope. The resurfacing of the telescope's primary reflector with aluminum panels and the installation of a high-powered transmitter operating at 2.38 GHz in the early 1970s, opened up new research areas in radio astronomy and greatly boosted Arecibo's sensitivity for radar studies of the terrestrial planets, satellites and asteroids and comets.

9.1 Computing and Data Acquisition

It was clear by 1965 that achieving many of the science objectives meant that the CDC 160A computer and the digital data acquisition system needed to be upgraded to allow faster processing of data and more flexibility in data acquisition and software control of external equipment. A CDC 3200 was rented from the Control Data Corporation and delivered to the observatory in mid-1965. It had 32K of 24-bit words of memory, four magnetic tape drives and two disk drives with removal platters that held 2M words or 6 Mbytes. Trivial by today's standards but a huge improvement in the observatory's computing capability. The CDC3200 had real time capability in that it could accept digital inputs from analog-to-digital converters similarly to the CDC 160A and it could also write to external devices such as

D. Campbell, *The Arecibo Observatory: A History of Innovation and Discovery*, Historical & Cultural Astronomy, https://doi.org/10.1007/978-3-031-75225-4_9

Fig. 9.1 The Arecibo CDC3300 computer in April 1972. By modern standards these were huge machines as they pre-dated integrated circuits using individual transistors on printed circuit boards. They were also expensive; the rent, including an on-site maintenance engineer, consumed about 10% of the observatory's budget. Credit: Cornell University/NSF

frequency synthesizers. Initial input was via punched paper tape. This was changed with the installation of a card reader for reading punched cards in late 1966. In early 1968, computer control of a frequency synthesizer was finally realized, greatly improving the radar observations of solar system bodies as it allowed the automatic adjustment of the local oscillator to track the changing radial velocity and distance to the target as the Earth rotates and the target moves in its orbit.

Gordon Pettengill and Miguel Feyjoo worked with the ADAGE corporation to build a special purpose data acquisition system with dual 7-bit analog-to-digital converters, each capable of sampling up to eight input channels sequentially at rates as fast as 3 μs. It also had a digital-to-analog capability that was very useful for displaying plots on oscilloscopes. The new interface, which cost about $65,000, was also installed in mid-1965. In one of the first major uses of the new ADAGE interface and the CDC3200, Ron Wand and Francis Perkins developed software that allowed spectra of the incoherent scatter echoes from a range of heights in the ionosphere to be obtained simultaneously. This ended the use of the analog RAYSPAN system (see Chap. 7) to no one's regret.

In early 1969 the CDC 3200 was changed for a CDC 3300 (Fig. 9.1), which had similar architecture and capabilities to the 3200 but resolved some reliability issues with this computer. One capability was a reliable foreground/background feature that allowed normal computations and data acquisition to take place at the same time. An interrupt system would stop the foreground computations whenever the ADAGE data acquisition system needed to transfer data to the computer.

Elliott Mayo joined the observatory's computer department in early 1968 working with Hiram Cabassa. After 6 years at the observatory, Cabassa left in the summer of 1969 and Mayo took over as head of the department. He was joined by Peter Watkins, a Canadian, in mid-1969, who in turn took over the computer department when Elliott Mayo left in July 1971 for what turned out to be for just a year. Watkins left in July 1972 and Mayo returned to his former position staying until December 1975.

9.2 Radio Astronomy: Three High Impact Areas

Radio source positions, size and structure: The approximate angular resolution of a telescope is given by the ratio of the wavelength used to the telescope's diameter. For an optical telescope with a diameter of, say, 5 m, the size of the California Institute of Technology's Palomar telescope, using a visible light wavelength of 0.7 μm, the nominal resolution is 0.03 arcseconds. In practice, turbulence in the Earth's atmosphere limits this to about 1 arcsecond although techniques have now been developed to overcome this. Because radio wavelengths are extremely long relative to those of visible light, a 5 m antenna operating at the wavelength of, say, 10 cm, would have an angular resolution of about 1 degree. Even Arecibo, with its 300 m diameter, had an angular resolution at 50 cm wavelength, its shortest operating wavelength in the 1960s, of about 6 arcminutes (0.1 degrees), not accurate enough to obtain positions of radio sources that can be used to find optical counterparts or to put meaningful upper limits on the size and structural details of sources with very small angular sizes.

Through much of the 1960s, three techniques were used to overcome these limitations on single dish radio telescopes for the study of small angular size radio sources, lunar occultations, interplanetary scintillations, and connected interferometers. As discussed in Chap. 7, in 1964 Marshall Cohen initiated a lunar occultation program at Arecibo consisting of a survey of the path of the moon over the sky looking for radio sources that the Moon would occult at a future date and then doing the occultation observation. Cohen left Arecibo about the time that Cyril Hazard arrived from Australia and Hazard vigorously pursued the occultation program. As recounted in Chap. 7, Hazard had used the Parkes 64 m telescope to determine the position of 3C273 using the lunar occultation technique, which led to the identification of the first known Quasar.

Scintillations In the early 1960s, Anthony Hewish and his group at Cambridge University discovered that fluctuations in the density of the solar wind close to the Sun caused small angular size radio sources to scintillate, a phenomenon equivalent at radio wavelengths to the twinkling of stars as their light passes through the Earth's atmosphere (Hewish et al. 1964). Again, it was Marshall Cohen who initiated a program at Arecibo utilizing the information contained in the scintillations to place upper limits on the angular size of the smallest component of radio sources. Scintillation measurements of the first quasar, 3C273, showed that 20% to 30% of its intensity came from a source within 3C273 with an angular size of less than 0.02 arcseconds (Cohen et al. 1966). At the distance of the quasar giving rise to the radio source this corresponds to less than about 90 light years. The resolution of 0.02 arc seconds was about five time better than could be achieved by connected interferometers at that time. The Arecibo group, primarily Marshall Cohen, Dan Harris and Ellen Gundermann, pursued a scintillation observation program until it became clear in the late 1960s that very long baseline interferometry (VLBI) was an even better and more versatile technique to study source size and structure.

Very Long Baseline Interferometry (VLBI) Two or more widely spaced radio telescopes which, when their outputs are combined, provide an angular resolution approximately equivalent to an antenna with a diameter equal to the largest spacing of the interferometer antennas. Connected interferometers, where the high frequency signal impinging on each antenna is converted using a common "local oscillator" to a much lower frequency without losing any of the inherent information, were in relatively common use from the late 1940s. For a two element interferometer, as the antennas scanned over the selected radio source, the product of the two signals would show a sinusoidal oscillation, referred to as fringes, if the source, or some component of it, has an angular size less than the equivalent resolution of an antenna with the same size as the length of the baseline, the distance between the two antennas. The baseline was typically a few hundred meters to a few km but no more than about 100 km, as it depended on being able to send the common local oscillator signal to each site.

As discussed in Chap. 8 of Kellerman et al. (2020), in 1965 two groups, one in Canada and one in the United States, started to investigate the possibility of using much longer baselines and independent local oscillators to improve the angular resolution, very long baseline interferometry (VLBI). One hurdle was the need for the independent local oscillators to have precisely the same frequencies and, related, knowing the time at each antenna with sufficient accuracy to know when the same signal from a radio source arrived at each antenna. These difficulties were overcome by using Rubidium frequency standards, which had a frequency stability of 1 part in 10^{11}, and associated clocks, plus a traveling very high precision clock to synchronize the times at each antenna. This was long before the Global Positioning System provided accurate time anywhere on Earth. The second hurdle was the need for high speed tape recorders and computers with sufficient power to combine the data from the two telescopes in an acceptable time. The Canadian group decided to use analogue recorders while the U.S. group chose digital tape recorders. The U.S. group consisted of Barry Clark and Ken Kellermann of NRAO and Marshall Cohen and David Jauncey, who had recently arrived from Australia, of Cornell University. Funding was split between both institutions (Kellerman et al. 2020). A system with a digital sampling rate of 720 kHz corresponding to a radio frequency bandwidth of

360 kHz was developed, the fastest data rate that could be written continuously to a magnetic tape at that time.

Perhaps given the Cornell connection and Arecibo's large sensitivity, the first serious U.S. attempt in January 1967 used the new 140 ft (43 m) telescope at NRAO's Green Bank, West Virginia site and Arecibo at a frequency of 611 MHz (50 cm wavelength), Arecibo's maximum operating frequency at that time. A "point" feed (see Fig. 8.4) must have been used as this predated the installation of Allen Love's 611 MHz flat line feed. No fringes were detected. A second try in February was equally unsuccessful. The equipment was shipped back to Green Bank and a test conducted between two antennas 650 m apart; it was successful. The next test on 8 May was from an 85 ft telescope at Green Bank to an 85 ft antenna of the Naval Research Laboratory at Maryland Point, a baseline of 220 km. It was also successful. See Kellerman et al. (2020) for a more detailed account of the early observations.

Progress was rapid. In August 1967 a transcontinental observation between the 140 ft telescope at Green Bank and an 85 ft telescope at Berkeley's Hat Creek observatory in northern California, a baseline of 3500 km, was successful giving an angular resolution of 0.01 arcseconds (Clark et al. 1967). Also in August, fringes were finally obtained between Arecibo and the Green Bank 140 ft telescope at 610 MHz (Jauncey et al. 1970). However, with the success of the intercontinental baseline observation and the move to shorter wavelengths, both of which gave improved resolution, there was little point in making additional observations with Arecibo with its 50 cm lower limit on the wavelength except where there was a need to use lower frequencies. One example was VLBI observations of the Crab Nebula pulsar at 111.5 and 196.5 MHz to study scattering of radio waves by the interstellar medium, a low frequency phenomenon. This was done by Vandenberg et al. (1973) in late 1969 and early 1970 utilizing Arecibo with the 92 m Green Bank telescope and with the 46 m Sugar Grove telescope about 50 km north-east of Green Bank. Arecibo's time as a significant player in VLBI science would come with the move to shorter wavelength capability after the resurfacing of the reflector in the early 1970s combined with its tremendous sensitivity.

Cornell's early involvement likely motivated Thomas Gold, CRSR Director, to take an interest and in a paper published in the journal *Science* in 1967 (Gold 1967) he pointed out that angular resolutions of better than 0.01 arcseconds would allow about 10 times better accuracy in measurements of the rotation period of the Earth. This was one of the papers that sparked a whole new discipline, Geodetic VLBI, the use of VLBI as a tool to monitor variations in the Earth's rotation period and movements of the Earth's tectonic plates among other applications. Marshall Cohen moved from Cornell to the University of California at San Diego in 1966 and, a few years later, to Caltech, where he continued to play a central role in the field of VLBI. David Jauncey remained at Cornell until late 1974, when he returned to Australia. He continued to play a central role in VLBI.

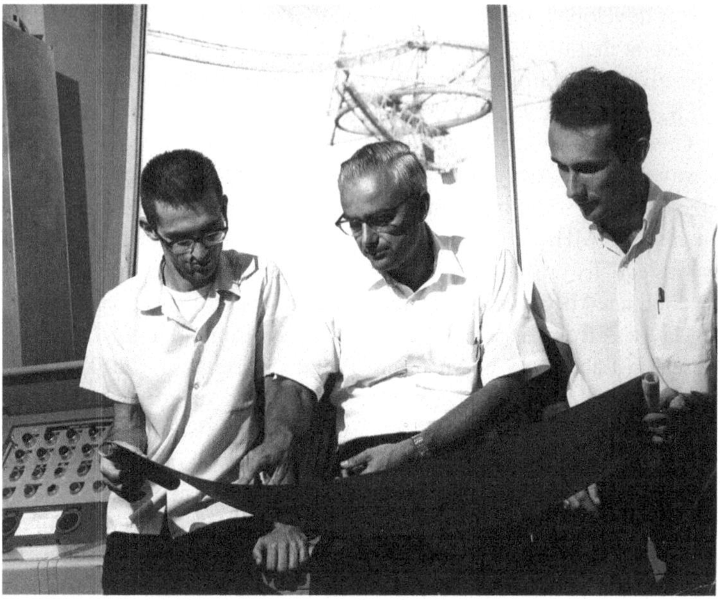

Fig. 9.2 Frank Drake (center) with then graduate students John Comella (left) and Harold Craft in the control room in 1968 looking at a strip chart recording. Credit: Cornell University

Pulsars With the discovery of pulsars by Jocelyn Bell, a graduate student at Cambridge University, in 1967, Arecibo jumped right in under the leadership of the then director, Frank Drake (Fig. 9.2). Arecibo was perfect for the study of these strange objects with its huge collecting area and capability to study pulses, whether reflected from the ionosphere or a planet or coming from a pulsar. At that time, most radio telescopes had very limited digital sampling systems as the maximum sampling rates required for typical observations were only a few samples per second. Because it was designed to study radar echo pulses, Arecibo had, for that time, high speed digital sampling systems with the rate limited by the writing speed of its magnetic tape recorders. This immediate advantage coupled with Arecibo's huge sensitivity and the fact that pulsar radio emissions are strongest at lower radio frequencies made Arecibo the ideal telescope to search for and study pulsars. It was the only telescope capable of studying the properties of single pulsar pulses, which are typically a few to tens of milliseconds long.

Drake recruited a number of Cornell graduate students to get involved with pulsar studies using Arecibo. John Comella and Harold (Hal) Craft were two of them (Fig. 9.2) and, of course, interested resident scientific staff became involved. At this time the origin of pulsar emission was still unknown. After the initial discovery by Jocelyn Bell of the first pulsar, CP 1919, there was talk of "little green men", but with the discovery of additional pulsars it was clear that there had to be a natural origin. But what was it? The most likely explanation was thought to be that they were related to pulsating stars, most probably pulsating white dwarf stars. These stars mark the final state of stars with masses less than a few times that of the Sun after they have exhausted all of their fusion derived energy and shed their outer envelopes with their core collapsing into an electron degenerate state with a diameter about the same as Earth and a mass about the same as the Sun. One critic of this explanation for pulsar emission was Thomas Gold, who proposed that pulsars were rotating neutron stars (as discussed above, the approximately 20 km diameter solar mass remnants left over from the supernova explosions of stars with masses above about ten solar masses) with very strong imbedded magnetic fields (Gold 1968). He pointed out that that the "great constancy of the intrinsic period suggests a massive object rather than some plasma physical configuration". He also suggested that most of the energy of the system would be in the rotational energy of the massive body and that a slight slowing down of the repetition rate of the pulsar may be expected as it radiated away its energy.

The first big pulsar science discovery at Arecibo came in November 1968 with the measurement of a very short period for a pulsar in the Crab Nebula, a remnant of the supernova explosion of a relatively nearby massive star in the year 1054, whose appearance was so bright that it was recorded in China. David Staelin and Edward Reifenstein had observed the Crab Nebula a few weeks earlier with the 92 m telescope at NRAO's Green Bank site looking for a pulsar. They observed four pulses over a 15 min interval but detected no periodicity from what they assumed was a pulsar, which they labeled NP 0532. Three Arecibo resident Cornell graduate students, John Comella, Halold Craft and David Richards plus resident scientist John Sutton and Len Tyler of Stanford University, observed the Crab nebula with high time resolution beginning on 9 November, 1968 and measured a pulse repetition period for NP 0532 that was the shortest of any pulsar to that date, 33 ms (Comella et al. 1969). This very short period gave considerable credence to Gold's theory as it was unlikely that a white dwarf star could exhibit a period this short. Further confirmation came a few weeks later with the detection at Arecibo of a slowdown in the rotation rate as predicted by Gold (Richards and Comella 1969).[1] In his autobiographical book Gold (2012) recounts his struggle to persuade his peers that pulsar emission originated from rotating neutron stars.

The Pop Culture Pulsar Because of Arecibo's ability to observe individual pulsar pulses, Harold Craft decided to base his PhD thesis research on studies of individual pulses with the aim of obtaining insight into the details of the pulse emission mechanism. As part of his research, he put together a sequence of pulse shapes for several pulsars including CP1919, the first pulsar discovered (Fig. 9.3). For still unknown reasons, a version of Fig. 9.3 appeared on the 1979 cover of the album

[1] A brief but interesting account of the discovery of the pulsar in the Crab Nebula can be found in Lovelace and Tyler (2012).

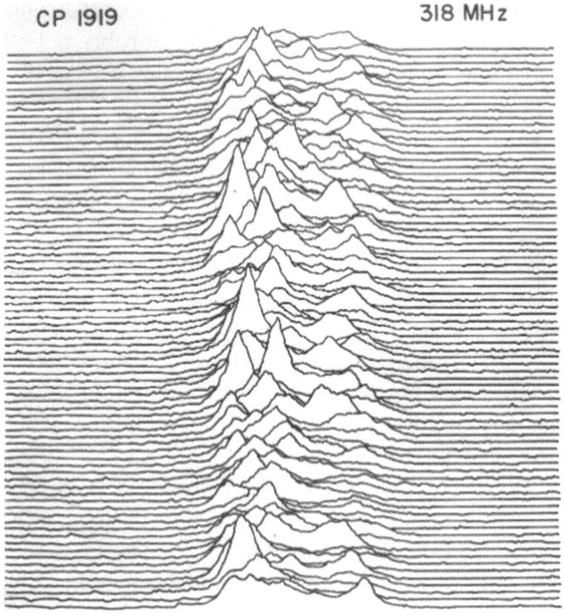

Fig. 9.3 The original image of the Pop Culture pulsar that appeared on the Joy Division band's album cover. It shows a series of successive pulse profiles for the pulsar CP1919 obtained with the Arecibo telescope. Credit: Harold D. Craft Jr.

Unknown Pleasures by the band Joy Division.[2] There was no other identifying information on the cover, just the sequence of pulsar pulses from CP1919.

Large Scale Surveys Merle LaLonde's and Allan Love's development of flat feeds at 318 and 611 MHz in the very late 1960s (see Chap. 8) opened the possibility of multi-beam systems for large scale sky surveys. Dave Jauncey, working with Cornell graduate students James (Jim) Condon, John Durdin and Dale Pleticha, initiated a 611 MHz multi-beam system, the final version of which consisted of ten 20 ft long flat line feeds designed and built by Merle LaLonde that were mounted in two rows of 5 along the sides of Carriage House 2 (Fig. 9.4). Michael Yerbury's Arecibo support group at Cornell designed and built the receiving system. The entire system was installed on the telescope in May 1972. Each line feed illuminated a 450 ft diameter area of the reflector giving a beam width of 13.5 arc min. The 10 beams on the sky were separated by 8.4 arc min. In use, the array was aligned north-south along the local meridian and scanned a 1.4-degree swath of the sky at whatever declination the telescope was pointing as the Earth rotated. A single pass over the entire sky visible with the telescope would take only about 650 h, the equivalent of 27 days. In practice, the swaths were separated by 0.7 degrees to give overlap doubling the time for a full survey. This was the first cm-wavelength multi-beam system on any large single dish radio telescope. However, by the time the system was installed on the telescope in May 1972 and began taking data later in the last

[2]An interesting article about the Pop Culture Pulsar and pulsars was written by Jen Christiansen and appeared in the October 2015 Scientific American. It can be found online under Pop Culture Pulsar.

Fig. 9.4 The 10 short flat feeds make up the 611 MHz multi beam system. The long 318 MHz feed behind the right middle 611 feed is mounted on the front center of the carriage house. Credit: Cornell University/NSF

quarter of the year it was too late to do a full survey. In the second quarter of 1974, all the feeds including the multi-beam system were removed from Carriage House 2 in preparation for replacing the carriage house with a much larger one that would support up to six line feeds and the new high powered planetary radar transmitter. This limited the time that the multi-beam system was in use. However, the survey was a precursor to many more surveys with the telescope over the following decades.

9.3 Planetary Radar Observations

Early Arecibo radar observations of the Moon and terrestrial planets concentrated on refining orbits and rotation periods and measuring the reflectivity and radar scattering properties of Mercury, Venus and Mars plus imaging observations of the surface of the Moon. One of the early surprising discoveries was that the rotation period of the planet Mercury was not 88 days, the same as its orbital period about the Sun, as long believed. Being close to the Sun, small and distant, Mercury was difficult to study with optical telescopes and all that was known was that it had bright and dark albedo features. Tracking of these albedo features seemed to confirm the belief that, like the Moon about the Earth, Mercury's rotation period was synchronous with its orbital period about the Sun. Using the same technique that had been used to measure the retrograde rotation period of Venus (see Chap. 7), Pettengill and Dyce

(1965) did a series of observations of Mercury during its 1965 close approach to Earth and showed that its rotation period was 59 ± 5 days, close to two thirds of its orbital period, a distinct surprise. The difference from the expected value of 88 days was quickly shown to be related to Mercury's highly eccentric orbit about the Sun and that the period was exactly two thirds of the orbital period, 58.6 days.

Venus was the real object of interest. Covered in dense white clouds of sulphuric acid droplets and with a dense carbon dioxide laden atmosphere driving an extreme greenhouse effect, the nature of its surface remained a mystery until slowly revealed by Earth based radar systems and then by both Soviet and US spacecraft born ones. Imaging a planetary surface with an Earth based radar requires measuring the echo power as a function of time delay, the echo from the point on the planet closest to Earth is received first and that from the "visible" edge or limb sometime later depending on the radius of the planet. The second dimension is obtained by measuring the echo power in each of the time delay intervals as a function of the Doppler shift again relative to that for the echo from the point closest to Earth. Due to the rotation of the planet,[3] the width of the Doppler spectrum increases as the distance from the center increases (Fig. 9.5).

Given the capabilities of Arecibo and the other Earth based radar systems in the US in the 1960s, the resolution of the radar images of Venus was 100–200 km, good enough to detect the presence of large features with very different reflection properties relative to their surrounding area but not good enough to identify what these features may represent in terms of the geology of the surface such as impact craters, volcanoes and mountain belts.

Looking at Fig. 9.5 it is clear that there is one serious problem with imaging the surface of Venus and other planets with Earth based radars, the annuli of constant delay and constant Doppler frequency cross at two locations north and south of the equator so that in a radar "image" the separate images of the northern and southern hemispheres are superimposed on each other, resulting in what is called the ambiguity problem. This north-south ambiguity was a serious complication for Earth based radar imaging of Venus and other planets. It was not a problem for the Moon. Arecibo's angular resolution at its transmitting frequency of 430 MHz, was close to 10 arc min, much smaller than the approximately 30 arc min angular diameter of the Moon. This meant that the telescope could be pointed at either the northern or southern hemisphere of the Moon with no power hitting the other hemisphere and, hence, no ambiguity problem as there is no radar echo from this hemisphere. Thomas (Tommy) Thomson, a Cornell University graduate student, mapped extensive areas on the Moon at about 30 km resolution for his PhD thesis work characterizing the

[3]The observed rotation period of a planet looking from Earth is made up of the sum of its intrinsic rotation and an apparent one due to the relative orbital motions of Earth and the planet. See Fig. 7.9 of Chap. 7, a plot of apparent rotation rates for Venus as a function of date around close approach of the planet in 1964.

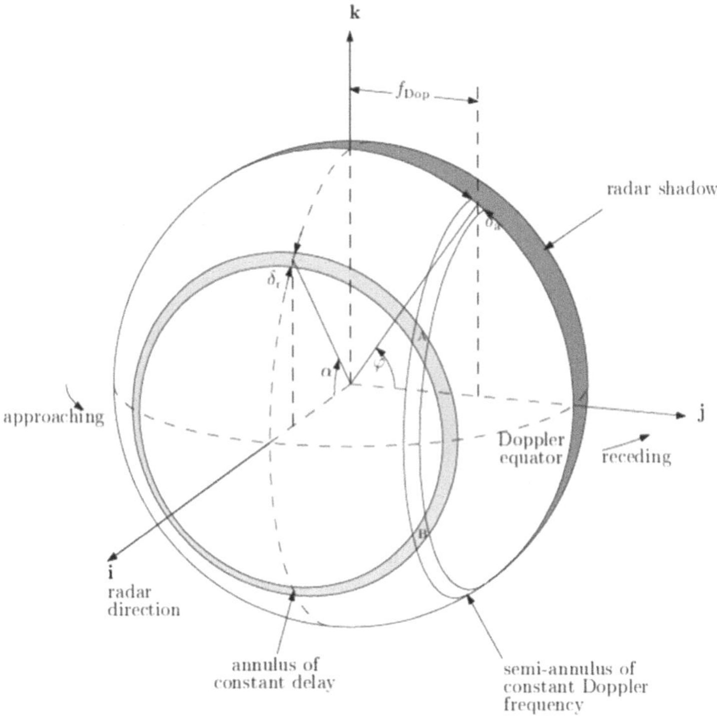

Fig. 9.5 The time delay-Doppler geometry for a spherical planet. Figure courtesy of Nicholas Stacy and Jean-Luc Margot

radar reflection properties of the dark Maria regions, the lunar highlands and large impact craters (Thompson and Dyce 1966).

Venus was first detected by radar during its close approach to Earth in 1961 using NASA's Deep Space Tracking Goldstone installation in California. This was the same close approach to Earth for which Arecibo was originally intended to be used to detect Venus had the original construction schedule been met. Close approaches to Earth occur every 19 months. For the next close approach of the planet in 1962, Richard Goldstein utilizing an antenna at NASA's Goldstone, California Deep Space Network detected two prominent radar-reflecting bright features on the planet's surface which he labeled α and β.[4] As discussed in Chap. 7, Arecibo was ready for the close approach of Venus in 1964 and detected the planet in February. Ranging measurements over the next few months refined our knowledge of the Astronomical Unit, the average distance between the Earth and the Sun, which sets the scale size of the solar system, and delay-Doppler measurements were used to refine the retrograde rotation period. Raymond Jurgens, a Cornell University graduate student, analyzed the delay-Doppler spectra and detected the two features that Goldstein had

[4]For a comprehensive discussion of the early radar imaging of Venus see Chap. 6 of Butrica (1996).

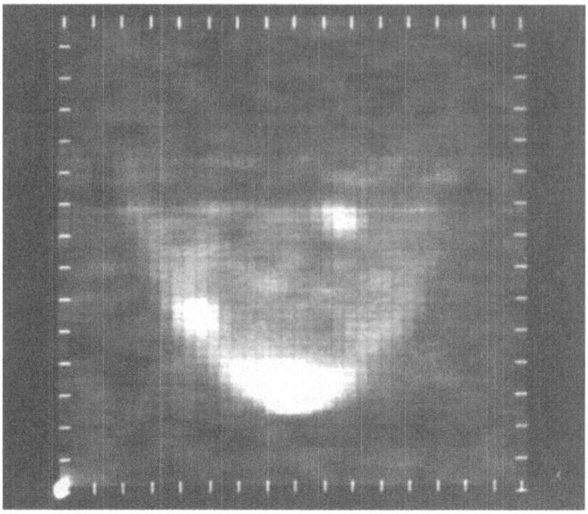

Fig. 9.6 A radar image of Venus made with the Arecibo radar system in 1967 showing two bright features, Gauss and Hertz (Goldstein's β) at lower left and Maxwell at upper right of center. The resolution of the image is about 150 km so the features are blurry and no details could be seen. The large bright region at lower center is the radar reflection from the center of the planet's disk, which is perpendicular to the line-of-sight so gives a bright mirror-like reflection. This delay-Doppler image is radar echo strength as a function of time delay (vertical) and Doppler shift (horizontal). Chap. 5, Fig. 20 from Butrica (1996)

discovered during the 1962 close approach. As recounted by Butrica (1996), Jurgens, encouraged by Thomas Gold, decided to name features on Venus after historical electromagnetic experts. Goldstein's feature α was named after Faraday, while the feature β, which had resolved into two closely spaced features, were named after Gauss and Hertz.

For the 1967 close approach, the next one that was observable with the Arecibo telescope,[5] Jurgens carried out a series of delay-Doppler observations of Venus with Arecibo as part of his PhD thesis work. He looked at latitudes higher than previous observations and discovered another very bright feature, which he named Maxwell after the originator of the famous Maxwell equations that describe the propagation of electromagnetic waves (Fig. 9.6).

Two new radar systems were completed in time for the 1967 close approach of Venus, the 37 m MIT Haystack radio/radar telescope completed in 1964 and the 64 m DSS14 antenna at NASA's Goldstone complex completed in 1966. Both radar systems operated at frequencies close to 8 GHz. Alan Rogers and Richard Ingalls (1970) developed an interferometer system based on the 37 m Haystack antenna

[5]Separated by 19 months, there are very close to five close approaches of Venus to Earth every 8 years. All of them have approximately the same hemisphere of the planet facing Earth, so only this hemisphere can be imaged with Earth based radars. Because of its limited declination coverage, Arecibo could observe Venus for only three out of the five close approaches.

transmitting and this antenna plus the nearby 18 m Westford antenna as an interferometer to receive the echo and resolve the north-south ambiguity problem. This was highly successful in identifying the hemispherical location of a number of features, but there was no improvement in the resolution. Following the lead of the NASA Goldstone and Arecibo groups, the Haystack group named their features Haystack I, Haystack II, etc. so that some features now had three names. This was finally resolved by the International Astronomical Union, which has the responsible to name features on planets and decided that all names on Venus should be female names, real people for craters, mythological figures for continental sized features, etc. However, in recognition of the Earth based imaging, three of the original names were preserved, Alfa Regio, Beta Regio and Maxwell, later Maxwell Montes, the highest mountain complex on Venus.

Given the success of the Haystack/Westford interferometer system in allowing imaging of Venus without the north-south ambiguity problem, Don Campbell, a Cornell graduate student at the time, set about organizing an interferometer for the next close approach of the planet visible with Arecibo in 1969 as his PhD thesis project. It was fortunate that an electronics engineer at the observatory, Sam Harris, who was famous in the amateur radio (Ham) community as the first Ham to communicate via bouncing radio waves off the Moon and back to Earth, had built a moon bounce antenna at his home in a secluded valley about 11 km north of the observatory. It was 50 ft (15 m) in size, a fixed parabolic antenna with a movable feed that allowed it to track up to 10 degrees from the zenith. With Sam Harris's agreement and engineering assistance, the antenna was expanded to 100 ft (Fig. 9.7) and a tower installed on an adjacent hill with antennas to receive a local oscillator signal from the main observatory and to transmit the received radar echo from Venus back.

Venus was successfully observed on a number of days near its closest approach in April 1969 resulting in images of the surface. The use of the interferometer to separate the radar echoes from the two hemispheres was successful although at the expense of some degradation of the sensitivity due to the use of the 100 ft antenna. While the resolution of the image was still about 200 km, it was clear that the interferometer technique was a success. Beta Regio, which contains Jurgen's Gauss and Hertz features, was in the correct (northern) hemisphere and Alfa Regio (Faraday) was also in the correct (southern) hemisphere (Campbell et al. 1970).

Installation of the new 430 MHz line feed in early 1972 gave an improved image (Fig. 9.8) but not significantly better resolution. This would have to wait on the resurfacing of the reflector and the installation of a 2.38 GHz 450 kW transmitter.

9.4 The Upper Atmosphere

In addition to the first altitude profile of electron density that extended well above the F2-region electron density peak at about 300 km, Chap. 7 discussed two of the very early results from the initial observations of the ionosphere, the presence of traveling

Fig. 9.7 The 100 ft moon bounce "Ham" antenna built by Sam Harris 11 km north of the Arecibo Observatory used with the 305 m telescope as an interferometer to image the surface of Venus. It could only steer 10° from the vertical so it was fortunate that the close approach of Venus in 1969 was observable. There was a slightly scary home made elevator up one of the vertical supports and a garbage can covered the receiving system but it all worked. Credit: photo by the author

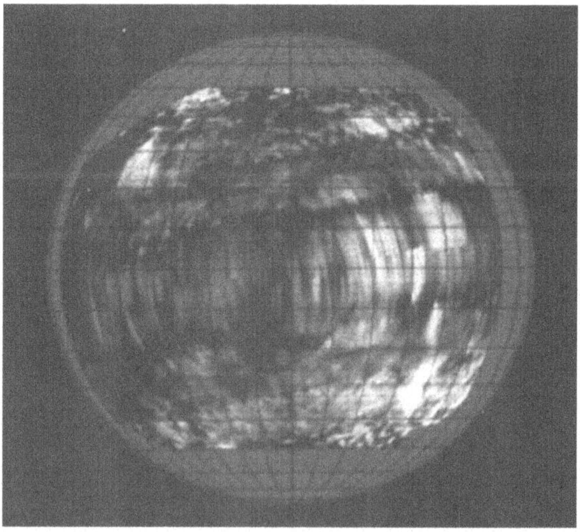

Fig. 9.8 A shaded radar image of the surface of Venus obtained with the 11 km baseline interferometer in 1972 after the installation of the new line feed. Alfa Regio is the bright feature in the southern hemisphere at lower right and Beta Regio can be seen at upper left. The dark almost horizontal lines are artifacts of the use of the interferometer. Credit: the author

ionospheric disturbances above the F2 electron density peak and the measurement of the ion composition at altitudes up to 1400 km showing the unexpected presence of helium ions in the 400 km to 600 km altitude range.

Over the following years, a wide variety of observations of the ionosphere and neutral atmosphere extending down to the stratosphere were undertaken. The 430 MHz incoherent scatter radar remained central to the program, but feeds and transmitters at other frequencies were introduced for specific experiments, remote receiving systems complemented the incoherent scatter measurements, optical instruments allowed studies of the neutral atmosphere at high altitudes, measurements in support of sounding rocket campaigns. Active heating experimentation with the ionosphere were pursued with a succession of transmitting systems at frequencies low enough that the power is absorbed, locally raising the temperature of the ionosphere and inducing a wide range of plasma physics effects. Not all these initiatives were successful. The first 1962 AFOSR operating budget included funds for a 40 MHz high powered radar system. It was installed in the operations building in 1965 with the power being sent to four yagi antennas located around the 430 MHz line feed via a 6-inch transmission line on the catwalk. For reasons that are still unclear it was never successfully used to study the ionosphere. It was used to obtain radar echoes from the Sun, conduct early ionospheric heating experiments, measure the electron density of the interplanetary medium between Earth and Venus and image the subsurface of the Moon. The transmitter and feed system were dismantled in mid-1971 in preparation for changing the frequency to 50 MHz using an amplifying tube obtained from ESSA (the Environmental Science Services Administration whose name was later changed to the National Oceans and Atmosphere Administration (NOAA)). However, this tube failed and the project was abandoned. The coaxial transmission line was repurposed to send the power to successive heating feeds before a dedicated transmitting facility was built in the late 1970s.

Review papers by Isham et al. (2000) and Mathews (2013) provide readable overviews of the history of the atmospheric science program at the Arecibo Observatory. This chapter will briefly describe the beginning of two long lasting research initiatives, the study of the high altitude neutral atmosphere above Arecibo as a complement to radar studies of the ionized component, and active experimentation in which a low frequency (historically designated High Frequency or HF), high powered transmitting facility "heats up" the ionosphere and the 430 MHz incoherent scatter radar is used to study the ionosphere's response. A third, shorter lived, program was the launching of sounding rockets into the lower ionosphere to both verify the temperature and electron density measurements obtained with the incoherent scatter radar and do experiments by releasing barium clouds that are then ionized by solar ultraviolet radiation.

The Neutral Atmosphere and Optical Instrumentation By definition the name ionosphere implies an ionized medium consisting of ions and electrons. However, the reality is that the ions and electrons comprise a very small component of the ionosphere, neutral un-ionized gas is by far the major constituent. At 300 km altitude the neutral species have a density of about 10^9 atoms per cm^3 while during daytime there are only about 10^6 ions per cm^3. Above about 500 km, the ion and neutral densities are so low that collisions between ions and neutral atoms or molecules are not important and the charged particles can be regarded as an independent plasma. Below about 500 km the ion and neutral densities are high enough that collisions can occur between ions and neutral species. This means that a full understanding of the Earth's atmosphere at high altitudes requires a means to study the neutral component in addition to the incoherent radar studies of the ionized portion.

An early discovery with the incoherent scatter radar was that in winter the ionosphere above Arecibo is heated hours before local sunrise by energetic electrons released by solar short wavelength UV radiation above Arecibo's magnetic conjugate point at Mercedes in south-eastern Argentina (i.e. the two locations are at the opposite ends of the same magnetic field line). Since in summer at Mercedes sunrise is earlier than in wintertime Arecibo, energetic electrons travel along the magnetic field lines and impinge on the ionosphere above Arecibo. Such an effect had been predicted, but this was the first confirmation (Carlson 1966). This discovery sparked interest in seeing if there was a pre-dawn enhancement of optical emissions (airglow) from the ionosphere over Arecibo due to the same energetic electrons escaping along the field lines from the magnetic conjugate point. As recounted by Craig Tepley in a talk he gave at the 40th anniversary of the completion of the telescope in 2003,[6] "In July 1965, Bill Gordon gave permission to Herb Carlson to purchase the required instrumentation, and a photometer that could measure airglow emissions at 6300, 5577, 5893, 5300 Å was acquired".

Faint optical emissions, airglow, results from a variety of chemical reactions occurring in the different parts of the atmosphere/ionosphere system as a consequence of de-excitation of excited atoms or molecules to the lower levels. One of the strongest airglow emissions, the redline at 630 nm wavelength, occurs at F-region altitudes near 300 km. It results from the dissociative recombination of O_2^+ (ionized oxygen molecules) with electrons that produces an oxygen atom in an excited state of oxygen atom O (1D). The spontaneous transition of this excited state to the ground state, results in the generation of a photon at 630 nm. Similarly, the generation of the oxygen green line at 557.7 nm wavelength is a two-stage process and peaks at around 94–96 km altitude. Monitoring these and other optical emissions provides an excellent method to probe the regions of the atmosphere where it originates. Since various airglow lines emanate at different altitude regions, optical instruments that record the time variation of the intensity/brightness of these emissions reveal properties of the medium/atmosphere at the height of a particular airglow layer under investigation. Waves passing through the atmosphere modulate the airglow intensity, and thus its temporal variation has been used to extract wave parameters like periods, wavelengths, speed etc.[7]

---

[6]Tepley, C.A. April 2004, Nearly four decades of optical measurements at Arecibo, National Astronomy and Ionosphere Center, Arecibo Observatory Newsletter, pages 10–11.

[7]Shikha Raisada, a long time member of the Arecibo observatory's scientific staff, contributed this discussion on airglow.

Over the following years, increasingly sophisticated optical equipment was installed in a small laboratory, commonly referred as the photometer shack, on the ridge above the operations building. A Fabry-Perot spectrometer was acquired with one early application being Leroy Cogger's and colleagues' comparison of electron temperatures in the F-region derived from incoherent scatter radar (ISR) measurements with optically derived ones by measuring the Doppler broadening of the atomic oxygen line, since at night the ion and electron temperatures are the same. Spacecraft-born probes measured electron temperatures that were substantially different from the ones measured by the ISRs. The optical measurements were in close agreement with the Arecibo ISR measurements confirming their validity (Cogger et al. 1970).

HF Heating Since the early days of the observatory, the ionospheric research community had a strong interest in studying non-linear effects in the ionosphere induced by transmitting a very low frequency, confusingly called high frequency or HF, radio wave at high power. The frequency of the wave is tuned to optimize its absorption at a specific height in the ionosphere, the height where the plasma frequency is close to the HF wave frequency. At the absorption height, the electrons are accelerated by the radio wave raising their temperature and, in turn, they transfer their energy to the ions via collisions raising the ion temperatures. As Gordon et al. (1971) stated in their paper reporting on the first results, "At F-region heights this ability should contribute to the thermal processes, the dynamics, and the ion temperature of the upper atmosphere and of some areas of plasma physics." If a HF transmitting capability could be installed on the Arecibo telescope, then the 430 MHz incoherent scatter radar would provide the perfect diagnostic capability to examine the effect of the heating.

A HF transmitter was obtained from the Institute of Telecommunication Sciences (ITS), which had built one of the first HF heater systems in Platteville, Colorado, and installed in the transmitter room of the operations building in October 1970. It was capable of producing 160 kW of power. To get started, a single frequency yagi feed tuned to about 6 MHz was built by Merle LaLonde and installed hanging from the center of the feed arm (Fig. 9.9). Transmitter power was fed to it via the 6 inch coaxial transmission line used for the 40 MHz transmitting system. To provide wider frequency coverage, a log-periodic feed covering 5 MHz to 12 MHz was designed and built, arriving in May 1971 (Fig. 9.10). As can be seen in the figure, these feeds were very large and limited Carriage House 1 with the 430 MHz line feed to zenith angles greater than about 4.5°. However, this did allow the ISR to interrogate about half the heated volume as the beamwidth of the HF feed was about 10°.

Bill Gordon had a strong interest in HF heating experiments, and he and his former and then current students at Rice University, Robert Showen, Ivan Kantor and Luis Diaz, along with Herb Carlson (Fig. 9.11), who now headed the atmospheric sciences group at the observatory, organized the early HF heating campaigns. The first measurements were made in the early morning hours of 26 October 1970 with the transmitter putting out 100 kW at a frequency of 5.62 MHz, very close to the plasma frequency of 5.7 MHz at the peak electron density of the F-region,

Fig. 9.9 The 6 MHz yagi HF feed in late 1970 suspended from the center of the feed arm. To its right is the original 430 MHz feed with the 40 MHz transmitting feed consisting of four yagi antennas wrapped around it. Credit: Cornell University/NSF

Fig. 9.10 The log-periodic HF feed on the telescope in July 1972. It was 160 ft on a side and weighed 8000 lb. When not in use the arms were folded down, the feed lowered and stored vertically at the base of Tower 8. The earliest experiments using the yagi feed were made using the original 430 MHz line feed. This photo shows the new 430 MHz line feed which was installed in early 1972. Credit: Cornell University/NSF

Fig. 9.11 Herbert Carlson, former graduate student of Bill Gordon and atmospheric sciences group head in the early 1970s. Credit: Cornell University

Fmax (Gordon et al. 1971). A 30% enhancement of the electron temperatures was seen at just below the 313 km height of Fmax, which declined again as soon as the heating was turned off. The first test of the HF system was a success.

Once the wide band log periodic feed was installed, HF heating could take place at a variety of frequencies providing information about "the thermal processes, the dynamics and the ion chemistry of the upper atmosphere, and the instabilities in plasmas." (Kantor 1974). However, thoughts were already turning to the possibility of building a separate HF transmitting facility that would provide greater flexibility and enhanced sensitivity.

Rocket campaigns: With three incoherent scatter radars (ISRs), Arecibo, MIT's Millstone Hill radar and the Jicamarca ISR in Peru, commencing measurements of the ionosphere at about the same time, there was strong interest in confirming their measurements of electron densities, electron temperatures and ion composition via in situ measurements with sounding rockets. In March 1968, there was a coordinated campaign of, primarily, electron temperature measurements over the height range of about 130 km to 300 km by the Arecibo ISR and rockets launched from the vicinity of the Tortuguero lagoon on the north coast of Puerto Rico about 40 km (25 miles) east of Arecibo. There was reasonably good agreement between the electron temperature measurements with height during the daytime launch, but the nighttime measurements were not in as good agreement (Brace et al. 1969). This was the first of a number of rocket campaigns over the years.

In May of 1968 a very different rocket campaign took place. This was driven by the observatory's funder, the Advanced Research Project Agency (ARPA) of the Department of Defense in conjunction with the Air Force Cambridge Research Laboratory (AFCRL), likely the only time that ARPA directly intervened in the observatory's research program. Their interest was in the effects of nuclear explosions in the upper atmosphere. Clearly, it was not possible to do the direct experiment, but releasing a barium cloud that would be partially ionized by solar UV radiation would at least partly simulate the event. A large number of sensors were deployed around Puerto Rico to observe the barium cloud, but Arecibo ISR measurements were the main reason that the campaign was mounted on the Island. As recounted by Frank Drake,[8] three successful early morning rocket launches took place with the Arecibo telescope pointed to the spot where the barium was to be released. Early morning so that that it was dark at ground level but the Sun had risen at the 200 km altitude of the release. The release was successful for all three but for the first two, the rocket's trajectory was not the one planned and the ISR measurements were not successful. Fortunately, the third one went where it was supposed to and good ISR data were acquired. Drake remembered that the barium clouds were spectacular, green for the non-ionized or neutral component and red-purple for the ionized one, which were separated due to different drift directions in the Earth's magnetic field. Measuring the motion of ionized and neutral components of the ionosphere was one of the major objectives of the study.

9.5 The First Upgrading Project

With the transfer of the observatory to the NSF on 1 October 1969 the stage was now set to proceed with the reflector surface upgrading project and the installation of a new high powered transmitter for planetary radar observations. William Brunk, the then head of NASA's planetary astronomy program, had indicated that NASA was interested in contributing to the project by funding the transmitter and some additional items to a total of $3M. NSF's FY 1973 budget for Arecibo, which started on 1 October 1972, included $5,654,600 for the reflector resurfacing and $1,545,000 of NASA funds for the S-band transmitter and other work. The NASA funds were transferred to the NSF and administered by the NSF to avoid multiple reporting requirements. For FY1974 the NASA amount was increased to the promised $3M and the NSF funding increased by a small amount to $5,760,000.

Based on the expectation of NASA funding, Cornell submitted a Proposal to NSF and NASA dated February 1970 for "Major Additions and Modifications of the Suspended Antenna Structure and Equipment of the Arecibo Observatory" with a

[8] A slightly humorous description of the rocket campaign is given by Frank Drake in his video interview, which can be viewed at the Cornell ecommons library site. https://ecommons.cornell.edu/handle/1813/40748

construction period from February 1971 to February 1973. It was signed by Thomas Gold, Assistant Vice-President for Research, Arecibo Affairs, and T.R. Rogers, Director, Cornell Office of Sponsored Research.[9] The proposal included new elevation rails, a larger Carriage House 2 to support up to six line feeds, a 900 kW, two klystron 2.38 GHz transmitter, three spare klystrons, two maser amplifiers and a dual polarization 2.38 GHz feed at a cost very close to $3M. It included a long discussion of the expected scientific dividends. What was not mentioned was the electrical power source for the transmitters. On site generators would be needed as the power provided by the electrical line from the Puerto Rican Water Resources Authority (renamed after 1979 to the Puerto Rico Electric Power Authority) was only adequate to power normal site operations including the 430 MHz transmitter. A two klystron 2.38 GHz transmitter putting out 900 kW would require about 2 MW of primary power as the amplifying klystrons were only about 50% efficient. A search for government surplus generators was initiated about the time the funding proposal was submitted.

Cornell issued a request for proposals in March 1971 for the resurfacing of the reflector. Proposals were submitted in July 1971 by the Rohr Corporation and the Ling-Temco-Vough (LTV) Corporation, a subsidiary of LTV Electrosystems (LTVE), which later became an independent company, E-Systems, in 1972. Technical and Business Evaluation Committees were formed with the technical committee consisting of William McGuire (chair), Donald Belcher, L. Merle LaLonde and J. H. Lancaster.

The Technical Evaluation Committee submitted its preliminary report[10] to Cooke on 30 September, "—the committee feels that the LTVE concept is clearly superior in overall technical merit" and "unless costs or other concerns not within the scope of the evaluation by the Technical Committee govern, it unanimously recommends that the Selection Board adopt the LTVE concept." The LTVE proposal was also the lowest, $3,557,699, and the final report of the Technical Committee did not change its preliminary recommendation resulting in the selection of LTVE's proposal. The recommendation to select LTVE was submitted to Cornell's Board of Trustees in November 1971 and approved. After negotiations the final cost of the proposal was $3,764,213.

About this time a search was begun for an Architectural and Engineering firm to design the required modifications to the suspended platform such as new elevation rails and the new Carriage House 2 needed to house the 2.38 GHz transmitter and multiple line feeds at other frequencies. The search resulted in the hiring of the engineering firm Ammann and Whitney (A&W), the beginning of a long association with the observatory. Othma Ammann was a Swiss-American civil engineer who designed many of the major bridges around New York city including the Verrazano-Narrows bridge. He founded A&W with Charles Whitney, a noted designer of

---

[9]Thomas Gold's files, Cornell University archives.

[10]Report of the Technical Evaluation Committee, Arecibo Reflector Upgrading Project, Ithaca NY, 15 Oct 1971. William McGuire files, Cornell University archives.

concrete structures, in 1946. For many years Joseph Vellozzi of A&W, a talented, always helpful, but somewhat acerbic engineer, dealt with the observatory's engineering needs.

On 1 November 1972, a request for proposals for the 2.38 GHz (S-Band) transmitter was sent to Continental Electronics Manufacturing Co., Energy Systems, RCA, The Raytheon Co., and Westinghouse Electric Corporation. To reduce costs and complexity, the transmitter power was reduced to 450 kW, which required only a single klystron. Three proposals were received by 13 December 1972 from RCA, Energy Systems, and Continental Electronics. On the basis of its $764,200 proposal, Continental, located in Dallas, Texas, was tentatively chosen based on cost but the Technical Evaluation committee chaired by Harold Craft specified a number of requirements that they must meet. On 6 March Continental delivered an alternate proposal for $978,481, a much improved proposal that was responsive to the input from the committee. RCA and Energy Systems did not significantly reduce their prices so Continental was selected. With NASA and NSF approval Continental was formally selected in early April 1973 and the contract awarded in late April at the cost of $978,481. The amplifying klystron and two spares were ordered separately from Varian Associates of Palo Alto, California.

LTV's (hereafter E-Systems) design for resurfacing the reflector to obtain a mean surface error of 3.2 mm allowing operation up to 5 GHz was to add 30 additional heavy lower cables, reducing the spacing from 100 ft to 25 ft and replacing the wire mesh surface with 38,778 aluminum panels. The additional cables would significantly reduce the changing sag with temperature of the upper light cables. The aluminum panels would be roughly 3 ft × 6 ft, to fit the approximately 3 ft spacing of the light upper cables and four would span the 25 ft spacing of the heavy, lower cables, with each panel consisting of a perforated aluminum skin supported by a backup structure of Z-shaped aluminum ribs. Each panel was shaped to match the required curvature of the reflector at its location; the panels are installed along E-W and N-S lines so the curvature relative to the orientation of the panels varies. Because their sheer volume made shipping difficult, a building was constructed on site to fabricate the panels (Fig. 9.12). Under the contract, at the end of the work, the building was to be transferred to the observatory for use as a warehouse. Robert Matyas, the head of the facilities office at Cornell, was the project manager ably assisted by José Maldonado and Harold Craft. Maldonado, the observatory's Head of Facilities, who was a native Arecibeño, was a talented engineer who had been in charge of the north western region for the Puerto Rico Electric Power Authority prior to joining the observatory staff. Harold Craft, a former graduate student working with Frank Drake, joined the observatory staff as Head of Technical Services and oversaw the transmitter procurement. In August 1973, he succeeded Tor Hagfors as Observatory Director.

The installation of the new large cables and repairs of some of the light cables were completed by February 1973, and installation of the panels begun (Figs. 9.13 and 9.14). This proceeded very rapidly and was completed on 27 November 1973 with the installation of the 38,778th panel (Fig. 9.15).

Fig. 9.12 Examining one of the first panel back structure z-struts in the new assembly building in April 1972. From left to right: Don Cooke, Cornell Vice President for Research, Merle LaLonde, Tor Hagfors, Observatory Director, José Maldonado, Head of Maintenance, Rolf Dyce and Frank Drake, NAIC Director. Credit: Cornell University

Once all the panels were installed the next task for E-Systems was to adjust the surface to the contracted mean accuracy of 3.2 mm. The corners of each set of four adjacent panels were attached to a small plate that was, in turn, attached to one of the light cables via a threaded rod allowing each plate and the panel corners to be moved vertically. Targets were placed on one of the panels next to each of the roughly 38,800 adjustment locations and their 3-D positions measured using three theodolites. Despite months of effort E-Systems did not succeed in bringing the surface to the required 3.2 mm accuracy and they finally gave up. José Maldonado took on the task and after extensive and careful work succeeded in bringing the surface to the required accuracy.

Ammann and Whitney designed the new Carriage House 2 that would support up to six line feeds and house the new S-Band transmitter. The shorter wavelength meant that it was not feasible to send the power to the platform using a waveguide as was the case for the 430 MHz transmitter. The feeds, including the 611 MHz multi-beam system, were removed from the original Carriage House 2 in the second quarter of 1974 and the Carriage House lowered. The new one was installed (Fig. 9.16) in the third quarter.

Continental Electronics completed the construction of the S-band transmitter and demonstrated it capabilities at their plant in Dallas, Texas. It was shipped to the observatory and installed in September, 1974 (Fig. 9.17). This was followed by

Fig. 9.13 Aerial photo from April 1973 showing the completion of the installation of the 30 new large cables and the start of the installation of the new aluminum panels. The new cables can't be seen but their locations are outlined by the rows of disturbed soil where the new tie down blocks are located that force the cables into the spherical shape. They are in groups of three with the blank spaces in between them the locations of the original large cables. Piles of fabricated panels can be seen at lower right stored outside of the panel fabrication building. The Rio Tanamá is at upper left showing where it goes underground for about 200 m. The beginning of the track from the observatory's upper road to the southern side of the river is at center left. Credit: Cornell University/NSF

acceptance tests which were successfully completed in October. The transmitter required approximately 1 MW of electrical power to operate and this plus the normal power requirements of the observatory exceeded the capabilities of the power provided by the local power authority. This meant that on site power generators were required for the transmitter. Two 750 kW refurbished turbine generators were acquired and installed in the third quarter of 1974. Merle LaLonde built a temporary linearly polarized S-band feed to use with the transmitter until a circularly polarized one could be designed and built. This gave great entertainment to the observatory staff as every time the transmitter was turned on flames would sprout from the top of the feed. The problem was finally solved by adding water cooling lines to the feed.

The entire project was completed in late 1974 (Fig. 9.18) with the dedication taking place on 16 November. Taking advantage of the newly installed 2.38 GHz transmitter, Frank Drake, Director of NAIC and well known for the first attempt to detect radio signals from extraterrestrial beings, Project Ozma in 1960, used the

Fig. 9.14 Panel installation in April, 1973. The concrete blocks at lower left are the tie down anchors for the new lower cables now at 25 ft spacing. The four corners of adjacent panels are connected to a plate which is attached to the upper cables via a threaded rod allowing the height of the plate holding the four corners to be adjusted. Credit: Cornell University/NSF

Fig. 9.15 Cornell President Dale Corson helping to install the 38,778th and last panel of the new reflector surface. Corson was an avid photographer and took many photos of the telescope. Tom Dickenson, the then new S-band transmitter engineer, is in the background. Credit: Cornell University/NSF

Fig. 9.16 The new Carriage House 2 being lifted to the feed arm. Credit: Cornell University/NSF. Photo by Harold D. Craft

Fig. 9.17 The new 450 kW S-band transmitter undergoing a tilt test at Continental Electronics plant in Dallas, Texas. All the piping and the white tubing are water cooling lines needed to dissipate up to 900 kW of heat via air cooled heat exchangers mounted on the top of the azimuth arm. Hard to believe that it will fit in the new Carriage House 2 of Fig. 13. Credit: Continental Electronics

Fig. 9.18 An aerial view of the resurfaced reflector and platform from late 1974 and an overall view of the site. The maintenance buildings are at lower left. At lower center are the operations building, the office building built in the late 1960s, the VSQ and the swimming pool. The road at lower right comes from the VSQ parking lot and its end on the left above the operations building is the site of the photometer laboratory. The end on the right with the cleared area is the future site of the LIDAR laboratory. The lower end of the road in the photo is the future site of the first VSQ expansion. Credit: Cornell University/ NSF

occasion to beam a message to the globular star cluster M13 to highlight the possibility of using powerful transmitting systems to contact extraterrestrial beings.[11] M13 is 25,000 light years from Earth so it would take 50,000 years for any reply to reach Earth. It was primarily a demonstration that the capability existed (Fig. 9.19).

9.6 The Discovery of the Binary Pulsar

One requirement of the resurfacing of the telescope's reflector was that the height of new aluminum surface be almost exactly the same as the old wire mesh surface. This meant that observations could continue as the new surface slowly replaced the old

[11] A video interview with Frank Drake can be found at the Cornell ecommons library site https://ecommons.cornell.edu/handle/1813/40748. In it he gives a vivid description of the content of the Arecibo message.

Fig. 9.19 Frank Drake watching Congressman Davis on the telephone to the observatory's control room requesting that they start sending the Arecibo message to the star cluster M13 using the newly installed S-band transmitter. Credit: Cornell University

one although calibration of the observations would be a problem due the variable gain as the wire mesh was replaced with the aluminum panels. For searches for new pulsars this did not matter and Joseph Taylor, then at the University of Massachusetts at Amherst, decided to take advantage of this to do a systematic search for new pulsars. There had been a number of previous searches with the Arecibo telescope, most of them targeting specific celestial objects in our galaxy such as the remnants of super nova explosions, utilizing the observatory's data acquisition system. Taylor, working with graduate student Russell Hulse, developed an independent data acquisition and analysis system (Fig. 9.20) to do a systemic untargeted pulsar search survey of the portion of the Milky Way visible with the Arecibo telescope. The sensitivity of this survey was about a factor of ten better than any previous pulsar survey at Arecibo or elsewhere (Hulse and Taylor 1974). The search got underway in early 1974. Through September, 30 new pulsars were discovered but they hit the jackpot in July 1974 when they realized that the arrival times of a newly discovered pulsar with a period of 59 ms, PSR 1913+16, were varying periodically by up to 3 s with a period of 7.75 h. This indicated that the pulsar, a neutron star, was in an eccentric orbit about another star with an orbital period of 7.75 h, the first discovery of a pulsar in a binary system (Hulse and Taylor 1975). It soon became clear that the pulsar was in orbit about another neutron star although one from which pulses were not detectable, perhaps because its beam did not cross the Earth.

Fig. 9.20 Russell Hulse in the control room of the Arecibo telescope in October 1974 adjusting the data acquisition system that he and Joseph Taylor used to find the binary pulsar, PSR 1913+16. Credit: Joseph Taylor

For this discovery, Hulse and Taylor were awarded the 1993 Nobel Prize in Physics with the citation "for the discovery of a new type of pulsar, a discovery that has opened up new possibilities for the study of gravitation". Gravitation was mentioned because observations of PSR 1913+16 over a number of years showed that its orbit is very slowly decaying due to the loss of orbital energy. The rate of energy loss is in precise agreement with Einstein's prediction of the energy loss for such a system due to the emission of gravitational waves (Fig. 9.21), the first, if indirect, confirmation of Einstein's prediction of the existence of such waves.[12] This was the best evidence for the existence of gravitational waves until their direct detection in September 2015 by the Laser Interferometer Gravitational Observatory (LIGO).

[12] A very readable discussion of the discovery and importance of the binary pulsar can be found in a *Scientific American* article by Weisberg et al. (1981).

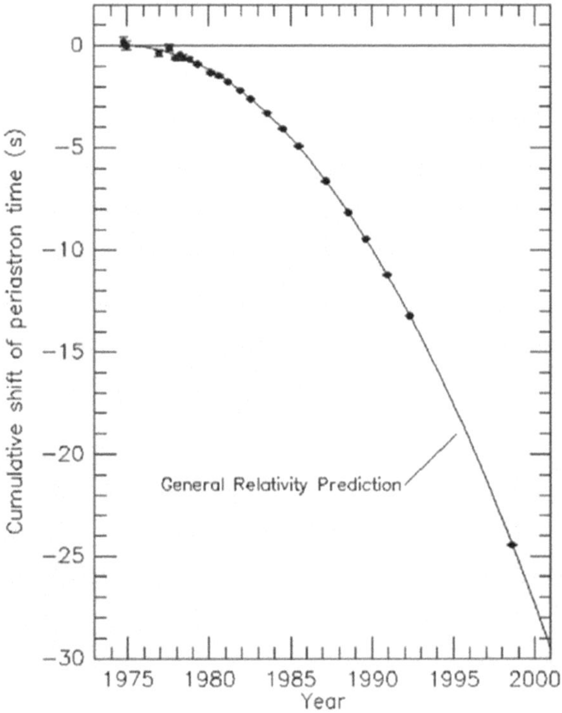

Fig. 9.21 The shrinkage of the orbit of PSR 1913+16 is only about 3.5 m per year, too small to measure. Instead, the time that the pulsar goes through periastron, the closest point to the center-of-mass of the system in its eccentric orbit, can be measured and goes as the square of the time from the first measurement. The dots are the measurements taken over about 15 years and the solid line is the prediction from Einstein's theory of General Relativity. They are in perfect agreement. Figure 21 from the October 1999 Arecibo Observatory/NAIC Newsletter (#28) courtesy of Joseph Taylor and Joel Weisberg

References

Brace, L.H., H.C. Carlson, and K.K. Mahajan. 1969. Radar Backscatter and Rocket Probe Measurements of Electron Temperature Above Arecibo. *Journal of Geophysical Research* 74: 1883–1885.

Butrica, A.J. 1996. *To See the Unseen: A History of Planetary Radar Astronomy*, NASA History Series. NASA SP-4218.

Campbell, D.B., R.F. Jurgens, R.B. Dyce, F.S. Harris, and G.H. Pettengill. 1970. Radar Interferometric Observations of Venus at 70-Centimeters Wavelength. *Science* 170: 1090–1092.

Carlson, H.C. 1966. Ionospheric Heating by Magnetic Conjugate-Point Photoelectrons. *Journal of Geophysical Research* 71: 195–199.

Clark, B.G., M.H. Cohen, and D.L. Jauncey. 1967. Angular Size of 3C 273B. *Astrophysical Journal* 149: L151–L152.

Cogger, L.L., G.J. Nelson, M.A. Biondi, R.D. Hake, and D.P. Sipler. 1970. Coincident F-Region Temperature Determinations from Incoherent Backscatter and Doppler Broadening of [OI] 6300 A. *Journal of Geophysical Research: Space Physics* 75: 4887–4889.

Cohen, M.H., E.J. Gundermann, H.E. Hardebeck, D.E. Harris, E.E. Salpeter, and L.E. Sharp. 1966. Radio Sources: Angular Size from Scintillation Studies. *Science* 153: 745–746.

Comella, J.M., H.D. Craft, R.V.E. Lovelace, J.M. Sutton, and G.L. Tyler. February, 1969. Crab Nebula Pulsar NP 052. *Nature* 221: 453–454.

Gold, T. 1967. Radio Method for the Precise Measurement of the Rotation Period of the Earth. *Science* 157: 302–304.

———. 1968. Rotating Neutron Stars as the Origin of the Pulsating Radio Sources. *Nature* 218: 731–732.

Gold, T. (ed. S. Mitton). 2012. *Taking the Back off the watch*, Springer.

Gordon, W.E., R. Showen, and H.C. Carlson. 1971. Ionospheric Heating at Arecibo: First Tests. *Journal of Geophysical Research* 76: 7808–7813.

Hewish, A., P.F. Scott, and D. Wills. 1964. Interplanetary Scintillations of Small Diameter Radio Sources. *Nature* 203: 1214.

Hulse, R.A., and J.H. Taylor. 1974. A High Sensitivity Pulsar Survey. *Astrophysical Journal* 191: L59–L61.

———. 1975. Discovery of a Pulsar in a Binary System. *Astrophysical Journal* 195: L51–L53.

Isham, B., C.A. Tepley, M.P. Sulzer, Q.H. Zhou, M.C. Kelly, J.S. Friedman, and S.A. Gonzalez. 2000. Upper Atmospheric Observations at the Arecibo Observatory: Examples Obtained Using New Capabilities. *Journal of Geophysical Research* 105: 18609–18637.

Jauncey, D.L., C.C. Bare, B.G. Clark, K.I. Kellerman, and M.H. Cohen. 1970. High Resolution Radio Interferometry at 610 MHz. *Astrophysical Journal* 160: 337–339.

Kantor, I.J. 1974. High Frequency Induced Enhancements of the Incoherent Scatter Spectrum at Arecibo, 2. *Journal of Geophysical Research* 79: 199–208.

Kellerman, K.I. et al. 2020. *Open Skies; The National Radio Astronomy Observatory and Its Impact on U. S. Radio Astronomy*, Springer Historical and Cultural Astronomy Book Series

Lovelace, R.V.E., and G.L. Tyler. 2012. On the Discovery of the Period of the Crab Nebula Pulsar. *The Observatory* 132: 186–188.

Mathews, J.D. 2013. A Short History of Geophysical Radar at Arecibo Observatory. *History of Geo- and Space Sciences* 4: 19–33.

Pettengill, G.H., and R.B. Dyce. 1965. A Radar Determination of the Rotation of the Planet Mercury. *Nature* 206: 1240.

Richards, D.W., and J.M. Comella. 1969. The Period of Pulsar NP 0532. *Nature* 222: 551–552.

Rogers, A.E.E., and R.P. Ingalls. 1970. Radar Mapping of Venus with Interferometric Resolution of the Range-Doppler Ambiguity. *Radio Science* 5: 425–433.

Thompson, T.W., and R.B. Dyce. 1966. Mapping Lunar Radar Reflectivity at 70 cm. *Journal of Geophysical Research* 71: 4843–4853.

Vandenberg, N.R., T.A. Clark, W.C. Erickson, G.M. Resch, J.J. Broderick, R.R. Payne, S.K. Knowles, and A.B. Youmans. 1973. VLBI Observations of the Crab Nebula Pulsar. *Astrophysical Journal* 180: L27–L29.

Weisberg, J.M., J.H. Taylor, and L.A. Fowler. 1981. Gravitational Waves from an Orbiting Pulsar. *Scientific American* 245: 74–82.

Chapter 10
A New Beginning

The large increase in frequency coverage resulting from the replacement of the telescope's reflector surface opened up many new opportunities in the fields of radio and radar astronomy. However, to take advantage of these opportunities new line feeds, especially circularly polarized ones, had to be designed and built and a wide band, multi-channel digital correlator-based spectrometer had also to be designed and built. Radar imaging of Venus with the new 2.38 GHz transmitter required the construction of a 100 ft (30.5 m) auxiliary interferometer antenna and the design and construction of a specialized correlator to "decode" the received radar echoes. Radar observations of Venus in the late 1970s utilizing the new interferometer antenna provided the first large scale image of the surface of Venus and observations of the icy satellites of Jupiter revealed the unusual radar scattering properties of icy surfaces in the solar system. A survey of atomic hydrogen in galaxies in the Pieces-Perseus super cluster of galaxies utilizing the new 1.4 GHz line feed plus a new receiver and a 1008 channel correlator revealed the filamentary structure of the distribution of galaxies in the local universe. In the pulsar area, the first milli-second pulsar was discovered as were the first planets around another star, a neutron star. In atmospheric sciences, a new ionospheric modification transmitting facility was built on the coast just east of Arecibo, a German group initiated a VHF radar program to study the lower ionosphere and the multi-instrument AIDA campaign took place in 1989.

10.1 The Benefits of the Upgrade

The completion of the first upgrading of the telescope resulted in an instrument with unmatched capabilities for research in radio astronomy and planetary science. Pulsars could be studied over a much larger range of frequencies and the capability to detect the 21 cm wavelength (1.4 GHz) radiation emitted by atomic hydrogen, the major gas constituent of the Milky Way and all other galaxies, opened up the

D. Campbell, *The Arecibo Observatory: A History of Innovation and Discovery*,
Historical & Cultural Astronomy, https://doi.org/10.1007/978-3-031-75225-4_10

possibility of studying the distribution of atomic hydrogen in our own galaxy and mapping the structure of the local universe. The big winner from the upgrade was the planetary radar program. An increase of the transmitter power by a factor of three and an improved antenna gain by a factor of 15 due to the higher frequency, plus lower additive noise in the receiver system gave an overall gain in the sensitivity for planetary radar observations of about 50. This allowed the imaging of the surface of Venus beneath its cloud cover at resolutions down to 2 km, fine enough to discern major geological features on the surface and the primary reason why NASA contributed funds to the upgrading project. Wernher von Braun, who played a critical role in the development of the U.S. rocket program in the 1950s and in the Apollo Moon landings, was rumored to have told NASA that providing about $3M in funding to enable the Arecibo telescope to map the surface of Venus was a bargain. The upgrading did not significantly improve the capabilities for ionospheric studies using the incoherent scatter radar system at 430 MHz but the program did benefit from the overall reinvigorating of the observatory due to the "new" telescope.

10.2 NAIC Management and the Budget

Tor Hagfors resigned as Observatory Director in the summer of 1973 to take a faculty position at the University of Trondheim in Norway and direct the construction of the European EISCAT incoherent scatter radar system located at several sites in northern Scandinavia. Harold Craft (Fig. 10.1) was appointed as the new

Fig. 10.1 Herb Carlson (left) and Hal Craft talking with José Maldonado (back to the camera) in the early 1970s. Credit: Cornell University

Observatory Director, the beginning of a much longer tenure in the position than the 2 years or less for the previous four directors. Rolf Dyce continued as the Associate Director.

In June 1981 Frank Drake ended his 10 year tenure as NAIC Director. He remained at Cornell University as Professor of Astronomy until 1984 when he was appointed Dean of Natural Science at the University of California at Santa Cruz. Harold Craft moved to the Cornell University campus as Interim NAIC Director while a search was conducted for a permanent one. After 7 years as Director of the European Incoherent Scatter Scientific Association (EISCAT), Tor Hagfors was interested in returning to Cornell and was appointed NAIC Director and Professor of both Astronomy and Electrical Engineering in July 1982. Hal Craft joined the Cornell administration becoming Head of Facilities and, eventually, Senior Vice President and Chief Financial Officer.

In Arecibo, Don Campbell, who had been appointed as Associate Director under Hal Craft after Rolf Dyce resigned in 1979, was appointed Observatory Director. In late 1987 he resigned and moved to the Cornell campus as Professor of Astronomy maintaining a close association with NAIC. Riccardo Giovanelli, who had been on the Observatory staff since 1978, replaced Campbell as site director for a year after which Michael Davis, another long term staff member and head of the observatory's radio astronomy group, succeeded him.

Budget: The FY 1971 operating budget starting on 1 October, 1970 was $2,456,000, which included $300,000 from ARPA, its last contribution to the operating budget. The budget increased to $4,068,000 in 1975 and to $5,229,000 in 1980. These numbers seem to represent a significant yearly increase, about 8%, but the decade of the 1970s was a period of very high inflation averaging about 7% over the decade.

As a result of its $3M contribution to the upgrading project, NASA contributed to the operating costs of the planetary radar program for the next 20 years. This funding started at about $150,000 per year in FY 1974 rising to about $500,000 at the end of this period.

VSQ expansion: With only six rooms in the original Visiting Science Quarters and no suitable hotels nearby or in Arecibo, there was a strong need to increase the observatory's on-site accommodation. Based on some small cottages built on the island of Culebra just off the east coast of Puerto Rico, Hal Craft had Cornell hire the same builder, Rudd Falconer, to construct four "bachelor units" each with a single bedroom, a small kitchen and a veranda, and two "family" units with two bedrooms, a living room/kitchen and a large veranda (Fig. 10.2). The units were built on the hill above the old VSQ (see Fig. 9.18) and were a very welcome improvement for anyone staying for some time at the observatory especially as they had kitchens and were away from the bustle surrounding the old VSQ and its cafeteria.

Fig. 10.2 One of the new VSQ family units, a very pleasant place to stay. A similar unit was featured in the 1997 movie Contact that was based on the book with the same name by Carl Sagan. Credit: Cornell University/NSF, photo by the author

10.3 Post-Upgrade Instrumentation

Line feeds: While the upgraded telescope opened up the prospect of major advances in astronomy and planetary science, these could not happen without major upgrades in the observatory's instrumentation. The first need was new line feeds for at least some of the six line feed mounting positions on the new Carriage House 2. Line feeds are intrinsically narrow band; they can operate over a very small range of frequencies requiring a new one to be designed and built for each frequency band, such as the S-band transmitter frequency of 2.38 GHz and the 1.4 GHz frequency of the emission of atomic hydrogen. As Merle LaLonde demonstrated (Chap. 8), single polarization flat feeds can be built relatively quickly but what was required for best performance were dual circular polarization feeds such as the new 430 MHz line feed installed in early 1972. It had been decided that a 12 m (40 ft) long feed, which illuminated 213 m (700 ft) of the reflector surface, was the best compromise between sensitivity and the illuminated area spilling over the edge of the reflector as the zenith angle increases. The reduced effective area due to the spillover matters, but a greater loss in sensitivity is due to the thermal emission (noise) from the hot (300°K, 27 °C) ground illuminated due to the spillover. This meant that Allan Love's cookbook as to how to build 29 m (96 ft) line feeds (Chap. 8) could not be used, and Merle LaLonde

Fig. 10.3 Merle LaLonde in 1976 at NAIC's Ithaca laboratory near the airport with the newly constructed 2.38 GHz circularly polarized feed for the S-band radar system. Credit: Cornell University/NSF

had to conduct a series of new measurements on short sections of a 12 m feed in order to create a new "cook book" for this length. As discussed in Chap. 9, Merle LaLonde built a 13 m flat feed at 2.38 GHz to transmit the Arecibo Message at the November 1974 dedication of the upgraded telescope and this feed was used for the early planetary radar observations. After also building a flat 1.4 GHz feed for initial observations of atomic hydrogen, LaLonde turned his attention to building a 2.38 GHz circularly polarized feed. This was completed in 1976 (Fig. 10.3) followed by circularly polarized feeds at 1.4 GHz and 1.65/1.67 GHz, the main emission frequencies of the hydroxyl molecule OH. Sadly, Merle LaLonde died very unexpectedly in April 1977, a great loss of someone who had made enormous contributions to the construction of the Arecibo telescope and to its subsequent success. Lynn Baker, who joined NAIC's staff in Ithaca in 1978, took on the design and production of future line feeds (Fig. 10.4).

Receivers: With the move to higher frequencies, it became critical to produce receivers, the first stage of signal amplification, with the minimum amount of additive noise. This meant cooling the amplifiers to about 20 degrees Kelvin (−253 °C). While Merle LaLonde was busy building line feeds, George Peter and staff at NAIC's Ithaca instrumentation laboratory produced a series of low noise receivers to be used with the feeds, initially at 1.4 GHz and 1.65/1.67 GHz. For the planetary radar system at 2.38 GHz, the Jet Propulsion Laboratory (JPL) provided two single channel maser amplifiers they were no longer using, one for the Arecibo telescope and one for the future interferometer antenna needed for imaging the

Fig. 10.4 Lynn Baker at the observatory installing Kevlar guy cables to a new line feed in the early 1980s. Credit: Cornell University

surface of Venus. Maser amplifiers, which operate at liquid helium temperatures of about 4°K (−269 °C), were, at that time, the Rolls Royce of amplifiers due to their very low additive noise. A short time later, Lynn Baker, as his first major project as an engineer working for NAIC on the Cornell campus, built two dual channel maser amplifiers with assistance from JPL, one for Arecibo and one for the interferometer antenna, that could receive both polarizations from a circularly polarized line feed or horn.

Digital correlator: To study the line emission from atoms or molecules a device, a spectrometer, needed to be designed and built. Jon Hagen, a newly appointed Arecibo scientific staff member, who had built digital equipment as part of his doctoral thesis under Donald Farley of Cornell's School of Electrical Engineering, took on the task of designing and building a 1008 channel digital autocorrelation spectrometer (Fig. 10.5). While, in theory, analogue filters can be used, digital systems have great flexibility, fine filter widths over a small range of frequencies, the bandwidth, or coarse ones over a much larger range. Observations of atomic hydrogen in external galaxies required spectral widths of about 40 kHz, so 1008 channels would cover about 40 MHz bandwidth, very close to the total bandwidth of the line feed at 1.4 GHz. Once the spectrometer was in place, Robert Duquet, who

Fig. 10.5 Jon Hagen in late 1976 connecting up the 1008 channel correlator he designed and built to allow observations of spectral lines from atomic hydrogen and other atoms and molecules. Credit: Jon Hagen

had succeeded Elliott Mayo as the head of the observatory's computer department, rapidly wrote the necessary software to make use of it for observations. This first correlator was followed in 1986 with a clone of a new Berkeley correlator: 2048 channels and a maximum clock rate of 40-MHz with either 1-bit or 1.6-bit (3-level) operation.

In 1993 NAIC entered into a collaboration with John Canaris, who was working at the NASA-sponsored Space Electronics Research Center, then at the University of New Mexico, to design a chip with 1024 1.6-bit correlator channels and a clock rate of 100 MHz. Jon Hagen worked closely with Canaris on the design ensuring that it was tailored to Arecibo's future correlator requirements. The chips were fabricated by Hewlett Packard and, after some problems on the initial run, a working chip was manufactured. Funding was provided by NAIC, NRAO and the Naval Research Laboratory. The chip was used for the 50 MHz bandwidth so called Interim Correlator and, very shortly after, for the WAPP correlators in preparation for the post-Gregorian upgrade requirements (see Chap. 12). NRAO used it in the correlator for the newly constructed 100 m Green Bank Telescope.

10.4 The Arecibo Electronics and Computer Departments

Electronics: While the cooled receivers, line feeds and some other instruments were built at the Ithaca instrumentation laboratory, it was the observatory's electronics department that had the responsibility for keeping all the equipment operating including multiple high power transmitters. They were also responsible for developing almost all of the needed electronic infrastructure such as local oscillator systems, intermediate amplifiers, filters and digital sampling systems (Fig. 10.6). In the early 1980s after about 20 years in the position Miguel Feyjoo stepped down as head of the observatory's electronics department. He was succeeded by Jon Hagen. Figure 10.7 shows most of the electronics staff in June of 1983. After a few years Jon Hagen moved to Ithaca to head the receiver section of the instrumentation laboratory and was, in turn, succeeded by Edgar Castro, a longtime member of the electronics department (Fig. 10.8). Eddie Castro stayed in the position for many years contributing greatly to the observatory's successes.

Fig. 10.6 The receiver room in the late 1970s. At the far end are computer controlled frequency synthesizers, the racks to the left of the technician, Carlos Hernandez, hold IF amplifiers, filters, mixers, and the rack to the right of Hernandez is the ADAGE digital interface to the CDC 3300 computer. The white panels are 12-bit A/D converters, an upgrade from the 7-bit ones installed in 1965. Credit: Cornell University/NSF

Fig. 10.7 Most of the Electronics department staff in June 1983. These were the people who kept the observatory's electronics and transmitters operating smoothly in support of the observational programs. Credit: Cornell University

Fig. 10.8 Eddie Castro in the cable car on the way to the suspended platform. Credit: Cornell University

Computing:[1] During the 1970s and 1980s the computer department was headed successively by Robert Duquet, Peter Shames and Richard Murphy. In 1983 Phillip Perillat joined the department followed by Arun Venkataraman in 1986, both of whom remained with the observatory until its closing in 2023. Richard Murphy left the observatory about mid-1986 and Phil Perillat took over temporarily. Arun Venkataraman was appointed head of the department in late 1988 remaining in the position until 2023. At this time, Harris 100 and 800 mainframe computers were used for data acquisition and a Harris 1000 for data analysis. The Harris 100 was soon phased out. However, Digital Equipment Corporation (DEC) VAX computers were preferred for data analysis by many astronomers so a DEC VAX 11/750 was installed. While the VAX was not extensively used for data analysis, it was interfaced to a high resolution graphics display allowing high quality displays of Venus and other astronomical data sets.

Starting in 1989, Arun Venkataraman organized a series of much needed upgrades; the first Sun workstations running the UNIX operating system were installed and, later, color Sun work stations. Single board computers utilizing VME based hardware were introduced to control pointing and data acquisition allowing the phasing out of the Harris H800 computer. Bill Zisk of the electronics department built the VME based systems to replace the ADAGE digital sampling system that had been installed in 1965 and the connections to other instruments. Phil Perillat created the needed software. Data recording and analysis transitioned after almost 30 years of relying on large 9-track magnetic tapes to, initially, 8-mm Exabyte tape drives and, later, to disks. A capability to display and process images was also installed. This was simpler to use and more versatile than the VAX based system. Workstations replaced the use of the Harris 1000 for data analysis especially after ethernet was made available in the offices in the very early 1990s, initially via coaxial cables and, later, via fiber optic cables allowing higher data rates. This meant that every scientist and engineer had their own workstation. After over 30 years of relying on mainframe computers, the Harris 800 and Harris 1000 computers were retired in the mid-1990s.

Up to the late 1990s, there was no general purpose user interface to the telescope for astronomical observations; separate data acquisition programs were used depending on which instrument was being utilized. It was not until shortly after the 1990s Gregorian upgrade was completed that the development of a general purpose interface, CIMA, was initiated by Chris Salter and Paul Goldsmith and created by, primarily, Duncan Lorimer and Jeffrey Hagen.

Phil Perillat, a Stanford University physics graduate, became involved with all aspects of the telescope's performance ranging from software systems, to monitoring receiver performance to using lasers to measure the catenary sag in the telescope's support cables to derive their tensions. Not long before the telescope's collapse in 2020, he was developing a technique to survey the shape of the primary reflector using a high powered laser system and, after the collapse, working with others to get

[1] Arun Venkataraman and Phillip Perillat contributed most of the details in this section.

the 12 m auxiliary antenna and receiving system working so that there would at least be a working radio astronomy antenna on the site that could be used for some pulsar timing observations.

10.5 New Opportunities in Radio Astronomy

Figure 10.9 shows a number of the astronomy scientific staff in the control room in 1978.

The 21 cm line and the structure of the local Universe: One of the scientific drivers for the upgrading was the prospect of being able to detect the emission from atomic hydrogen (HI). As a result of the Universe's origin in the "Big Bang", on the large scale all galaxies are receding from each other in what is called the Hubble flow with the velocity of this recession increasing with distance by an amount dictated by what is known as the Hubble constant. Consequently, if the recessional velocities of galaxies, whose positions on the sky are known, can be accurately measured then their distances can be approximated and the 3-dimensional structure of the Universe mapped. The 21 cm radiation of atomic hydrogen in galaxies is Doppler shifted to longer wavelengths by an amount that depends on their recessional velocities, so a

Fig. 10.9 Members of the astronomy scientific staff in the telescope's control room in 1978. Back row from the left, John Harmon, Dale Frail, Martha Haynes and Ken Turner; front row from left, Tim Hankins, Mike Davis and Riccardo Giovanelli. Credit: Cornell University

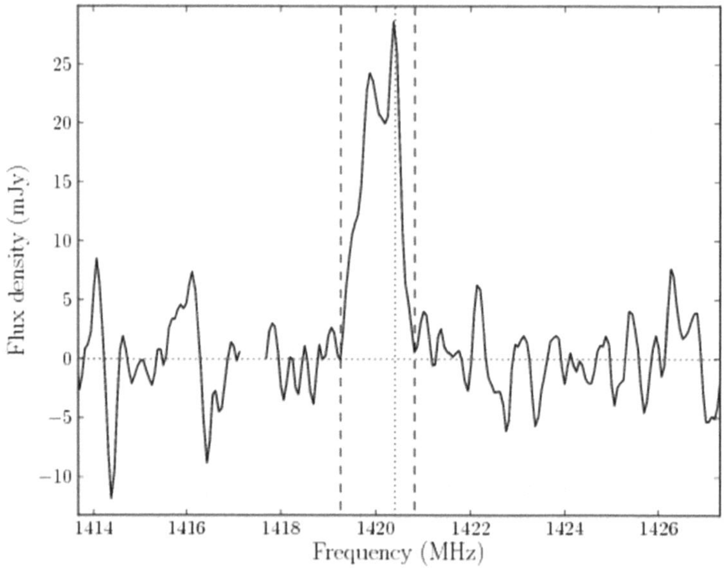

Fig. 10.10 A spectrum of the HI emission from a nearby galaxy. Its width is determined by the size, rotation rate and orientation of the galaxy relative to the line-of-sight from Earth. The vertical black dotted line shows where the spectrum would be centered if it was not moving away from our galaxy. Credit: Martha Haynes and Riccardo Giovanelli

measurement of the Doppler shift provides a direct measurement of the recessional velocity and, hence, the galaxy's distance (Fig. 10.10). By systematically observing the entire sky, or some fraction of it, visible with the Arecibo telescope, it is possible to detect all galaxies containing significant amounts of atomic hydrogen out to a distance determined by the sensitivity of the survey, and build up the 3-D image.

Two former graduate students at the University of Indiana, Riccardo Giovanelli and Martha Haynes, who both joined the staff of the Arecibo Observatory in 1978, took up the challenge of mapping the 3-D distribution of galaxies in about half of the sky observable with the Arecibo telescope, a region that encompasses what is known as the Pieces-Perseus supercluster of galaxies. Using a recent catalog of positions of spiral galaxies on the sky, which, similarly to the Milky Way, have significant amounts of atomic hydrogen in their interstellar mediums, they plotted the positions and saw that they were concentrated in a long band across the sky. This was a two dimensional plot with no information as to distances. Arecibo, with its tremendous sensitivity, was the best radio telescope to measure the Doppler shifts (Redshift as it is called because at optical wavelengths the recessional velocity of galaxies moves their optical emissions towards the red end of the spectrum) of the atomic hydrogen in these galaxies and, from this, obtain their distances using the Redshift-Distance relationship. In 1983 building on previous work, Giovanelli and Haynes set about measuring the redshifts of several thousand of these galaxies to obtain their

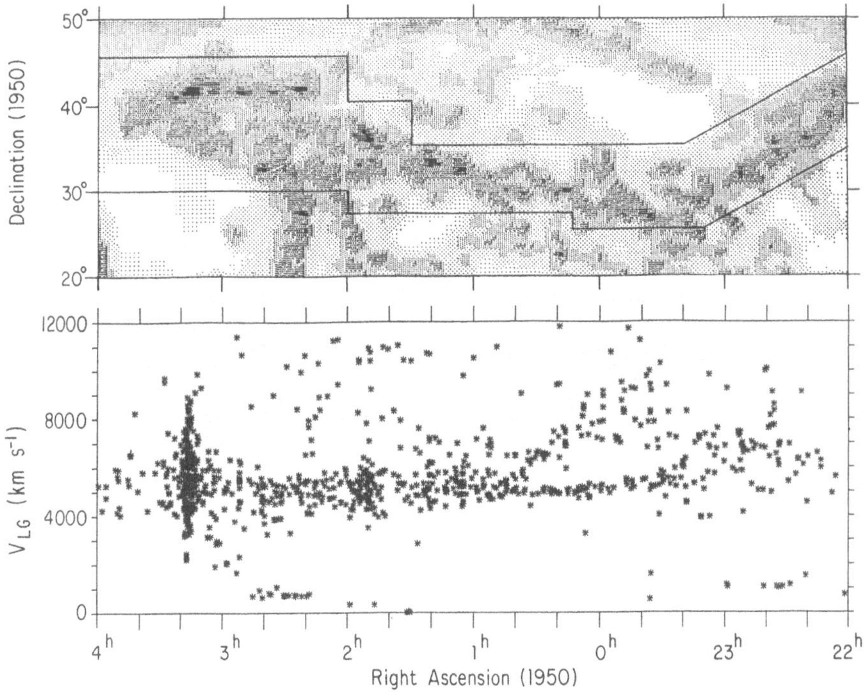

Fig. 10.11 Upper: The plane of sky densities in right ascension and declination of galaxies in the Pisces-Perseus region of the sky with the major ridge outlined. Lower: Measured redshifts in the outlined area of the upper plot as a function of right ascension and recessional velocity in km/s. From Giovanelli et al. (1986)

distances. They had to observe each galaxy for about 15 min to get an accurate measurement requiring a lot of telescope time, over 1000 h. Fortunately, the Pisces-Perseus section of the sky was not in great demand by other observers so time was available. After several years of work, they found that the galaxies were almost all at similar distances and were clustered in a filament that extended over a vast region of space, the first 3-D view of the large-scale structure of the Universe and the first realization of the filamentary distribution of galaxies (Fig. 10.11). For this discovery in 1989 Riccardo Giovanelli and Martha Haynes were awarded the Henry Draper medal of the U.S. National Academy of Sciences (Fig. 10.12).

OH Megamasers: In 1982 Willem Baan of the observatory's scientific staff, Perry Wood and Aubrey Haschick (1982) used the Arecibo telescope to detect OH emission from the galaxy Arp 220 (IC 4553), the first time that OH emission had been detected from an external galaxy. They discovered intense maser emission—the radio equivalent of lasers. Subsequent interferometric observations with the NRAO's Very Large Array (VLA) showed that the OH emission was coincident with the strong continuum source in IC 4553 and likely originated in a molecular

Fig. 10.12 Riccardo Giovanelli and Martha Haynes receiving the Henry Draper medal at a ceremony at the National Academy of Sciences in 1989 for "The first three-dimensional view of some of the remarkable large-scale filamentary structures of our visible universe." Credit: National Academy of Sciences

disk surrounding the continuum source (Baan and Haschick 1984). This was the first detection of an OH Megamaser with many more following.

Pulsars: Pulsar science remained a key research area, Arecibo was still the most sensitive radio telescope and it now had much larger frequency coverage. As of the end of 1973 about 100 pulsars had been discovered, primarily by the Australian Molongo telescope, the Jodrell Bank telescope and the Green Bank telescopes. Searches at Arecibo had had limited success until Joseph Taylor of the University of Massachusetts at Amherst decided to build a standalone search instrument for Arecibo. As recounted in Chap. 9, Taylor and Russell Hulse, a graduate student working with him, built the search system, took it to Arecibo and between 1974 and 1975 discovered 40 new pulsars including the first binary pulsar, PSR 1913+16.

On 25 September 1982 a pulsar was discovered by Donald Backer and colleagues that rotated about 600 times per second (1.558 milliseconds rotation period), vastly higher than the then know fastest rotator, the very young Crab Nebulae pulsar, at about 30 times per second. Unlike the Hulse–Taylor binary pulsar, which was discovered via a systematic survey of the sky, the discovery of this new 1.6 millisecond period pulsar was the result of detective work. It had been known for some time that the radio source named 4C21.53 scintillated—the radio wave equivalent of the twinkling of stars—indicating that at least parts of it had angular diameters of less than 1 arc second (1/3600 of a degree). Its brightness dropped very strongly with increasing radio frequency, a behavior similar to that of pulsars but no pulsar could be detected. Donald Backer, a former Cornell graduate student then at the University of California Berkeley, began a campaign to unravel the mystery using radio interferometers to delineate the structure of 3C21.53 and Arecibo to do further searches for a pulsating source. These searches, conducted by then Berkeley graduate student Shri Kulkarni, initially looked for a pulsar with a rotation period greater than 10 milliseconds since the Crab Nebulae, with its 33 milliseconds rotation period, is a very young pulsar and it was assumed that 10 milliseconds was likely the shortest period for a newborn pulsar. Success finally came with a search at Arecibo for even shorter periods resulting in the discovery of a new class of pulsars, ones with millisecond rotation periods (Backer et al. 1982). It was finally realized that these were "reborn" pulsars, normal pulsars (neutron stars) that had been in a binary orbit with another star. As the orbit distance between the star and the neutron star decreased due to the emission of gravitational waves, the gravitational force due to the neutron star on the outer gas layers of the star became stronger than the gravity of the star resulting in the outer atmosphere of the star being stripped away and transferring its angular momentum to the neutron star speeding up its rotation. If the inclination of the neutron stars rotation axis was favorable, then it would be a visible pulsar on Earth with a very short period. Millisecond pulsars turned out to have very stable rotation rates, almost perfect clocks, making them ideal for future efforts to use the arrival times of multiple millisecond pulsars distributed over the sky to detect the gravitational waves predicted by Einstein.

Another major discovery using the Arecibo telescope was the detection of terrestrial planet sized bodies in orbit about a pulsar, the first discovery of planets about a star other than our sun, albeit a neutron star (Wolszczan and Frail 1992; Fig. 10.13). For this discovery Alex Wolszczan, an Arecibo staff scientist at the time, received the Beatrice Tinsley prize of the American Astronomical Society in 1996 for his "creative and innovative analysis of binary and relativistic pulsars culminating in the discovery of a system of planets outside our solar system". Given his Polish background Alex Wolszczan received a number of awards for his work from Polish institutions culminating in sharing a 2001 Polish stamp set (Fig. 10.14) that included such distinguished Polish figures as former president Lech Waleza and Pope John Paul II.

10.6 Solar System Studies

Venus: The first post upgrade close approach of Venus to the Earth would be in September 1975, only 10 months after the upgrading completion and there was a lot to be done to get ready. An interferometer system was still needed to do the Venus imaging requiring a new 100 ft (30.5 m) diameter semi-steerable antenna to replace the 100 ft wire mesh reflector antenna at the Los Canos site 11 km north of the observatory (see Chap. 9). Rolf Dyce organized its design and construction and it was built at a cost of only $300,000, a real bargain (Fig. 10.15). It was one of the largest parabolic radio astronomical antennas in the United States at the time.

Because of the high data rate required for the high resolution imaging of the surface of Venus, a special purpose digital processor was needed as computers in the

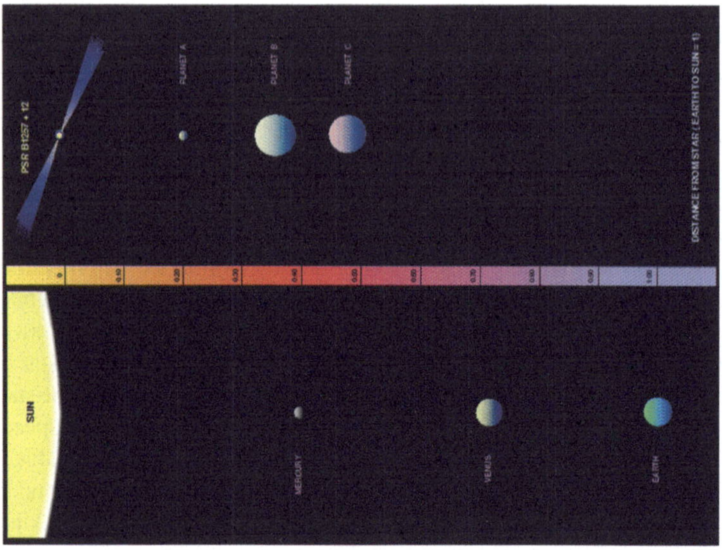

Fig. 10.13 Top: The location of the three planets in orbit about the pulsar PSR B1257+12 with the distances from the neutron star as shown in the horizontal scale in AU with 1 AU being the mean distance between the Sun and Earth. Bottom: The distances of Mercury, Venus and Earth from the Sun in AU for comparison. The sizes of the planets in both images are indicative of their relative masses. Credit: Alex Wolszczan

Fig. 10.14 Alex Wolszczan's photo on a 2001 Polish stamp honoring him for his many achievements including the discovery of planets around a star in 1992. He is in very good company; the left side of the stamp has an image of Copernicus

mid-1970s were not fast enough to do the required computations in real time and the rate at which you could store data on the then used magnetic tapes for later processing was not high enough. Valentin Boriakoff, a former graduate student of Frank Drake, who had built a digital pulsar processor as part of his doctoral thesis,

Fig. 10.15 The newly constructed 100 ft (30.5 m) Los Canos antenna in 1975. Its sky coverage was the same as that of the Arecibo telescope. The 2.38 GHz horn and the sub-reflector were provided by the Jet Propulsion Laboratory. The design and construction of this telescope was organized by Rolf Dyce. Credit: Cornell University/NSF

designed and built the real time processor. Unfortunately, it was not quite ready for the September 1975 observations limiting the resolution in the radar images to about 15 km due to the magnetic tape writing speed limitation (Fig. 10.16).

In April 1977 Venus made another close approach to Earth providing another opportunity to image the surface with the Arecibo radar system. Combining the results from 1975 and 1977 gave an image covering about 25% of the planet's surface (Fig. 10.17) the first large scale view of this intriguing planet (Campbell and Burns 1980). Impact craters, probable volcanoes and coronae can be seen as well as features interpreted as mountain chains and at least one rift valley extending through Beta Regio and further south. Perhaps the most interesting result was the paucity of large impact craters compared with Mercury, the Moon and Mars. Using estimates of the impact rate of asteroids and comets on a planet over the age of the solar system, the density of impact craters can be used to estimate the average age of the surface. Barbara Burns, a Cornell graduate student working with Don Campbell, analyzed the number and size distribution of the craters in the 25% of the surface imaged and estimated that this surface had an average age of about 600 million years, much younger than the several billion year average ages of the surfaces of Mercury, the Moon and Mars (Burns and Campbell 1985). The implication is that Venus is resurfaced via volcanic/tectonic processes on a timescale of about 600 million years. The Magellan orbiting radar about Venus in the early 1990s confirmed this result.

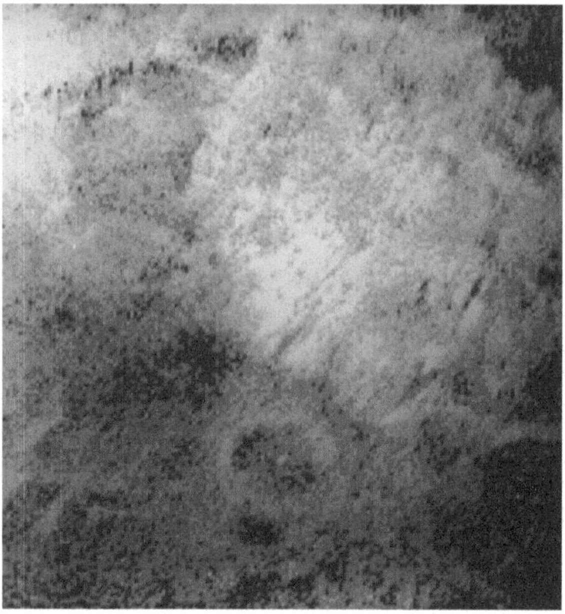

Fig. 10.16 The very first image obtained from the 1975 observations. It is of Alpha Regio, one of the first features discovered on the surface of Venus in the observations of the early 1960s. The approximately 200 km diameter circular structure just south of Alpha Regio was initially thought to be an impact crater but its characteristics, especially the small central feature, are not consistent with an impact structure of this size. It turns out that it was a new type of planetary feature displaying both volcanic and tectonic structures and thought to be the surface expression of sub-surface upwelling of mantle material. Many of these features are present on Venus and they were given the name Corona. The Corona below Alpha Regio was named Eve as the central bright spot was used to initially define the position of zero longitude. Since the planet's rotation period was still not known with adequate precision, it was important to have a physical feature defining the origin of longitude. The image is approximately 20° in longitude by 20° in latitude and was made using an early Versatec electrostatic printer, the only way to make a digital image at the observatory at that time. Credit: Barbara Burns and Don Campbell

Subsequent Venus imaging benefitted from much better data acquisition and processing capabilities and realized the approximately 2 km resolution that the radar system was designed to achieve (Fig. 10.18).

Icy surfaces in the solar system: A major discovery was the unusual radar reflection properties of icy surfaces in the solar system. All four of the large Galilean satellites of Jupiter, Io, Europa, Ganymede and Callisto were detected following the installation of the new 2.38 GHz radar system. What was strange was that the satellites with icy surfaces, Europa, Ganymede and Callisto (Io is very volcanic due to tidal torques exerted by Jupiter) all acted as retro-reflectors lighting up under the radar similarly to a highway sign that is illuminated by a car's headlights. The returned echoes were very strong and, equally unusual, most of the signal was in the same sense of circular polarization as was transmitted instead, as for normal surfaces, in the opposite sense (Campbell et al. 1978). This phenomenon is still not fully

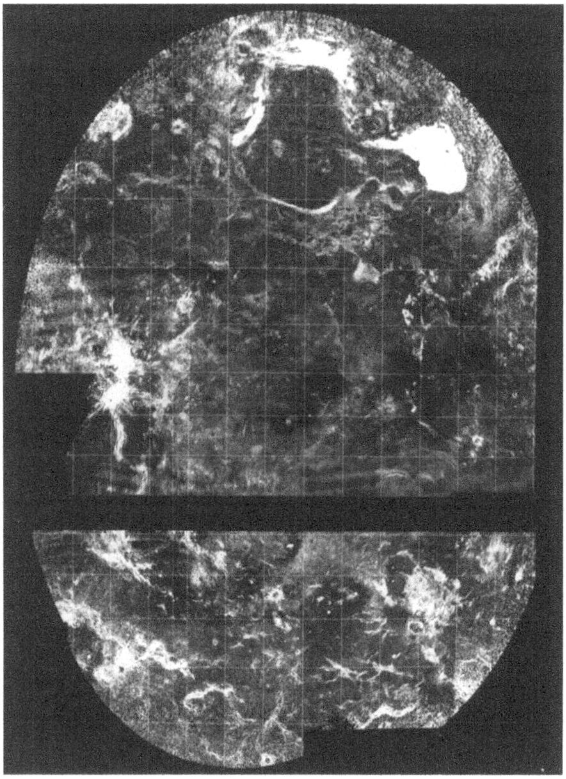

Fig. 10.17 A radar image of Venus based on data from 1975 and 1977, the first large scale image of the planet's surface with sufficient resolution to identify the nature of individual features. Grid lines are spaced 10 deg. apart in latitude and longitude. The resolution varies within the image from 5 km to 20 km. Maxwell Montes is at upper right, Beta Regio at center left and Alfa Regio at lower right. The roughly horizontal bands are an artifact resulting from using an interferometer to resolve the north-south delay-Doppler ambiguity. Figure 2 from Campbell and Burns (1980)

understood but is related to the propagation of the radar signal in the sub-surface; low temperature water ice is very transparent to radio waves.

This discovery provided a new technique for detecting icy surfaces in the solar system including ones covered by a thin layer of dusty material making the ice invisible in optical images; the radar wave can penetrate such a layer. This led to the discovery of ice at the poles of Mercury, very surprising since Mercury is the closest planet to the Sun. However, Mercury's rotation axis is almost perfectly perpendicular to its orbital plane about the Sun. If you are standing at one of the poles of Mercury, the limb of the Sun barely rises above the horizon. If you are in a depression such as the bottom of an impact crater the Sun will never shine on the surface, it will be extremely cold as there is no atmosphere to transmit heat, and ice formed from water delivered by, say, cometary impacts would be stable over the life of the solar system. John Harmon, a staff scientist at Arecibo, used the radar system to co-discover apparent water ice deposits at the poles (Harmon and Slade 1992) and, shortly later, map out the distribution of ice deposits in the polar regions of Mercury finding that they were all inside impact craters at least for the section of the polar regions imaged by the Mariner-10 spacecraft as it flew by Mercury (Fig. 10.19, Harmon et al. 1994).

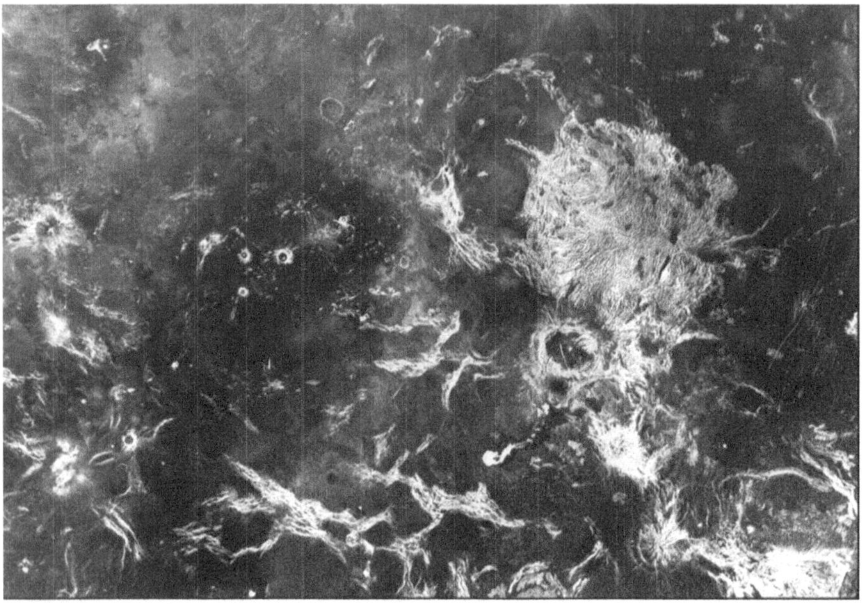

Fig. 10.18 An image of Alfa Regio and its surrounding area on Venus made in 1988 with a resolution close to 2 km. There are three volcanoes in a vertical line at far left. Several impact craters—referred to as the crater farm—at center left, and three coronae south of Alfa Regio. Credit: Don Campbell

Fig. 10.19 An Arecibo radar image of the north polar cap of Mercury from observations in 1991 and 1992 showing the distribution of ice deposits associated with impact craters. Figure 1a of Harmon et al. (1994)

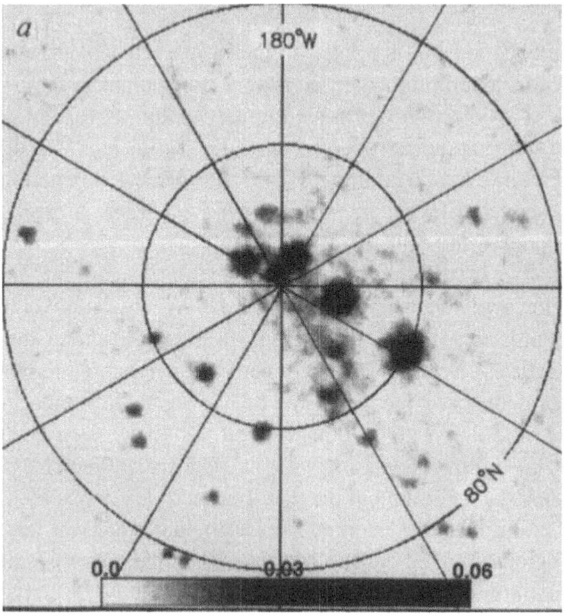

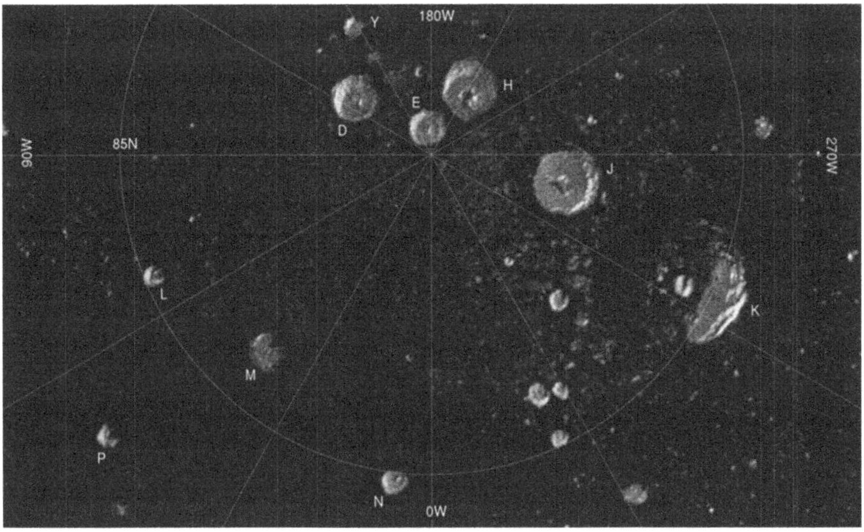

Fig. 10.20 A higher resolution Arecibo radar image of the north polar cap of Mercury from observations in 1999 and 2004 showing the distribution of ice deposits in the permanently shadowed areas of impact craters. For craters away from the pole, the area of the crater floor permanently shadowed by the crater rim is the part farthest from the pole. Figure 3a from Harmon et al. (2011)

Higher resolution imagery of Mercury's polar areas was obtained after the 1990s Gregorian upgrade of the telescope (see Chap. 11). This clearly showed that the bright radar echoes came from the permanently shadowed area of the craters in the polar regions confirming that the bright radar echoes were almost certainly due to water ice (Fig. 10.20, Harmon et al. 2011).

If there is ice at the poles of Mercury, then why not at the poles of the Moon since the Moon's rotation axis is very close to perpendicular to the Earth-Moon orbit plane about the Sun leaving polar impact craters in permanent shadow as on Mercury? There had been a suggestion based on observations from the Clementine lunar orbiter that there may be large deposits of ice in the polar regions of the Moon. However, a search carried out by Nicholas Stacy, a Cornell graduate student working with Don Campbell, using the Arecibo radar system did not detect any signature of ice equivalent to that found for Mercury (Stacy et al. 1997). It is clear from other remote sensing techniques from orbiting spacecraft and, in 2020, the aircraft mounted Sofia mid-infrared telescope, that there is water at low concentrations in the lunar soil either in the form of water ice, bound up with other minerals or incorporated in glass beads but not in large concentrated deposits as seems to be the case for Mercury.

Mars Viking mission support: 1976 was a big year for Mars exploration with two Viking mission orbiters each carrying a lander, which were going to touch down on the planet with the primary aim of detecting signs of past life. The Viking 1 lander was scheduled to touch down on July 4, American Independence Day. While the selected site looked good in images from the Viking 1 orbiter, the resolution of these images was not good enough to determine if the site was smooth or rough at scales important for landing Viking 1. Radar measurements could provide information on this averaged over about a degree of longitude as well as estimates of the surface density provided that the site was at the sub-Earth location on Mars—i.e. at the center of the disk of the planet.[2] In late May 1976 Arecibo was used to observe the proposed Viking 1 site and found it to be rough whereas nearby sites were much smoother. Many of the site selection team wanted to proceed with the landing anyway as changing sites would miss the July 4 date and also entail considerable re-planning. Higher levels in NASA decided that they could not ignore the Arecibo data, and a formal request was made that Arecibo be used on July 2, 3 and 4 to assess another site. Unfortunately, this site also appeared to be rough and unacceptable. However, previous Arecibo observations has revealed a quite smooth area further west. Photos from the orbiter were consistent with this and also indicated that the area met the aims of the mission. It was selected and on 20 July 1976 Viking 1 touched down safely in the Chryse Planitia area on Mars. The Viking 2 landing was too far north on Mars for the site to be assessed with Arecibo. It too touched down safely on 3 September. Given the stakes, Arecibo more than repaid the $3M that NASA had contributed to the upgrading and new transmitter on the telescope.

Asteroids and Comets: The tremendous increase in the sensitivity of the radar system for solar system studies with the resurfaced reflector and new transmitter opened up a whole new area, the study of small bodies, asteroids and comets. Asteroids fall into two classes, near-Earth ones (NEAs) that primarily orbit in the inner solar system inside the orbit of Mars and those in the belt of asteroids between the orbits of Mars and Jupiter known as main belt asteroids (MBAs). Prior to 1975 only two asteroids, both NEAs, had been detected by radar and none with Arecibo. This changed in January of 1975 when Arecibo detected the large NEA Eros with the 430 MHz radar as the new 2.38 GHz one was not yet working. The first with the new radar system was Betulia, a 2.9 km NEA which was detected in May 1976. This was followed by Ceres, the largest MBA at about 950 km in diameter, in March/April of 1977 and Vesta at about 525 km in diameter, in November 1979. Between 1980 and early 1992, the beginning of the Gregorian upgrade of the telescope, the number of asteroids observed each year ranged from five to a high of 13 in 1985 with slightly more MBAs than NEAs.[3] The number of NEAs was limited by their discovery rate with optical telescopes. As of early 1990 only 140 NEAs were known and not all of these came close enough to Earth to be detectable with the Arecibo radar.

Round-trip time delay and Doppler shift measurements provide precise distances and line-of-sight velocities for NEAs which, when combined with plane-of-sky optical positional measurements, greatly refine our knowledge of their orbits

---

[2] A description of the measurements and their interpretation can be found in Tyler et al. (1976).

[3] A discussion of radar observations of asteroids and a list of their detections by Arecibo and the radar system on NASA's 70 m Deep Space Network antenna at Goldstone in California can be found at https://echo.jpl.nasa.gov/asteroids/index.html

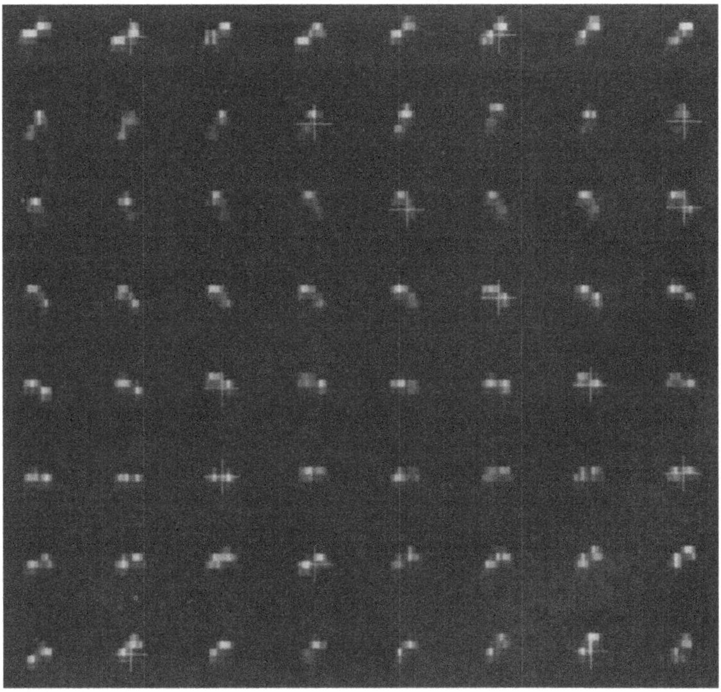

Fig. 10.21 64 Arecibo radar time delay-Doppler images of the NEA 4769 Castalia taken over about a 2.5 h period on August 22, 1989. In time the images go sequentially from left to right starting with the top row. Time delay is in the vertical or y-axis direction and Doppler frequency is along the x-axis. The effective location of the radar system is looking down from the top. Figure 3 from Ostro et al. (1990)

compared with just plane-of-sky positional measurements. This allows their orbits to be predicted far into the future allowing a determination as to whether a specific NEA may pose a serious threat to Earth in the near or far future. Being able to distinguish between those NEAs that may pose a threat to Earth based on uncertain orbits and those that are genuine threats based on more precise orbit predictions is a tremendous benefit.

The first good 2-D delay-Doppler images of an NEA were obtained in August 1989 of 4769 Castalia (initially known as 1989 PB) which came within 0.037 AU of the Earth on August 22 while still within Arecibo's declination range. Its closest approach to Earth was on August 25 at 0.027 AU but, unfortunately, not observable with Arecibo. The NASA Goldstone 70 m antenna could have observed it then but was occupied with the Voyager 2 spacecraft's close approach to Neptune so could not observe 4769 Castalia until August 30 at a distance of 0.057 AU. Sixty four delay-Doppler images were obtained with Arecibo on August 19 (Fig. 10.21) over about 2.5 h of tracking time. The time delay resolution was 300 m and the resolution in the Doppler frequency direction was approximately 150 m. The images clearly show two small bodies rotating about each other over a little more than half a turn

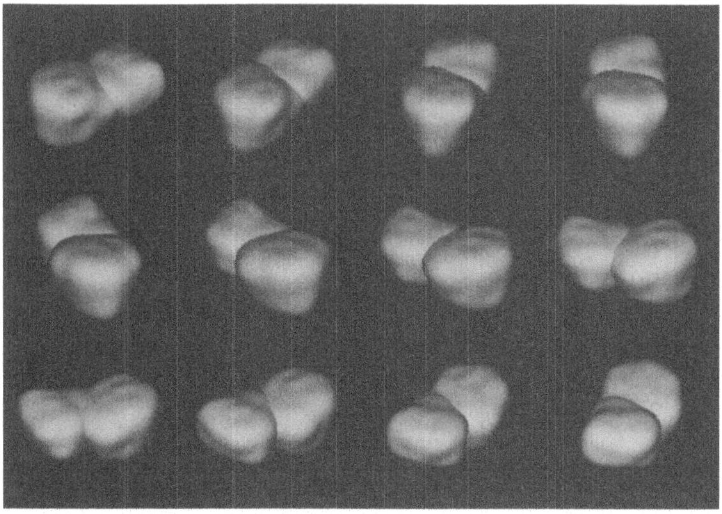

Fig. 10.22 A shape reconstruction model of 4769 Castalia by Scott Hudson based on the 64 delay-Doppler images of Fig. 10.20. The model clearly shows the contact binary structure of the asteroid. Figure 3 from Hudson and Ostro (1994)

corresponding to a rotation period of 4.07 h. They are either attached to each other or, as stated in Ostro et al. (1990), possibly separated by less than 100 m. The overall dimensions of the system are about 1.0 km × 1.7 km. Castalia was the first clear case of a contact binary asteroid possibly formed by a low velocity collision of two small bodies.

Scott Hudson, then a graduate student at the California Institute of Technology (Caltech) working with Steve Ostro, developed inversion techniques to obtain 3-dimensional shape models of 4769 Castalia from the 64 delay-Doppler images of Fig. 10.21. The results showed that Castalia has dimensions of 0.7 × 1.0 × 1.6 km with a 100 to 150 m deep waist strongly indicating that it is a contact binary resulting from the low velocity collision of two bodies (Fig. 10.22, Hudson and Ostro 1994). Scott Hudson's development of the shape reconstruction technique based on a series of delay-Doppler images made as the asteroid rotates was a major contribution to the study of asteroids allowing 3-D images to be obtained for numerous asteroids and contributing to the planning and success of the Japanese Hayabusa mission to the NEA 25143 Itokawa and the US OSIRIS-Rex mission to the NEA 101955 Bennu.

Over the succeeding years and, especially, after both the Gregorian upgrade of the Arecibo telescope and the increase in the NEA discovery rate, many more contact and true binary NEAs were discovered via imaging with the Arecibo and NASA Goldstone radar systems (Chap. 12). No currently existing optical telescope has the resolution to image a 1 km sized NEA at the typical distances of their closest approach to Earth.

In November 1980, comet 2P/Encke was the first comet detected with the upgraded radar system. Encke was followed by 26p/Grigg-Skjellerup in May

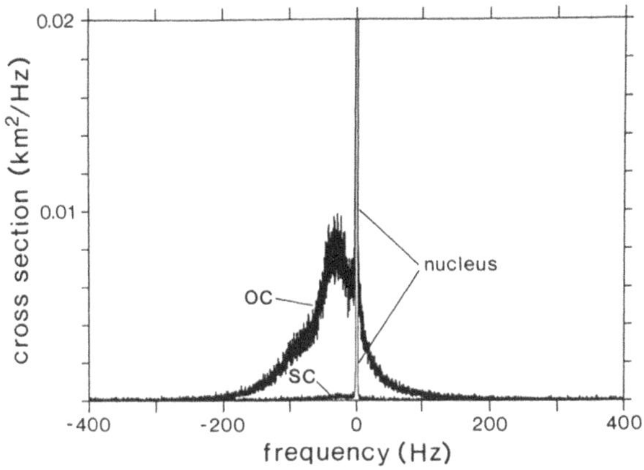

Fig. 10.23 The radar echo spectrum from Comet IRAS-Araki-Alcock. As indicated, the very narrow component at 0 Hz is from the nucleus and the broad component (skirt) offset from the nucleus echo is from a coma of cm or larger particles that extends up to ~1000 km from the nucleus. The OC component of the skirt is the echo power in the expected sense of received circular polarization while the SC one is in the opposite sense. If the particles are ice, then the very low level of the SC component indicates that the great majority of the radar echo is from particles with an upper limit of about 5 cm for their size. Figure 2a from Harmon et al. (1989)

1982. Grigg-Skjellerup was visited by the European Giotto probe in 1992 after Giotto's encounter with Comet Halley in 1986. The next comets observed were IRAS-Araki-Alcock in May 1983 and Sugano-Saigusa-Fujikawa in June 1983 followed by Halley's comet in November 1985.

By far the best opportunity to study a cometary nucleus and its environs was the very close approach to Earth of IRAS-Araki-Alcock at a distance of 0.037 AU. Unfortunately, it was moving so fast in declination that it was only in view of the Arecibo telescope for one 2.5 h period on 11 May 1983. Due to uncertainties in the comet's position and the very short observing time, only the spectrum of the radar echo was obtained, delay-Doppler imaging observations were not possible. The spectrum of the radar echo showed a very strong echo from the nucleus and a broad spectral component offset from the nucleus echo (Fig. 10.23). From the echo strength the inferred diameter of the nucleus was between 5 km (an ice surface) and 16 km (loosely packed snow). Other non-radar observations of the nucleus indicate about a 10 km diameter. Using approximately 10 km and the width of the nucleus spectrum the rotation period is in the 2–3 day range. This period is consistent with the lack of motion of features in the nucleus spectrum over the 2.5 h observing time (Harmon et al. 1989). However, the broad spectral component was the real interest as it indicated a coma similar to the optically observed dust coma but consisting of cm sized and larger particles. The offset spectrum is due to the large particles, like the dust, being carried off the surface of the comet on its sunlit side by gas drag due to the sublimation of surface volatiles such as ice. Modeling and comparison with results from the NASA Goldstone radar indicate that the particle cloud likely extended up to 1000 km from the nucleus. This was the first time that a large particle coma had been directly observed.

Halley's comet was observed in November 1985, when at a distance of 0.062 AU, almost twice the distance at which IRAS-Araki-Alcock was observed and, hence, a potentially significantly weaker echo given the inverse fourth power dependence of echo strength on distance. No clear echo from the nucleus of Halley was detected but there was a weak echo with a broad spectrum similar to that received from IRAS-Araki-Alcock indicating that Halley also had a coma of cm or larger sized particles. Unfortunately, Halley was not in view with Arecibo during its closest approach to Earth in 1985 at a distance of 0.042 AU.

10.7 New Initiatives for Studies of the Upper Atmosphere

The Chain of Radars and the World Day program: In the early 1980s, Herb Carlson (Fig. 10.1), the former graduate student of William Gordon, while in charge of the Upper Atmosphere Facilities section at the NSF, arranged for the move of SRI International's Incoherent Scatter Radar (ISR) in Alaska to Sondrestrom in Greenland to form a chain of four ISR radars, Jicamarca (Peru), Arecibo, Millstone Hill (west of Boston) and Sondrestrom (see Fig. 10.24), all at close to the same longitude and covering a very wide range of latitudes. With the chain in place, the World Day program was initiated in which the ISR radars measured a variety of ionospheric parameters at the same time. This occurred at roughly monthly intervals with the aim of developing a very long term record of ionospheric parameters as a function of latitude. Other ISR radar systems joined the program including the European Incoherent Scatter (EISCAT) radars in northern Scandinavia and Svalbard in the Artic (see Isham et al. 2000). In the early 1980s the World Day program in the U.S. became part of the NSF sponsored CEDAR—Coupling, Energetics, Dynamics of Atmospheric Regions—program. Approximately 500 h/year was devoted to the program, about one third of the total time that was scheduled for ionospheric radar observations at Arecibo. The program continued at Arecibo until the collapse of the telescope in December, 2020.

The New Heating Facility: Improving the Observatory's ionospheric modification capability was a priority in the mid-1970s resulting in the decision to build a separate transmitting facility at an isolated site near the coast at Islote just east of Arecibo. This was a major endeavor not helped by the site being a low lying marshy, mosquito infested area that had been drained via excavated water channels leading to a pumping station that kept the area about 3 ft (1 m) below sea level so that it could be used for sugar cane farming. By the 1970s it was no longer in use for farming. Rolf Dyce organized the construction assisted by Albert Veldhuis with Antonio Crespo as the field supervisor. The facility consisted of 32 inverted log periodic antennas that could radiate over the frequency range 2–12 MHz (i.e. the long 2 MHz dipole elements are at the top and the shortest 12 MHz ones at the base) supported from 75 m (246 ft) high towers which were separated by about 100 m (Fig. 10.25).

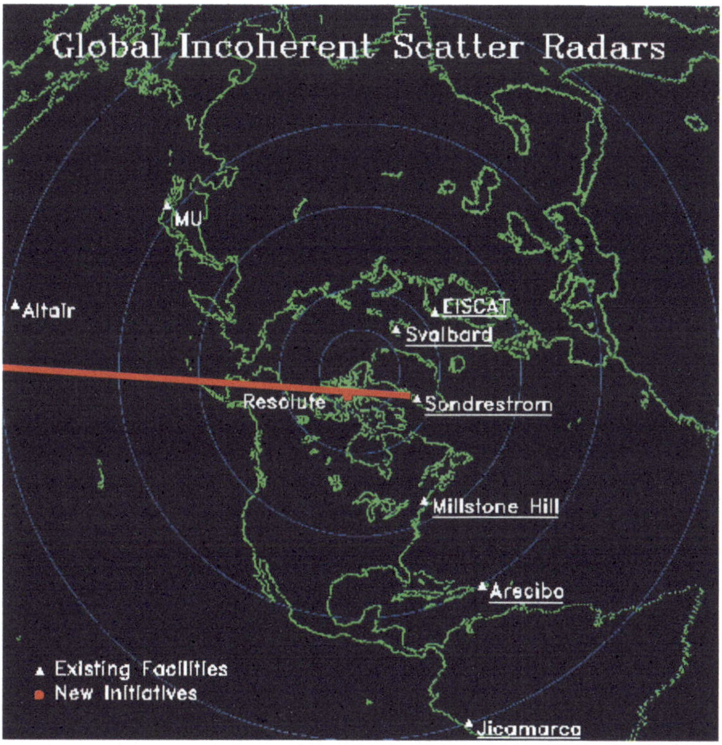

Fig. 10.24 The locations of the Chain of Radars sites including the EISCAT ones. The red bar indicates the move of the Sondrestrom radar from Alaska. Figure from: CEDAR 1983-2005 by Tim Killeen, page 23, CEDAR web site https://cedarscience.org/sites/default/files/2021-10/Killeen05.pdf

Four surplus 150 kW transmitters were acquired from the US Government with the electrical power being provided by four diesel generators. The array was in two 4x4 quadrants with two transmitters feeding the two polarizations of each quadrant. Transmitter power was fed to the antennas via, initially, pseudo coaxial lines, a ring of wires with a central conductor that mimicked a solid outer cylinder with a central conductor (Fig. 10.26).

The facility was designed and built by observatory staff including additional staff hired for the project. Mosquitoes were a major problem especially at the beginning. After 14 years with the observatory Rolf Dyce resigned in 1979 to join a startup radio communications company and Albert Veldhuis took over the project with José Maldonado organizing the construction aspects.

In the late 1990s the facility faced two threats which caused it to be closed down. There was pressure from environmental groups in Puerto Rico to return the land to its original marsh state by turning off the pumping system, which had kept the water level below sea level. This may have eventually caused the facility to close but the real end came abruptly due to Hurricane Georges, which passed directly over the

Fig. 10.25 The Islote HF facility showing the layout of the towers supporting the log periodic antennas. The antenna wires are so fine that it is barely possible to see them in the photo. Originally, four pseudo coaxial transmission lines brought the power from the transmitter building to the center of the array, two for each 4x4 sub-array and then with progressive splitting of the power to separately feed all the N-S and E-W antenna curtains radiating linear polarizations. The two polarizations could then be combined to give either left or right circular polarization. The apexes of the base of the log periodic antennas are at the center of each four towers where the transmissions lines end. The transmitter and other buildings at upper left give some idea of the huge size of the array. Due to arcing and other problems with the pseudo coaxial lines, in the early 1990s they were replaced by solid lines. This photo was taken after that change was made. Credit: Cornell University/NSF

Island and the Observatory on 21 September 1998. The damage to the HF facility was such that it was not worth repairing especially given the issues related to the pressures from the environmental groups. However, despite being a significant operational headache, the facility, operating in conjunction with the Observatory's incoherent scatter radar and optical sensing instrumentation, provided a wealth of data on non-linear effects in a sparse plasma.

The SOUSY-VHF radar system: In the late 1970s, a group at the Max Planck Institute for Aeronomy (MPAe) in Germany developed a Very High Frequency (VHF, 30 to 300 MHz) SOUSY (SOUding SYstem) radar to study the Earth's mesosphere, troposphere and stratosphere (see Fig.1.3 of Chap. 1). On a visit to Germany, James (Jim) Walker, the head of the atmospheric sciences group at Arecibo, encouraged Jürgen Röttger and his group to bring a VHF radar to Arecibo as its observations would complement those of the observatory's 430 MHz incoherent scatter radar. To do this, the initial need was to identify a suitable frequency that was as free of interference as possible and which did not interfere with other users. A

Fig. 10.26 The transition from the four solid coaxial lines coming from the transmitter building to the pseudo coaxial wire lines. Only two of the pseudo coaxial lines were in place when this photo was taken. Arcing of the pseudo coaxial lines under the high transmitter power was a significant problem leading to their being replaced by solid coaxial lines in the early 1990s. The corner of the building housing the generators is at lower right. Credit: Cornell University/NSF

wide band feed (Fig. 10.27) was built and sent to Arecibo for this purpose. A frequency of 46.8 MHz was found to be satisfactory. Initial tests with a 4 kW transmitter gave encouraging results; on 3 May 1980 good echoes were obtained from the mesosphere at a height of 80 km. This led to the decision to construct a 144 kW transmitter and data acquisition system in a trailer and transport it to Arecibo by plane. After arriving in New York for transshipment to Puerto Rico, the trailer was lost, ending up in Miami, trucked to New York and finally flown to San Juan. There was considerable damage to the radar system but this was repairable. Once operational the SOUSY radar had a very productive campaign studying waves and turbulence in the mesosphere. The system was returned to Germany in late 1981 or early 1982, where it was refurbished and sent to Norway.

In September 1983, while Jürgen Röttger and Tor Hagfors were on a hiking trip in the French Alps, Hagfors persuaded Röttger to return to the Arecibo Observatory for an extended stay. Röttger arrived back in Arecibo in late 1984 staying for several years and serving as Acting Head of the Atmospheric Sciences Group. Röttger's first task well before he returned to Arecibo was to participate in a proposal to the NSF for the installation of a permanent VHF radar system on the Arecibo telescope. This was approved and a VHF radar system with a transmitter power of 50 kW and operating at 46.8 MHz was installed in early 1985. A very readable account of the earlier campaign and Röttger's Arecibo experiences can be found in Rottger (2013).

Fig. 10.27 The SOUSY wideband feed on the Arecibo telescope. From Rottger et al. (1981)

The AIDA campaign: In 1987 the then head of the Atmospheric Sciences group, Colin Hines, started planning for the Arecibo Initiative in the Dynamics of the Atmosphere (AIDA). This would, in part, be a response to the development of the NSF supported program Coupling, Energetics and Dynamics of Atmospheric Regions (CEDAR). As outlined in Hines' Preface to a 1993 issue of the Journal of Atmospheric and Terrestrial Physics (Hines 1993), the AIDA campaign originated in the idea of acquiring an Imaging Doppler Interferometer (IDI) for Arecibo and utilizing the Islote HF system as a transmitter, aimed at studies of the mesosphere and stratosphere. As a test, Gene Adams of Utah State University and John Brosnahan, then of Tycho Industries, were invited to bring their IDI system to Arecibo and set it up at the heating facility site. In order to have independent measurements for comparison with the IDI results, Chet Gardner of the University of Illinois was invited to bring his CEDAR related lidar system to Arecibo to obtain a vertical profile through the mesosphere. This would also give the Arecibo staff an introduction to lidar systems in preparation for acquiring one for the observatory. As another comparison, the Observatory acquired a Spaced Antenna Radar (SAR) system from Bob Vincent of the University of Adelaide, Australia.

When combined with the observatory's incoherent scatter system and airglow instruments, the array of instruments was impressive. This led to a general invitation to the CEDAR community to bring additional instruments to Puerto Rico to participate in a large campaign in 1989. Many of these instruments were placed at various

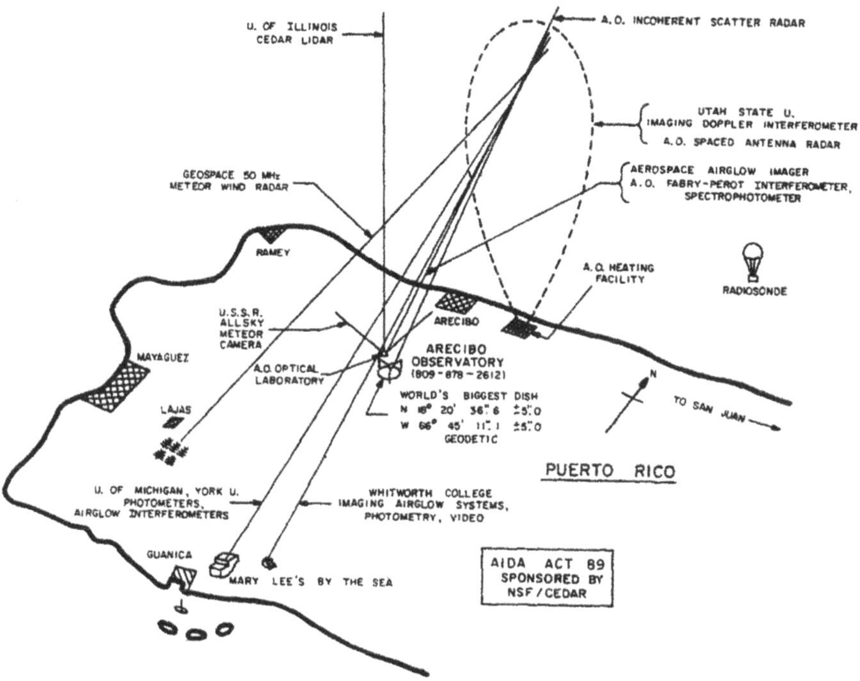

Fig. 10.28 The locations and type of instruments participating in the 1989 AIDA campaign. From Hines (1993)

locations on the island (see Fig. 10.28). The campaign took place over three periods of 8, 14 and 8 days, centered on the new moon to optimize airglow results, in March, April and May of 1989. The campaign was a big success and about 20 papers discussing the results were published in a March 1993 special issue of the Journal of Atmospheric and Terrestrial Physics (Vol. 55, No. 3), which included the Preface by Colin Hines. The results were also disseminated through special sessions at meetings of the American Geophysical Union and a dedicated two day workshop at the University of Wisconsin in October 1990. Three significant outcomes of this campaign amongst many others were (S. Raizada, private communication):

- A spectacular 6 day (5–11 April, 1989) continuous run of the various radar systems that allowed in-depth analysis and characterization of tides in the lower ionospheric regions. The results showed that between the 70—110 km altitude region, diurnal tides are the most dominant feature over Arecibo. The zonal and meridional amplitudes were observed to be ~50 m/s and ~25 m/s respectively, while the vertical winds were found to be in the range of tens of centimeters per second (Roper et al. 1993).
- Validation of neutral winds both at mesosphere and thermosphere altitudes using multiple techniques in the mesosphere (Djuth and Elder 1993; Bird et al. 1993). This campaign resulted in the acquisition of radar hardware, development of new

analysis techniques, installation of imaging Doppler Interferometer (IDI) radar that operated at medium frequency (Brosnahan and Adams 1993).

- Comparisons between Na lidar and airglow emissions revealed a good agreement between the wave parameters derived from active and passive optical techniques (Hecht et al. 1993). This study showed that the integrated Na abundance between 86 and 88 km correlated well with the OH emission lines, while O2 atmospheric bands were shown to peak between 90 and 94 km based on the column abundance of Na lidar data at these altitudes.

The Airglow Program: The incoherent scatter radar (ISR) could make very precise measurements of the properties of the ionized plasma in the Earth's upper atmosphere but it could only infer the properties of the neutral upper atmosphere. Optical airglow observations could do direct measurements complementing the ISR ones. As recounted in Chap. 9, the first optical airglow instrument, a rotating filter photometer, was borrowed from NOAA and installed in a tiny "shack" near the operations building in preparation for observations in the last dark of the moon period in the winter of 1966. This was initiated by Herb Carlson, who had joined the Arecibo scientific staff after finishing his PhD working with William Gordon. The primary objective was to verify the early morning heating of the ionosphere due to high energy electrons emanating from Arecibo's magnetic conjugate location in Argentina.

The success of these initial airglow observations led to the construction of a more substantial photometer "shack" on the hill above the west side of the operations building (see Fig. 9.18). Over the next few years photometers, an Ebert-Fastie spectrometer and a Fabry-Perot interferometer were installed aimed at measuring emission line spectra in order to obtain temperatures and winds for the neutral atmosphere. Quoting Herb Carlson (private communication, September 2020); "Biondi and Sipher brought their instrumentation to Arecibo to launch a visiting research program that lasted through much of Arecibo's first decade. A graduate student, Vince Wickwar, from Rice University did his PhD applying these airglow observations to determining chemical recombination rates in the ionosphere over Arecibo (Wickwar et al. 1974). A Canadian postdoc (Leroy Cogger) came to apply ISR plus optical observations to initiate a new experimental capability for the study of neutral density variations (Cogger et al. 1974). This launched a new research field based on combined ISR and optical emissions from the neutral atmosphere to learn the role of this key coupling in shaping Earth's upper atmospheric environment."

Starting in the early-1970s, the airglow program was driven successively by staff scientists John Meriwether, Leroy Cogger, Roger Burnside and Craig Tepley. Based on a talk[4] that Craig Tepley gave at the 2003 40th anniversary celebration of the observatory, the decade 1973–1983 was the period when spectroscopy dominated the airglow program. During this period, measurements of 7320 Å twilight emissions were made and imagers were first installed by visitors at the observatory. In the

[4]Page 10 of the National Astronomy and Ionosphere Center/Arecibo Observatory Newsletter #37, April 2004.

Fig. 10.29 Dedication event for the upgraded photometer building in early 1980s. The power pole is the termination of the 13 kV power line from the sub-station about 3 miles (5 km) south-east of the observatory. Credit: Cornell University/NSF

early 1980s the photometer building was expanded to two rooms, the observers' room and the instrument room (Fig. 10.29).

The decade 1983–1993 was marked with the development of interferometery. Roger Burnside used interferometric methods on the 6300 Å emission line of atomic oxygen to measure thermospheric winds. In 1986, Robert Kerr and his colleagues made observations of the Hα line that originates from the geocorona. This work was followed by Kerr and Tepley (1988) inferring hydrogen concentrations in the exosphere. A readable review of Arecibo's optical upper atmosphere program can be found in Isham et al. (2000).

References

Baan, W.A., and A.D. Haschick. 1984. The Peculiar Galaxy IC4553: VLA-A Observations of the OH Megamaser. *Astrophysical Journal* 279: 541–549.

Baan, Willem A., Perry A.D. Wood, and Aubrey D. Haschick. 1982. Broad Hydroxyl Emission in IC443. *Astrophysical Journal* 260: L49–L52.

Backer, D.C., S.R. Kulkarni, C. Heiles, M.M. Davis, and W.M. Goss. 1982. A Millisecond Pulsar. *Nature* 300: 615–618.

Bird, J.C., G.G. Shepherd, and C.A. Tepley. 1993. Comparison of Lower Thermospheric Winds Measured by a Polarizing Michelson Interferometer and a Fabry-Perot Spectrometer During the AIDA Campaign. *Journal of Atmospheric and Terrestrial Physics* 55 (3): 313–324.

Brosnahan, J.W., and G.W. Adams. 1993. The MAPSTAR Imaging Doppler Interferometer (IDI) Radar: Description and First Results. *Journal of Atmospheric and Terrestrial Physics* 55 (3): 203–228.

Burns, B.A., and D.B. Campbell. 1985. Radar Evidence for Cratering on Venus. *Journal of Geophysical Research* 90: 3037–3047.

Campbell, D.B., and B.A. Burns. 1980. Earth Based Radar Imagery of Venus. *Journal of Geophysical Research* 85: 8271–8281.

Campbell, D.B., J.F. Chandler, S.J. Ostro, G.H. Pettengill, and I.I. Shapiro. 1978. Galilean Satellites: 1976 Radar Results. *Icarus* 34: 254–267.

Cogger, L.L., V.B. Wickwar, and H.C. Carlson. 1974. Combined Airglow and Incoherent Scatter Observations as a Technique for Studying Neutral Atmospheric Variations. *Radio Science* 9: 205–210.

Djuth, F.T., and J.H. Elder. 1993. The VHF Meteor Radar System Used During the Arecibo Initiative in Dynamics of the Atmosphere (AIDA) campaign '89. *Journal of Atmospheric and Terrestrial Physics* 55 (3): 229–239.

Giovanelli, R., M.P. Haynes, and G.L. Chincarini. 1986. Morphological Segregation in the Pisces-Perseus Supercluster. *Astrophysical Journal* 300: 77–92.

Harmon, J.K., and M.A. Slade. 1992. Radar Mapping of Mercury: Full-Disk Images and Polar Anomalies. *Science* 258: 640–643.

Harmon, J.K., D.B. Campbell, A.A. Hine, I.I. Shapiro, and B.G. Marsden. 1989. Radar Observations of Comet IRAS-Araki-Alcock 1983d. *Astrophysical Journal* 338: 1071–1093.

Harmon, J.K., M.A. Slade, R.A. Velez, A. Crespo, M.J. Dryer, and J.M. Johnson. 1994. Radar Mapping of Mercury's Polar Anomalies. *Nature* 369: 213–215.

Harmon, J.K., M.A. Slade, and M. Rice. 2011. Radar Imagery of Mercury's Putative Polar Ice: 1999 – 2005 Arecibo Results. *Icarus* 211: 37–50.

Hecht, J.H., T.J. Kane, R.L. Walterscheid, C.S. Gardner, and C.A. Teplky. 1993. Simultaneous Nightglow and Na Lidar Observations at Arecibo During the AIDA-89 Campaign. *Journal of Atmospheric and Terrestrial Physics* 55 (3): 409–423.

Hines, C.O. 1993. Arecibo Initiative in Dynamics of the Atmosphere, Aida Act 89. *Preface, Journal of Atmospheric and Terrestrial Physics* 55 (3): 197–199.

Hudson, R.S., and S.J. Ostro. 1994. Shape of Asteroid 4769 Castalia (1989 PB) from Inversion of Radar Images. *Science* 263: 940–943.

Isham, B., et al. 2000. Upper Atmosphere Observations at the Arecibo Observatory: Examples Obtained Using New Capabilities. *Journal of Geophysical Research* 105: 18609–19637.

Kerr, R.B., and C.A. Tepley. 1988. Ground Based Measurements of Exospheric Hydrogen Density. *Geophysical Research Letters* 15: 1329–1332

Ostro, S.J., J.F. Chandler, A.A. Hine, K.D. Rosema, I.I. Shapiro, and D.K. Yeomans. 1990. Radar Images of Asteroid 1989 PB. *Science* 248: 1523–1528.

Roper, R.G., G.W. Adams, and J.W. Brosnahan. 1993. Tidal Winds at Mesopause Altitudes Over Arecibo (18 deg N, 67 deg W), 5–11 April 1989 (AIDA '89). *Journal of Atmospheric and Terrestrial Physics* 55 (3): 289–312.

Rottger, J. 2013. My Time with the Arecibo Observatory: An Exotic Experimental, Scientific and Personal Challenge. *URSI Radio Science Bulletin* 2013: 17–27.

Rottger, J., P. Czechowsky, and G. Schmidt. 1981. First Low-Power VHF Radar Observations of Tropospheric, Stratospheric and Mesospheric Winds and Turbulence at the Arecibo Observatory. *Journal of Atmospheric and Terrestrial Physics* 43: 789–800.

Stacy, N.J.S., D.B. Campbell, and P.G. Ford. 1997. Arecibo Radar Mapping of the Lunar Poles: A Search for Ice Deposits. *Science* 276: 1527–1530.

Tyler, G.L., D.B. Campbell, G.S. Downs, R.R. Green, and H.J. Moore. 1976. Radar Characteristics of Viking 1 Landing Sites. *Science* 193: 812–815.

Wickwar, V.B., L.L. Cogger, and H.C. Carlson. 1974. The 6300 Å O(D-1) Airglow and Dissociative Recombination. *Planetary and Space Science* 22: 709–724.

Wolszczan, A., and D.A. Frail. 1992. A Planetary System Around the Millisecond Pulsar 1215+12. *Nature* 355: 145–147.

Chapter 11
The Gregorian Upgrade and Angel Ramos Foundation Visitor Center

While the new reflector surface opened up many opportunities for new research initiatives as described in the previous chapter, utilizing its full capabilities was still hampered by the need to use line feeds with their limited bandwidths. This prompted Frank Drake in the late 1970s to begin looking into the possibility of changing to a Gregorian secondary reflector coupled with a third reflector to correct for the spherical aberration of the primary reflector. This in turn led to an extended effort under the new NAIC director, Tor Hagfors, to develop the concept, obtain the funding and initiate a major upgrading of the telescope and planetary radar system that would lead to a much more versatile and sensitive telescope for radio and radar astronomy and ionospheric observations. Hagfors resigned in early 1992 during the very early stages of the construction work. He was succeeded by Paul Goldsmith in early 1993 with Don Campbell filling in for the intervening year. Contracts were let for the structural modification to the telescope including the Gregorian dome, the reflectors, and a new 2.38 GHz transmitter. The "new" telescope came on line in 1997. At the same time, the Angel Ramos Foundation Visitor Center was designed, built, equipped with displays and completed also in 1997.

11.1 The Initial Idea

While the resurfacing of the reflector allowed the telescope to be used at much higher frequencies, its use was still dependent on building line feeds tuned to whatever frequency was required (Fig. 11.1). Apart from their limited bandwidth, line feeds had one other problem, they could not efficiently illuminate along the direction of their major axis leaving an approximately 300 ft (100 m) under-illuminated area at the center of their illumination pattern resulting in a loss of sensitivity and relatively high near-in sidelobes. Perhaps remembering Bill Gordon's 1964 plan to mount a Gregorian reflector on the original Carriage House 2 that would focus the incoming radar waves over a wide range of frequencies (see Chap. 7), in the late 1970s Frank

D. Campbell, *The Arecibo Observatory: A History of Innovation and Discovery*, Historical & Cultural Astronomy, https://doi.org/10.1007/978-3-031-75225-4_11

Fig. 11.1 Five line feeds and a point feed on Carriage House 2 in the late 1980s. Carriage House 1 (right) has the 96 ft long 430 MHz feed plus a 46 MHz yagi feed around it near the top. Credit: Cornell University/ NSF

Drake started to look into the feasibility of a Gregorian reflector system for Arecibo. While a single reflector can bring all the rays from the primary spherical reflector to a focus it has two drawbacks: the illumination of the prime reflector has an inverse taper, that is the outer edges of the reflector are illuminated more efficiently than the central areas rather than vice-versa as is the case for line feeds, and the focal point is far below the Gregorian reflector and so inaccessible for installing receivers much less a transmitter. As discussed by Drake in a 1980 report,[1] a third reflector would be the solution to both these problems, correcting the illumination issue and bringing the focus to an accessible location not far below the azimuth arm. Drake also proposed a design where the illuminated area is offset from one centered on the direction the telescope is pointing. This has the big advantage of delaying the spillover of the illuminated area over the edge of the main reflector as the zenith angle increases.

As recounted in Chap. 10, Frank Drake stepped down as NAIC Director in July 1981 and, after a search, Tor Hagfors, who had been the Arecibo Observatory Director from 1971 to 1973, was appointed the new NAIC director as of July

[1]Gregorian Feed Concepts for the Arecibo Telescope, Frank D. Drake, November 1980, NAIC Report 147. Cornell University archives under Arecibo Ionospheric Observatory.

1982. Harold Craft served as interim director for the year. In the intervening years Hagfors had been the director of the European Incoherent Scatter Scientific Association (EISCAT). Very shortly after being appointed, Hagfors, likely based on Drake's 1980 report, asked Sebastian von Hoerner, a highly respected antenna expert at the National Radio Astronomy Observatory's site in Greenbank, West Virginia, to do a further investigation of a two reflector system to replace the line feeds. For simplicity, von Hoerner chose to model a system with a secondary reflector shaped to bring the rays reflected from the primary reflector to a focus and a tertiary reflector below that to translate the focus to a convenient location. This did not solve the illumination problem but it did allow him to investigate secondary reflector sizes as a function of the degree of offset of the illumination pattern to delay the zenith angle at which spillover occurs.[2] Von Hoerner's modeling assumed a 700 ft (213 m) diameter illuminated area, the same as for the standard 40 ft lines feeds then in use, and he modeled two offsets of the area relative to one symmetrically located around the direction the telescope is pointing, 50 ft and 125 ft The 50 ft offset would delay the spillover of the 700 ft illuminated area from about 11 degrees zenith angle to close to 15 degrees. A 125 ft offset would avoid any spillover out to 20 degrees zenith angle.

11.2 Working Out the Details

Based on von Hoerner's 1983 report, in 1984 Tor Hagfors initiated work on both the electromagnetic design of the reflector system and the structural modifications to the telescope that would be required to support the significant additional weight of this system. Lynn Baker did the initial design of the reflector system using a geometric optics approach and assuming a 50 ft (15 m) offset in the illuminated area. A larger offset would require a secondary reflector that would be too large. Unlike von Hoerner, Baker did not have the secondary reflector bring the rays to a focus thus allowing a normal illumination pattern with the edges tapered relative to the center. The approximate sizes of the secondary and tertiary reflectors were 65 ft (20 m) and 35 ft (11 m). Per-Simon Kildal, then at the Norwegian Institute of Technology, who had designed the large cylindrical reflector of EISCAT's 224 MHz incoherent scatter radar, was recruited to do a full physical optics analysis of the reflector system. Hagfors would have been well acquainted with Kildal from his time as director of EISCAT. The physical optics analysis did reveal a problem with the low frequency performance the solution of which required the vertical distance between the secondary reflector and the "paraxial" focus, the focus of a paraboloid tangent to the spherical primary reflector, to be increased by about 5 ft (1.5 m) (Kildal and

[2] Arecibo Three-Mirror System: Optimizing the Optics, Sebastian von Hoerner, Engineering Division Internal Report No. 112, National Radio Astronomy Observatory, Greenbank, West Virginia, February 1983. Cornell University archives under Arecibo Ionospheric Observatory.

Skyttemyr 1989). Solving this would require raising the telescope's suspended platform by this amount, something that could be relatively easily done.

Paul Stetson was hired to do the initial structural and mechanical design study of the modifications to the telescope required to support the dual Gregorian system with the issue of feasibility being paramount. Stetson had been a longtime engineer working on antenna issues with MIT's Lincoln Laboratory and Hagfors would have met him when he worked for Lincoln Laboratory at their Millstone Hill site west of Boston in the 1960s. Paul Stetson's work, while little celebrated, was crucial to the success of the Gregorian project.

What was the Gregorian project? It evolved over time to the following components:

- The design and implementation of a dual Gregorian reflector system that would be usable for frequencies between 300 MHz and 8 GHz. The illuminated area would be offset by 50 ft (15 m) to delay the zenith angle at which spillover occurs from 11° to about 15°. To maximize the sensitivity, the illuminated area would be elliptically shaped, 700 ft (213 m) × 765 ft (234 m), with the long axis tangent to the rim of the primary reflector to avoid any additional spillover. The reflectors would be enclosed in an 80 ft (24 m) diameter spaceframe structure—the Gregorian Dome—covered by aluminum panels to shield the reflectors and horns from the wind and radio frequency interference coming from the horizon direction. A rotary turret in a separate cabin would bring multiple receiving and transmitting horns to the tertiary's focal location (Fig. 11.2).

- The installation of a Ground Screen, initially a 60 ft (18 m) high wire mesh screen around the periphery of the telescope reflector to shield the receiving systems from ground thermal radiation at zenith angles larger than about 15° (Fig. 11.3). The design was later changed to 50 ft (16 m) as a cost saving measure.

- The installation of a 1.0 MW 2.38 GHz CW transmitter, over twice the power of the one installed in 1974.
- The installation of waveguide on the Gregorian reflector side of the azimuth arm to enable a 430 MHz transmitting capability in the dome that houses the Gregorian reflectors.
- The readjustment of the primary reflector surface to bring it to 2.0 mm rms allowing good aperture efficiency at 8 GHz. Between 1984 and 1986 there was a large effort to repair and reset the reflector surface organized by Lynn Baker and José Maldonado at a cost of about $700,000. They achieved 2.3 mm rms, a very significant improvement over the approximately 3.4 mm rms prior to the resetting.

What was being proposed was perhaps the most radical modification of a large astronomical telescope that had ever been attempted. If successful, the payoff would be very significant in terms of sensitivity for both radio and radar observations, larger frequency coverage, lower sidelobes, higher power for planetary radar observations,

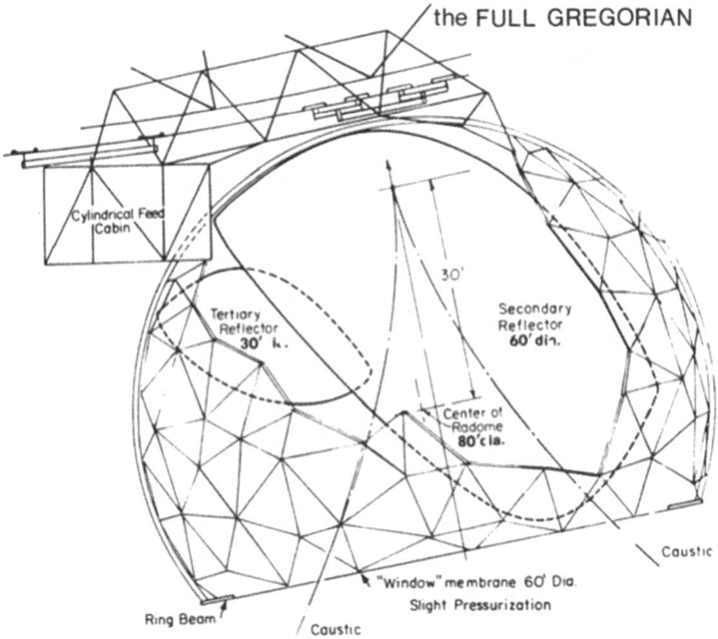

Fig. 11.2 The initial design of the Gregorian system showing the two reflectors inside an 80 ft diameter radome and a separate cylindrical feed cabin housing the feed horns on a rotary floor and also the 1.0 MW 2.38 GHz transmitter. Prior to the final structural design, Ben Hooghoudt redesigned the system to incorporate the rotary floor that brings the horns to the tertiary reflector's focal location and the transmitter into the now 83 ft diameter dome, dispensing with the feed cabin. This lowered the distance from the feed horns to the tertiary, significantly reducing the size of the horns. Credit: Cornell University/NSF

and a two beam ionospheric incoherent scatter system utilizing Carriage House 1 and the Dome.

Radio astronomy: As mentioned earlier, due to line feeds inability to radiate along their axis, there is an approximately 300 ft (92 m) "hole" at the center of their illumination patterns. This reduces the sensitivity and leads to near-in side lobe levels of only a factor of 10 (−10 db) below that of the main beam. For observations at frequencies above 1 GHz, the increase in sensitivity due to the increased illuminated area, higher aperture efficiency and reduced system noise temperatures due to the ground screen was expected to be about 2.4 at zenith angles out to 11° increasing to approximately a factor of 8 at 20° zenith angle. The first sidelobe level would decrease significantly to about a factor of 75 (−19 db) lower than the main beam.

Radar astronomy: As with the early 1970s upgrade, the big "winner" from the upgrading plans was planetary radar observations. Apart from the doubling of the transmitter power, the two-way nature of radar means that it benefited from the increased aperture size and efficiency twice. For observations of Venus and Mercury, the average increase in sensitivity was expected to be a factor of about 7. The high temperature of Venus reduces the sensitivity gain when receiving the echo. For

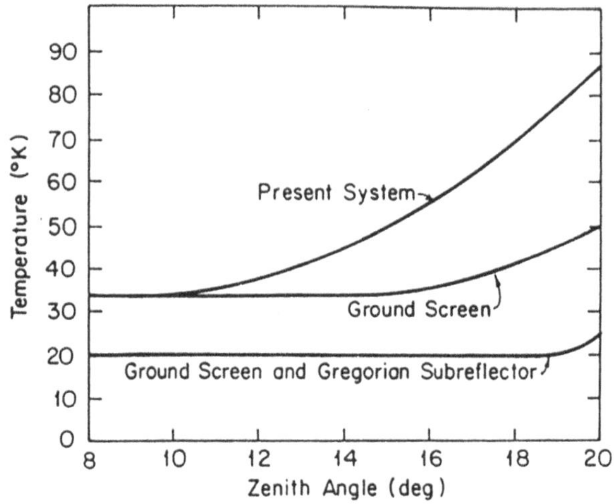

Fig. 11.3 The predicted improvement in system noise temperatures (Tsys) with the installation of a ground screen and then the full Gregorian system. The "present system" plot shows the very large increase in Tsys as the zenith angle of the receiving line feed increases to 20° for frequencies above 1 GHz. The loss in sensitivity is a factor of 2.5 compared with zenith angles below 11°. With the ground screen in place the Tsys at 20° zenith angle is reduced by almost a factor of 2. With the Gregorian reflectors installed with their offset illumination the Tsys is almost constant with zenith angle. The prediction of a Tsys of 20°K was optimistic, in practice it was closer to 24°. Credit: Cornell University

Mars and the Jupiter system an average increase of about 13. For asteroids and comets, an average increase of a factor of 22. For the Saturn system, which can only be observed at high zenith angles due to its approximately 2.25 h round-trip-light time, the increase would be almost a factor of 40.

On 13–15 October 1986 a meeting attended by about 40 scientists was held at Cornell University to discuss the scientific benefits of an upgraded Arecibo telescope. Brief discussions of the technical aspects of the upgrading were followed by talks covering many of the possible observational programs enabled by the Gregorian system.[3]

11.3 Finding the Funding

The October 1986 meeting at Cornell University was attended by Claud Kellett, Program Manager for NAIC in the Division of Astronomical Sciences at the NSF, and Richard Behnke, Program Manager for Upper Atmosphere Research in the

[3]Proceedings of the Arecibo Upgrading Workshop, Eds. Joseph H. Taylor and Michael M. Davis, National Astronomy and Ionosphere Center, October 1987. Available in the Cornell archives.

Geosciences Directorate of the NSF and a former Arecibo staff member in the ionospheric group. Their presence reflected the ongoing discussions with the NSF, primarily the Division of Astronomical Sciences, about the funding for the Ground Screen and the Gregorian system. NASA expressed interest in possibly funding a more powerful transmitter, 1 MW at 2.38 GHz, over twice the power of the then current one. However, NASA was also interested in SETI and had had a low level program since 1975 funded at a few hundred thousand dollars per year. Starting in 1981 this ramped up to between $1M and $2M per year and increased to over $2M per year in 1987 likely in preparation for the formal start of funding in FY 1991 for the $100M over 10 years High Resolution Microwave Survey (HRMS) search for extraterrestrial intelligence. Arecibo's role was a targeted search, that is, looking at specific stars, with this part of the project managed by the NASA Ames Research Center at Moffett Field in Mountain View, California. In parallel, an all sky survey was to be conducted using an antenna at NASA's Deep Space Network site in California to scan the entire Milky Way. A proposal dated 11 March 1991 submitted to the Arecibo Observatory by Jill Tarter of the SETI Institute on behalf of the International Working Group for SETI requesting 2600 h of observing time gave a good idea of the plans for the targeted search. Of the 773 targeted solar type stars that had been identified, 245 were observable with Arecibo and the plan was to observe them over the frequency range 1.0 to 3.0 GHz at frequency resolutions as fine as 1 Hz. This was a problem since the line feeds had a bandwidth of only about 40 MHz. Moving the feeds radially could extend this bandwidth but only by a small factor. Since feeds only existed to cover the HI band near 1.4 GHz, the hydroxyl OH band near 1665 MHz and the 2.38 GHz radar frequency, a number of new lines feeds, perhaps 23, along with receiver systems would need to be built. Clearly, the proposed two reflector Gregorian system would provide higher sensitivity and make new line feeds unnecessary. It was suggested that instead of building new line feeds and receiver systems that the HRMS project contribute funds for the Gregorian upgrade project. With an estimated cost of $100,000 for each line feed a contribution of $2.3M was proposed with the condition that NAIC guaranteed 2500 h for the HRMS survey after the completion of the Gregorian upgrade.

The formal start of the HRMS project was in early October 1992 (US Fiscal Year 1993) with a budget of $12M. However, in September of 1993 an amendment was proposed in the US Senate to the FY1994 US budget to cancel the HRMS project. It was passed ending the NASA sponsored project. By this time NASA had committed the $2.3M to the Gregorian project. A new privately funded SETI program was formed based on the SETI Institute, a non-profit corporation located near the Ames Research Center and founded in late 1984 starting operations in February 1985.[4] Project Phoenix as it was called was limited to the targeted search part of the HRMS program utilizing the Arecibo telescope. Despite the change, NAIC did honor the 2500 h of telescope time after the Gregorian upgrade was completed. A detailed discussion of the history of the HRMS program can be found in Garber (1999).

[4] See the History of the SETI Institute available on-line at seti.org/history-seti-institute

Based on the studies by Paul Stetson, Lynn Baker and Per-Simon Kildal, a proposal for the construction of a 60 ft Ground Screen was submitted to the NSF in March of 1986 at a cost of $1,560,020. This was followed by a proposal to both the NSF and NASA for the Gregorian feed system and the 1 MW transmitter in December 1987. The proposal was for $17,900,000, $11,841,000 for the Gregorian feed system including the needed structural reinforcement and $6,059,000 for the transmitter including the power generator.

On 30 September 1988 there was a meeting at NASA Headquarters attended by Laura (Pat) Bautz (Director of the NSF Division of Astronomical Sciences), Kurt Riegal (NSF Astronomy Centers Section), Claud Kellett (NAIC NSF Program Manager), Geoffrey Briggs (Director, NASA Solar System Exploration Division), Henry Brinton (Manager, NASA Planetary Astronomy and Atmospheres program), Jurgen Rahe (Discipline Scientist, NASA Planetary Astronomy Program and responsible for NASA's support of the Arecibo planetary radar program), Lynn Griffiths (SETI Program, NASA Life Sciences), Tor Hagfors (Director of NAIC) and Don Campbell (NAIC). The objective was to decide how to split the costs of the Gregorian upgrading project between the NSF and NASA. Much to everyone's surprise, Geoff Briggs took control of the meeting by stating that NASA would pay for the Ground Screen ($2M in 1991 to 1993), the new 1.0 MW transmitter plus the power generating units ($6M in 1993/94), and would contribute the SETI $2.3M (from NASA Life Sciences in 1990/91) to the costs of the Gregorian reflector system. The NSF would be responsible for the remainder of the Gregorian system's costs including any increases ($9.7M). This was clearly a larger contribution from NASA than the NSF had expected leaving both agencies satisfied with the outcome and the Cornell group very relieved. Pat Bautz did point out that the NSF could not budget this far in advance.

A memorandum of Agreement (MOA) between the NSF and NASA covering the two agencies' responsibilities for the funding of the Gregorian upgrade went through a number of iterations with the final one being signed by Richard Truly, the NASA Administrator on 4 May 1990 and by Erich Bloch, the Director of the NSF, on 1 June 1990. By this time the estimated total cost was $22.7M. NASA's costs had increased to $11.3M due to an increase in the transmitter cost to $7M. NSF's cost for the Gregorian system increased to $11.4M. On 1 February 1991 the Cornell University Board of Trustees approved the expenditure of $22.8M for the Gregorian upgrade. The reason for the $100,000 discrepancy with the NSF/NASA MOA is not clear.

The collapse of the National Radio Astronomy Observatory's 300 ft telescope at its Green Bank site in West Virginia on 15 November 1988 put the funding for the Arecibo upgrade in doubt. The US senators from West Virginia, especially Robert Byrd, wanted the NSF to rebuild the telescope at an estimated cost of about $50M, and the concern at Cornell was that this would seriously affect the NSF's ability to fund its share of the upgrade. Would Puerto Rico be able to help out? An approach was made to the Puerto Rican government arranged by Wallace Gonzalez Oliver, a Cornell law school graduate and a former Secretary of the Treasury in Puerto Rico, who was highly regarded across the political spectrum in Puerto Rico. This eventually resulted in a 24 August 1989 letter from Sila Calderon, the Secretary of State in

Puerto Rico, to Erich Bloch, the NSF Director, offering to contribute $6M over 3 years starting in FY 1991 to the upgrading project via a proposal to the local legislature. However, for reasons that are not clear, there was no follow up on the offer.

On 5 June 1990 Don Campbell, then acting director of NAIC while Tor Hagfors was on sabbatic leave for a year, received a letter from Frank Giovane, Program Manager for NAIC in the Division of Astronomical Sciences at the NSF, saying the National Science Board, the oversight board for the NSF, had approved funding for both the Ground Screen and the Gregorian Upgrade proposals. Great news!

11.4 Organizing the Management of the Project

Tor Hagfors as Director of NAIC was the Project Executive. On 31 March 1992 Michael Davis stepped down from his position as the Arecibo director to take on the task of Project Scientist. Daniel Altschuler, on leave from a faculty position at the University of Puerto Rico and the current Associate Director, succeeded him as the observatory director. John Kiefer, a long time member of Cornell's facilities department, was appointed the Project Manager for the Ground Screen with José Maldonado, the head of the facilities department at the observatory, as the on-site construction manager. After a wide search, Ben Hooghoudt, a Dutchman with long experience in designing and building radio astronomy antennas, was appointed the overall Project Manager for the Gregorian project taking up the position in February 1991. José Maldonado was to be the Arecibo upgrading resident engineer in addition to his other duties with Aida Carrasquillo, a longtime observatory staff member, appointed as his Administrative Assistant. In late July 1992, Kurt Samuelson, a civil engineer from the Ithaca area, was appointed Assistant Project Manager and Pam Sieradski the Project Administrator. In mid-1993 Evan Kurtz was appointed as the project manager for the procurement of the transmitter.

In mid-1994 Ben Hooghoudt had to resign as project manager and return to his home in Leiden, The Netherlands, due to a serious illness. He was succeeded as project manager by Kurt Samuelson. Sadly, Ben Hooghoudt died on 8 February 1995 at the age of 70.

Tor Hagfors resigned as the NAIC Director effective 31 January 1992 to take up a position as a director of the Max Planck Institute for Aeronomy in Katlenburg-Lindau, Germany, now the Institute for Solar System Research in Göttingen. Paul Goldsmith, Professor of Astronomy at the University of Massachusetts at Amherst, was appointed NAIC Director as of early 1993 with Don Campbell as interim director for the intervening year and the new Project Executive.

11.5 The Mini-Gregorian Demonstration Project

To overcome any skepticism within the relevant research community and the funding agencies as to the viability of the offset Gregorian design, the decision was made to construct a quarter scale demonstration system with a 14 ft × 16 ft secondary reflector and an 8 ft tertiary. Lynn Baker was given the responsibility for its design, construction and testing. Planning and design work started in late 1985 with the initial work concentrating on how to fabricate the two reflectors and how to test them. The best option was to use the Large Optical Generator at the University of Arizona, the same machine that was used to make molds for large optical mirrors. Testing would require the construction of a near-field measurement range. A geodesic dome would support the two reflectors and the 5 GHz test feed horn and the whole system would be housed in an inflatable dome next to the NAIC development laboratory near the Ithaca airport.

During the summer and fall (autumn) of 1986, reflector mold blanks were prepared, trucked to the University of Arizona in Tucson, figured on the Large Optical Generator and trucked back to Ithaca. After the mold surfaces were polished and their shape checked, the reflectors were fabricated using carbon fiber with a foam core composite. Both reflectors were then metallized, installed and aligned in the geodesic dome. A positioning table was installed in the inflatable building on which the mini-Gregorian system—the reflectors mounted in the geodesic dome with a radiating 5 GHz horn at the focus—would be mounted for the near field measurements. An arch was constructed over the entire system to hold the near-field receiving horn.

Over the winter, spring and summer of 1988, the emphasis was on the near field measurement system, both the hardware and, especially, the analysis software. This involved a collaboration with the US National Bureau of Standards. After months of effort the system was perfected with the results indicating that the Gregorian system was meeting it design goals. The mini-Gregorian was then disassembled and prepared for shipment to the Arecibo Observatory, where it was installed on the telescope (Fig. 11.4).

After correcting some alignment and other problems, the mini-Gregorian system performed up to expectations. It illuminated 350 ft (105 m) of the primary reflector with a 50 ft (15 m) offset and was the largest single dish radio astronomical system in the world for frequencies above 5 GHz since Arecibo was not equipped with line feeds at these frequencies. It was later equipped with receivers at 3 GHz and 8 GHz and used as a regular observing system until it was decommissioned in 1993, the beginning of the upgrading work. However, its most important contribution was the verification of the dual reflector feed concept putting to rest any doubts as to the viability of the plans for the Gregorian upgrading of the telescope's performance.

Fig. 11.4 The mini-Gregorian mounted on the Arecibo telescope. It is on a trolley towed by Carriage House 1, which supports the 430 MHz line feed. The two reflectors can be clearly seen with the 5 GHz horn just inside the geodesic dome at upper left. The illuminated area is 350 ft (105 m) and, unlike the full Gregorian, the 50 ft (15 m) offset of the illuminated area was uphill towards the rim of the main reflector. This was done to avoid the reduction in sensitivity due to blockage of the incoming radiation by the large suspended structure of the telescope. Given the relatively small size of the illuminated area, it did not spill over the primary reflector even at 20° zenith angle. Credit: Lynn Baker and Cornell University/NSF

11.6 The Ground Screen

Specifications for the Ground Screen were: (1) A 60 ft (18 m) high wire mesh "fence" around the approximately 1.0 km periphery of the 305 m primary reflector to prevent the area illuminated by the dual reflector Gregorian system from spilling over the top of the ground screen at zenith angles less than 20°; (2) A tilt of the screen away from the reflector at an angle of 25° to the vertical combined with a 2° maximum deviation of the plane of the screen to minimize the reflection of ground radiation from the surrounding hills; (3) A mesh with galvanized wires spaced 0.25 inches (6 mm) apart to minimize transmission of ground radiation at frequencies below 8 GHz.

On 8 January 1991, the NSF authorized Cornell to proceed with the construction of the Ground Screen. A request for proposals was issued in May 1991 with the lowest bid received from Contemporary Constructors Inc. (CCI) of San Antonio, Texas. It was for $2.8M, significantly higher than the budgeted $2M, which included management costs. The initial solution was to lower the height of the screen to 50 ft

Fig. 11.5 One section of the ground screen installed as of 8 May 1993. Multiple sections were erected simultaneously. Seven panels covered the 50 ft height of the screen. Their width varied depending on their position around the dish. Credit: Cornell University/NSF

(15 m) and to leave a 60° gap on the western side. The west side was chosen due to the desire to do radar studies of the Saturn system, the rings and larger satellites. A round trip light time of about 2.25 h to Saturn combined with the limited tracking time meant that radar transmission would occur for up to 25 min after Saturn was observable and reception of the echo would be for the last 25 min of tracking time. The missing fence on the west side would not affect transmissions but would significantly affect reception due to the increased system noise level if it was on the east side. Jurgen Rahe at NASA was willing to accept the 50 ft height but not the gap, Cornell had promised a full 360° screen. After protracted discussions amongst all the parties, a solution was finally arrived at; CCI quoted $2.4M for a 50 ft high, 360° screen, NASA contributed the promised $2M and Jurgen Rahe[5] allowed NAIC to use $270,000 in carryover funds from its yearly $420,000 NASA funding for the planetary astronomy program. The NSF also allowed NAIC to use $250,000 from its NSF operating budget. This gave a total of $2,520,00, $2.4M for the CCI contract and $120,000 in management costs. On 12 February 1992, the NSF gave approval for Cornell to enter into a construction contract with CCI for the Ground Screen at a cost of $2,404,760.

The CCI contract was a design-build one requiring CCI to do the detailed design of the ground screen. They chose to use a relatively light weight 50 ft high steel structure that was supported by beams to the top of concrete pillars, concrete mounting blocks in the cliffs when that was convenient or large concrete pads on the ground where the area outside of the telescope's reflector was sufficiently wide

[5]Jurgen Rahe, Director for Solar System Exploration in NASA's Office of Space Science, was a very considerate and helpful person and a pleasure to work with. Sadly, he tragically died in a freak accident in June 1997 when a tree fell on his car during a storm while he was driving home.

Fig. 11.6 A view from the back of the screen showing one concrete support pillar and cliff anchorages taken on 17 June 1993. Credit: Cornell University/NSF

(Figs. 11.5, 11.6 and 1.7). There was a requirement to maintain a 12 ft wide road around the periphery of the reflector rim wall. Seven 7 ft wide wire mesh covered panels with lengths between 20 and 38 ft would form the screen. National Steel Erectors (NSE), which had experience erected communications towers in Latin America and had worked with CCI previously, was to do the construction work. On 29 and 30 April 1992, John Kiefer, Paul Stetson and William McGuire visited NSE's plant in Oklahoma to inspect a test panel. Structurally it was fine, but the half inch welded wire mesh did not meet the corrosion resistance specification. However, the decision had already been made to purchase stainless steel mesh, which was made by a firm in India mainly for use in continuous baking ovens. The cost would be similar to the cost of wire mesh in the United States and, since no firm in the United States produced such mesh in quantity, it could be imported duty free. Almost four acres of mesh was required for the ground screen.

NSE set up a production facility for the panels in a rented space in Arecibo. When the first stainless steel mesh arrived, it came in very tightly wound rolls making it difficult to flatten to the specifications on the panel frame, a problem that was eventually solved by developing a method to stretch the mesh more tightly and keep it from pulling back. A number of panels had already been installed and these had to be modified. Erection of the first panel support frames started in late February 1993, and the first panels were installed on 4 March. Work proceeded very rapidly and the entire ground screen was completed in August 1993 (Fig. 11.7).

CCI and NSE did a very good job, the ground screen lived up to expectations and remained essentially maintenance free for many years. There was a claim against Cornell by CCI due to cost overruns; it was eventually settled without litigation.

Fig. 11.7 The ground screen completed in August, 1993. Credit: Cornell University/NSF

11.7 The Structural Upgrading and Dome

With approval to proceed with the Gregorian upgrading, the first task was a detailed engineering design for the needed structural upgrading of the cables and suspended structure of the existing telescope to support the additional weight of the dome and its contents. The dome but not the reflectors and their backup structures, was included. A contract was let to NAIC's long term engineering consulting firm, Ammann & Whitney (A&W) with Joseph Vellozzi the lead engineer. Ben Hooghoudt made a major contribution to the project by re-designing the dome so that the feed turret and transmitter were inside the dome rather than having them in a separate feed cabin. This brought the final focus location much closer to the tertiary reflector significantly reducing the size of the feed horns (Fig. 11.8). It was this dome that A&W incorporated in their design work.

With the dome redesign it was possible to finalize the specifications and expected performance of the reflector system[6] taking into account the constraints imposed by the limit on the weight of the dome, 90 tons, and its radius set by wind loading considerations, 83 ft. This was later increased to 84 ft. The resulting specifications were: (1) an elliptical aperture size of 700 ft × 776 ft with the long dimension tangent to the reflector rim; (2) a downhill offset of the illuminated area of

[6]The specifications and performance were outlined in a series of memos written between March and May of 1992 by Lynn Baker, Michael Davis, Martin Johansson and Per-Simon Kildal. They are summarized in Appendix B of the minutes of the 12th meeting on 19 May 1992 of the Planning and Design Task Group for the Gregorian Upgrade, which are in the Cornell University library archives under Arecibo Ionospheric Observatory.

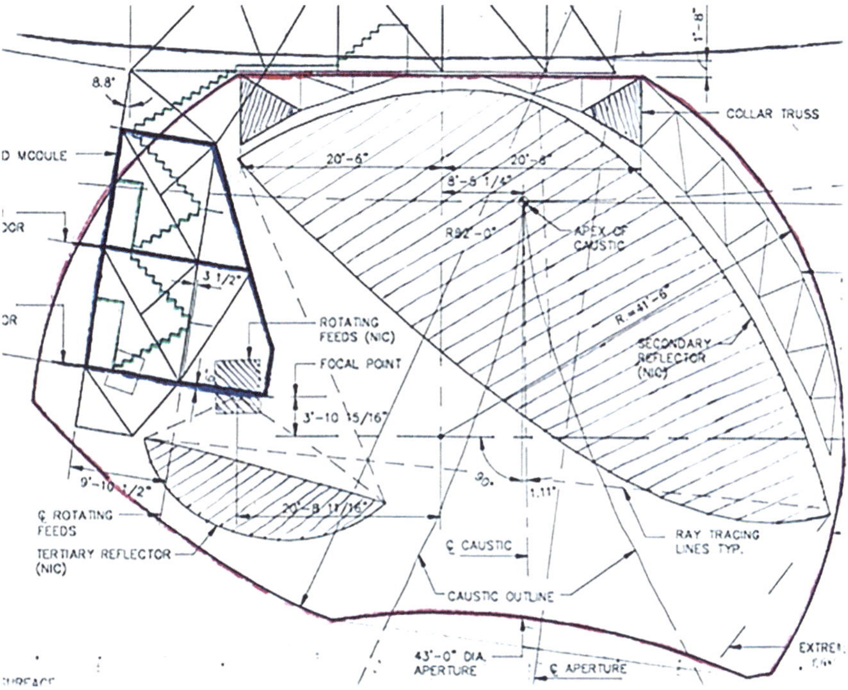

Fig. 11.8 A&W's design drawing of the re-designed dome with some colored outlines by Lynn Baker to make it clearer. The overall inside diameter is 84 feet (26 m) with the secondary reflector at upper right and the tertiary reflector at lower left. A 43 ft (13 m) aperture at the bottom of the dome allows ingress of the radio waves reflected from the primary reflector. The two-story feed cabin at left has the rotary floor bringing the feed horns to the focal point at the bottom and the 2.38 GHz transmitter at the top. The jagged lines are stair cases. From A&W drawing number S33 dated 12 October 1992. Cornell University archives

47 ft; (3) a paraxial clearance, the vertical distance from the paraxial surface (the dashed line passing through the Apex of Caustic in Fig. 11.8) to the secondary reflector surface, of 10 ft. An innovative design of the dome's elevation trolleys by A&W kept the top of the dome close to the bottom of the feed arm, but a 4 ft lift of the entire suspended structure would still be required. The close proximity of the rotary feed platform and the tertiary reflector (Fig. 11.8) allowed for small horns with full opening angles of 120°, an enormous improvement over the very large horns required for the original concept of a separate feed cabin.

The aperture illumination function selected was very flat with a sharp roll off to 20% amplitude at the edge. This minimizes the effects of blockage by the suspended structure and the side lobe level is primarily determined by the central dip caused by this blockage. The predicted aperture efficiency was excellent, 79% at 250 MHz, 83% at 500 MHz, 84% at 1.0 GHz and 68% at 5 GHz. The reduction at 5 GHz is due to the then 2.3 mm rms of the primary reflector surface. These are impressive numbers compared with a standard parabolic reflector. The illuminated area, equivalent to a 225 m circular aperture, would provide a sensitivity of 11°K/Jansky at 1.4 GHz, significantly better than the approximately 8°K/Jy of the 40 ft line feeds at 1.4 GHz. Estimates of the additive noise temperature at 1.4 GHz due to cosmic and atmospheric contributions, tertiary spillover, blockage and vignetting were 14°K to 16°K depending on zenith angle. This did not include additive noise in the receiver which, if kept low, would allow system noise temperatures close to 20°K.

On 14 February1992, the NSF approved the issuance of a request for proposals (RFP) for the structural work. A&W's design work was largely completed by the end of June; some of the details of the dome were not completed until October, allowing the RFP to be sent to 17 selected engineering firms. A pre-proposal conference was held at the Arecibo Country Club, a social club, on 16 and 17 September 1992. Of the 17 firms; 14 attended. Proposals were due 23 October 1992. Separate technical and financial committees reviewed the proposals resulting in the decision to enter into contract talks with the Universal Antennas Division of Radiation Systems Inc. (UAD/RSI). After extensive negotiations between Cornell and UAD/RSI over technical details and pricing, in February 1993 the NSF gave approval for Cornell to enter into a contract for $8,416,690 with UAD/RSI for the new cables, structural modifications and the dome needed to house the Gregorian reflectors. The contract did not include the reflectors and their backup structures or the dome's drive system. The contract was signed in early April, and a kick-off meeting at Cornell took place on 21 April with a follow up meeting at A&W in New York the next day. RSI was already contracted to build the replacement for the collapsed 300 ft Green Bank telescope of NRAO, which, at 100 m (330 ft) diameter, would be the world's largest fully steerable radio telescope. Adding the upgrading of the world's largest radio telescope must have appealed to Richard (Dick) Thomas, the founder and Chief Executive of RSI.

Considerable structural reinforcement of the platform and feed arm was needed to support the estimated 90 ton load of the Gregorian dome and its contents. This added weight plus that of the dome and additional drive systems would increase the overall weight of the suspended structure by about 300 tons over the original 600 tons. This, in turn, meant that additional main and backstay cables would be needed. Despite the planned addition of 20 tons of lead ballast at the far end of the feed arm away from the dome, the dome would still be an unbalanced load as Carriage House 1 weighed only about 17 tons. To stop the platform tipping as the dome moved in zenith angle, vertical cable tie downs from the corners of the triangular platform to large concrete blocks below the reflector equipped with hydraulic motors to adjust the tension were planned. Six laser range finders, Distomats, with targets on the three apexes of the platform would be used to measure the height and tilt of the platform providing input to the tie down control system. Booms extending from the apexes of the triangular platform and stayed by vertical masts would be needed for the vertical tie downs in order for them to clear the feed arm and dome.

Civil work: Before significant load was added to the suspended structure the two auxiliary main and two auxiliary backstay cables for each tower had to be installed. This required large anchor blocks to be constructed behind the existing ones and the addition of steel collar trusses on the top of the towers to hold the new auxiliary

Fig. 11.9 Preparing the new anchor block at Tower 4 for the pouring of concrete. The large steel rods in the foreground and at the other end support the load of the two new auxiliary backstay cables. At upper right is the dark back of the original anchor block. Credit: Cornell University/NSF

cables. Very large concrete blocks were also needed below the primary reflector to take the load on the vertical tie down cables. Nogama, a construction firm in Puerto Rico, was contracted by RSI to do the needed civil work. This was primarily excavating large areas behind the three existing cable anchor blocks to provide room for the new anchor blocks and installing the concrete blocks and cable anchorages (Fig. 11.9). Work on the A12 auxiliary anchor block excavation started in May 1993. Access to the back of the T4 anchor required a substantial road to be constructed. For the T8 auxiliary anchor, it was finally agreed that RSI could access it from the front via a narrow road between the original anchor and the cliff. This would require a crane to lift loads over the original backstay cables as the cliff prevented it swinging the other way. Damage to these cables was a concern for Cornell.

RSI's initial schedule had the new anchors, tower top work and the 12 new auxiliary cables installed and preliminarily tensioned by January 1994. The dome installation was to start in August and the entire project completed by December 1994. Despite quickly extending the completion date by 3 months, the schedule was highly optimistic. However, the three anchor blocks were completed in early March 1994, the tower top work in September and the large vertical tie down concrete blocks were finished in mid-October. This ended Nogama's civil work, a good milestone but far from finishing the overall project.

Cable installation: Because of the complexity of adding cables at the apexes of the suspended platform, it was decided to attach them about 2/3 the distance along the bottom chord of the platform (see Fig. 11.10). As can be seen from Fig. 11.10, this means that the two auxiliary cables from a pair of towers cross at the same height

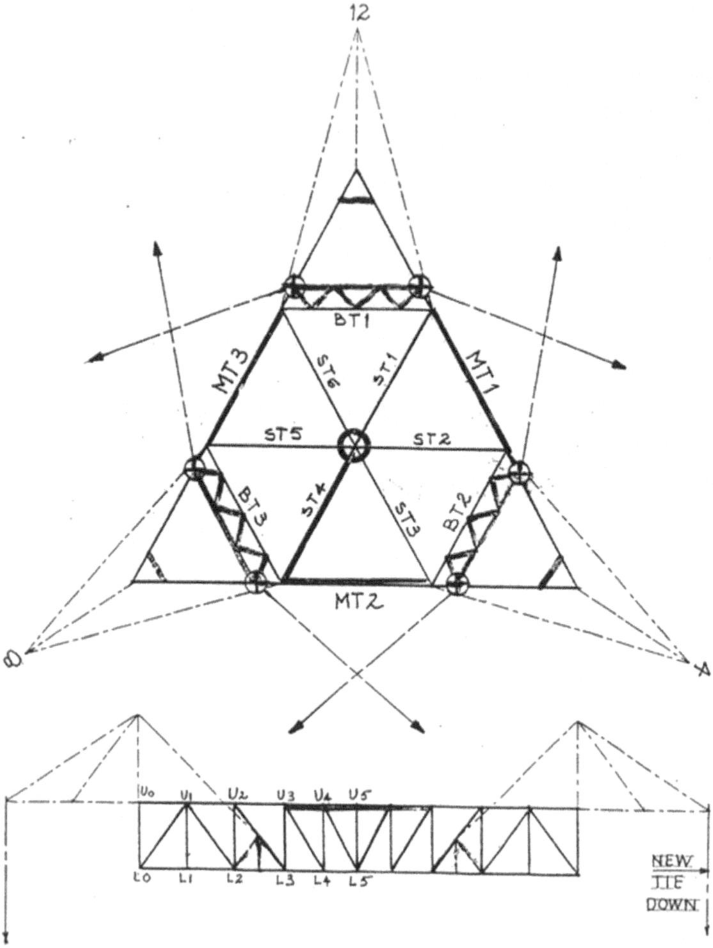

Fig. 11.10 Sketches showing the locations of the attachment points for the auxiliary main cables coming from the towers and the masts and booms needed to hold the new vertical tie down cables so that they cleared the azimuth arm and dome

requiring steel rubbing plates to be installed. It also meant that before the cables could be installed, steel attachment structures had to be installed plus diagonal cable trusses to balance the perpendicular loads on the two opposite platform lower cord attachment points (BT1, 2 and 3 of Fig. 11.10). North American Tower/Tower King (NATK), a Texas firm, was contracted to install the cables, a far from simple task (Fig. 11.11). However, this work went very smoothly with the six auxiliary backstay cables installed by mid-November 1994 and the six auxiliary main cables by mid-February 1995. All of the cables had their initial tensioning completed by mid-March.

Fig. 11.11 Installation of one of the auxiliary main cables (upper left) showing the (yellow) jacks for tensioning the cable. At bottom right is the truss cable connecting the two opposite auxiliary cable connection points allowing the perpendicular loads to be balanced. This cable also had to be tensioned as shown by the yellow jacks. The old azimuth trolley is at bottom left. It will be replaced by four trolleys connected to two outriggers extending from the feed arm to distribute the increased load over a larger area of the azimuth girder. Credit: Cornell University/NSF

Platform strengthening: With the auxiliary cables installed, work proceeded on the strengthening of the platform and feed arm to take the load of the dome, and with the installation of outriggers on the feed arm to support additional azimuth trolleys to distribute the load on the azimuth girder. The strengthening involved reinforcing or changing virtually every beam on the azimuth arm and reinforcing virtually all the vertical and diagonal trusses between the top and bottom chords of the platform. This work also included adding the masts and booms at the apexes of the triangular platform needed for the vertical tie downs to the large concrete blocks below the primary reflector.

By the end of 1995, the azimuth arm and platform structural modifications were complete (Fig. 11.12), the azimuth outriggers (Fig. 11.13) were in place and the eight new azimuth trolleys connected and taking the load of the azimuth arm. Vertex, the firm responsible for the new drive systems, had provided temporary drives to allow azimuth motion. The realignment of the elevation rails, needed as the platform would be raised by about 4 ft, was in progress. Still to be done were the replacement of the azimuth rails and the installation of about 20 tons of lead ballast at the end of the azimuth arm away from the dome to provide an additional partial counterweight for the dome apart from the 17 ton Carriage House 2. Installation of the tie down masts and booms would take place after the dome was lifted and attached to the

Fig. 11.12 Work on the azimuth arm on 6 June 1995. Virtually every beam was either replaced or strengthened. The large cross beams in the bottom center will take the load on the outriggers that will be attached to the four new azimuth trolleys at each end of the arm. Credit: Cornell University/NSF

azimuth arm. All that remained before the dome could be lifted into place was the completion of the dome assembly, the installation of the secondary reflector panels and their initial alignment.

The Dome: The Temcor company was selected by RSI to build the dome. RSI's plant in Mexia, Texas built the collar truss and the secondary reflector backup structure. A successful trial assembly was done at the Mexia site in May and June 1995. At the same time, a 100 × 100 ft platform was constructed above the center of the primary reflector as a base for the dome and a central scaffold tower to support the collar truss was installed followed by an erection crane (Fig. 11.14).

Secondary reflector: Once the structural/dome contract was in place, attention turned to the 22 m secondary reflector's panels and backup structure. Eight firms were invited to submit proposals for the construction of the panels with seven compliant proposals received. The Cornell review panel met on 20 April 1994 and recommended further discussion with ESSCO and RSI, the two lowest bidders. A contract was eventually let to RSI for 414 low weight secondary reflector panels at a price of $775,000. One advantage that RSI had was its capability at its Sterling, Virginia, Technical Products Division to produce panels curved in two dimensions. With the panel sizes and shapes finalized by 30 January 1995 by Lynn Baker with checking by Per-Simon Kildal, Baker then developed the basic geometry of the secondary reflector backup structure with input from RSI. Ammann & Whitney did

Fig. 11.13 Two of the new outriggers projecting from the side of the azimuth arm each of which support two azimuth trolleys. The trolleys distribute the greatly increased load of the azimuth arm and dome along the azimuth girder. The old azimuth trolley is at upper center and was removed. At the top is the cable truss which counteracts the perpendicular loads on the platform from the new main auxiliary cables. Credit: Cornell University/NSF

the detailed structural design and RSI implemented it. The secondary reflector was installed in the dome prior to its lift to the azimuth arm of the telescope (Fig. 11.15). Before the lift, the panels were positioned as accurately as possible with the final setting done once the dome was in place (Fig. 11.16).

Dome lift: Assembly of the dome including the secondary reflector backup structure and panels was completed in early May 1996. In order to install the bottom cap of the dome and the bottom reflector panels, the rigging for the eventual lift was installed and the dome lifted off from the platform in late March by enough to allow this work to proceed. The final lift was on 16 May 1996 (Figs. 11.17 and 11.18).

Finishing up the structural work: With the installation of the dome, the remaining RSI platform tasks were the replacement of the azimuth rails, completion of the resetting of the elevation rails, the installation of the tie down masts and booms, the addition of the 20 ton counter weight and the final tensioning of the cables including the original main and anchor cables to balance the loads and raise the platform by approximately 4 ft. Apart from the final cable tensioning, the installation of the masts and booms dictated the schedule (Fig. 11.19).

Preparations started in late 1996 for the tensioning of the cables with jacks installed on all the cables (Fig. 11.20). The tensioning of the backstay cables was completed in January 1997. For the tensioning of the original main cables restraining nuts on the U-bolts behind the steel blocks in which the cables were socketed (see

Fig. 11.14 Assembly of the dome showing the central scaffolding tower supporting the collar truss to which all the dome ribs are attached. The people on the top of the dome are installing the top cover plates. Clearly a hot day as indicated by the umbrella. There is a great deal of debris lying on the reflector, which is covered with algae. The algae, while unsightly, did not affect the performance of the reflector. Credit: Cornell University/NSF

Fig. 11.19) had to be moved back. This was a problem for the cable attached to the T8 apex that was spliced during the original constructions as Bethlehem Steel had made the cable too short (see Chap. 5). As a result of the splicing the cable ended up slightly longer than the other main cables. This was not a problem but it meant that the restraining nut had to be removed to allow the block to move back the required distance. In removing the nut on 31 January 1997, the RSI employee ended up cutting into one of the arms of the U-bolt in two locations, one of which was close to 0.5 inches (1.25 cm) deep. An emergency telephone conference was held on Saturday, 1 February between Joe Velozzi of A&W, RSI and Cornell people in Ithaca and Arecibo. Was the bolt in danger of failing? Joe Vellozzi stated that there was a concern. To alleviate this concern a 4 ft end section of the 23 ft long U-bolt was removed and taken to Lehigh University in Pennsylvania to be load tested in the University's testing facility after identical cuts were made. The threaded rod passed the test and the tensioning proceeded finishing the structural work much to the relief of everyone involved with the project (Fig. 11.21). After the completion of the upgrading project José Maldonado added two additional spools with U-bolts to reduce the load on the damaged U-bolt.

Tertiary reflector: Once the secondary reflector's construction was finalized, attention turned to the 9 m diameter tertiary reflector. This was contracted with RSI as an add on to the secondary reflector contract. Again, Lynn Baker did the sub-divisions and shapes of the 60 panels plus the central 8 ft one with checking by

Fig. 11.15 The inside of the dome during construction showing the backup structure of the secondary reflector supported from the collar truss. The dome's outside panels are partially installed. The red stairs give access to the top of the scaffolding tower that supports the load of the dome. Credit: Cornell University/NSF

Per-Simon Kildal. Baker suggested a space frame of nodes and struts as the backup structure and this was eventually accepted by RSI, which sub-contracted the node and strut fabrication to the German firm Mero. The contract specified that RSI would supply the panels including the central round plate machined out of a solid block of aluminum, plus all the nodes and struts. Cornell would assemble the sub-reflector and install it in the dome after the dome was lifted to the azimuth arm.

Assembly and initial alignment of the panels took place on a cleared area next to the rim wall of the primary reflector just below Tower 8. It was an "all hands" effort by the local staff including some of the scientific staff plus some NAIC staff from Ithaca (Fig. 11.22). José Maldonado organized the very complicated 30 September 1996 lift through the 43 ft (13 m) aperture of the dome and into place below the feed cabin (Figs. 11.23 and 11.24). A long process of alignment of the panels followed.

The Elevation and Azimuth Drive Systems: To avoid problems related to contractor interactions, it was decided to sole source the Carriage House 1 and Dome elevation drive systems and the azimuth one with RSI, which had a Precision Controls Division with the relevant experience. The overall specifications were developed at an 11 August 1994 meeting at the observatory with RSI personnel including Ralph White, the President of the Controls Division, José Maldonado and other observatory technical people plus Kurt Samuelson, Gene Bartell and Don

Fig. 11.16 Targets were placed at each corner of every panel of the secondary reflector and video photogrammetry used to set the 414 panels plus the central circular one before the dome was lifted to the structure. It was repeated after the dome was put in place on the suspended structure. The two-floor instrumentation cabin can be seen below the reflector. The bottom one houses the rotary feed floor allowing the selected receiver and horn to be moved to the focal location. The S-band transmitter will be in the top one. The walls were attached after the dome was lifted into place. Credit: Lynn Baker and Cornell University/NSF

Fig. 11.17 The dome on its way to the platform, 16 May, 1996. Credit: Cornell University/NSF

Fig. 11.18 After the lift; connecting the azimuth trolleys—the four white structures that everyone is looking at—to the dome attachment points. There are four more trolleys on the other side of the azimuth arm. Two of the four vertical multi-part lines that were used for the lift can be seen left of center and at the right. The frames at bottom center support the heat exchanges for the new S-band transmitter. Credit: Cornell University/NSF

Fig. 11.19 The tie down masts and booms being installed on 3 December, 1996. The one on the right at the T8 apex of the platform is finished and the T12 one is almost finished. The two vertical tie down cables span the four large main cables. The booms were installed vertically then lowered down into position. Credit: Cornell University/NSF

Fig. 11.20 Jacks in place on the four main cables at the T8 apex of the platform in readiness for the final tensioning of all the cables. The steel blocks into which the cables are socketed were constrained by nuts on the threaded U-bolts at the front and back of the block, and the nuts at the back of the block had to be moved back before the tensioning as shown in the figure. The far cable was slightly longer than the other three, which meant that its nuts had to be removed as just winding them back did not leave enough room to tension the cable. It was during this process that the hairpin rod ended up with two cuts, one of which was approximately 0.5 inches (1.25 cm) deep. Credit: Cornell University/NSF

Campbell from the Cornell campus. Robert Dickman, NAIC's Program Officer at the NSF also attended. This resulted in a conceptual design from RSI and, inexplicitly, a cost estimate that was several times what Cornell had budgeted. There was a strong feeling at Cornell that RSI was attempting to make up for its anticipated losses on the structural/dome contract. The decision was made to go out for competitive bids.

A request for proposals was issued to five firms including RSI on 30 January 1995 with proposals due 7 March. The Duisburg, Germany based Vertex Antennentechnik GmbH (VA), was selected based on its technical proposal and price. VA was an independent company within the Vertex Group headquartered in Kilgore east of Dallas, Texas. Vertex grew out of a company building satellite ground stations that was co-founded by J. Rex Vardeman in 1973. Vardeman had been the site supervisor for E-Systems during the early 1970s resurfacing of the Arecibo reflector and left in the middle of that project to co-found the company that, later, became Vertex. Awarding the contract to Vertex led to a difficult conversation with RSI Chief Executive Richard Thomas. Given the circumstances, on the Cornell side there was concern about contentious contractor interactions from working in

Fig. 11.21 Two of those very relieved people, Don Campbell (left) and NAIC Director Paul Goldsmith after the completion of the installation of the tie-down system and the final tensioning of the cables, the last major tasks of RSI's structural contract. One of the tie-down masts and booms is behind Campbell and Goldsmith. Credit: Cornell University

Fig. 11.22 The tertiary reflector fully assembled near the edge of the telescope's primary reflector on 18 September 1996. Lynn Baker is trying to retrieve a sunshade. Credit: Cornell University/NSF

Fig. 11.23 The tertiary reflector lifted through the 43 ft aperture of the dome on 30 September 1996. It was attached to controllable actuators mounted on the sides of the feed cabin to allow focus positioning. Credit: Cornell University/NSF

Fig. 11.24 The interior of the dome showing the 22 m secondary reflector, the instrumentation tower and the 9 m tertiary reflector mounted below it. Credit: Cornell University/NSF

close proximity, but this turned out not to be a significant problem. VA delivered very capable azimuth, dome and Carriage House 1 drive systems plus their operating system that served the observatory well for 23 years.

11.8 The 2.38 GHz Transmitter

Having provided the original 450 kW 2.38 GHz transmitter in 1974, Continental Electronics of Dallas, Texas was commissioned to do a design study and cost estimate for the new 1.0 MW 2.38 GHz transmitter to be installed in the dome. The Varian company of Palo Alto, California was to provide a cost estimate for the needed klystrons. In their report "S-band Tx Upgrade Study for The Arecibo Observatory" dated January 1990, Continental and Varian estimated the transmitter cost at $5.5M including the two klystrons with an additional $400k for two spare klystrons. In a letter dated 19 January 1990 to Hal Theiss, Manager of Deep Space Network Systems in NASA Office of Space Operations, Campbell provided these estimates plus an estimate of the cost of the power generator for a total of $7.7M including a 10% management fee. The Office of Space Operations, NASA code TN, had insisted that they would fund the transmitter as its primary use would be the study of asteroids and comets, a NASA priority related to both the coming CRAF and Cassini missions.

A Request for Proposals for the transmitter was sent to seven firms on 26 March 1993, and a pre-bid conference was held on 15 April with four pre-qualified firms in attendance including Continental Electronics. Three proposals were received. Both the business and technical review committees recommended that Cornell enter into negotiations with Continental. A meeting between Cornell and Continental was held on 30 June to discuss, primarily, technical issues. After receiving NSF permission on 28 September, a contract with Continental was signed in early October with delivery within 18 months. Under the contract, Cornell agreed to provide any items from the original Continental built 2.38 GHz transmitter and power supply that would be useful for the new one. The main items were the waveguide switches, which Cornell had refurbished, two 0.5 MW loads and the large inductrol high voltage regulator, which was also refurbished. Because of the long lead time for building klystrons, a purchase order was issued by Cornell in August 1993 to Varian for delivery of three klystrons, two to Continental as customer provided equipment and one to the observatory, at a cost of $270,000 each. Their bandwidths would be at least 25 MHz with 30 MHz being a goal. Maximizing the bandwidth was important as it determined the range resolution for imaging near-Earth asteroids and comets. A 25 MHz bandwidth would allow 6 m range resolution.

An initial design review was held at Continental's plant in Dallas on 19 October 1993 with a UAD/RSI representative also in attendance. There were myriad small details that needed to be resolved related to the transmitter's installation in the dome and who, RSI, Continental or Cornell, was responsible for the work and cost. A major issue was the proposed mounting of the transmitter's two heat exchanges on the dome. With a critical design review between RSI and Temcor, the builder of the dome, scheduled for mid-December, RSI needed to know if the load of the heat exchanges required a redesign of some of the dome's trusses.

On 15/16 December 1993, an intermediate design review was held at Continental's plant in Dallas, where most of the details of the transmitter, its installation and operational issues were discussed and largely settled. One issue was getting the power to the transmitting horn on the rotary floor of the receiver room which was below the transmitter room. This would be done via a waveguide run down the rotation axis of the floor. However, this required a rotary coupler due to the floor rotation but which only had to allow transmitter power to pass when the transmitting feed was at the focal location. Jon Hagen, then heading NAIC's Ithaca based instrumentation laboratory, designed the rather complex coupler and had it built. It worked for the life of the transmitter.

A final design review was held in May 1994 at Continental's plant in Dallas. Varian delivered two klystrons to Continental and one to Arecibo in late summer. By the end of March, the transmitter was almost fully assembled at Continental and acceptance tests were expected at the end of May. However, it was discovered that the new 1 MW loads were defective preventing full power testing. The 0.5 MW ones provided by Cornell from the old transmitter allowed testing to half the designed output power, 0.5 MW. A new firm was contracted to manufacture the 1 MW loads but, despite promises of early delivery, they were not received until early March of 1996. Tony Crespo and Evan Kurtz went to the Continental plant in Dallas in mid-March and the acceptance tests, also attended by Jon Hagen and Don Campbell, took place in late April. The tests were satisfactory but resulted in a significant number of small details that Continental had to address. This held up shipping of the transmitter until July. By early October the transmitter was installed except for the final waveguide run to the transmitting feed under the receiver cabin rotary floor, which, along with the receiving feed, had been constructed by Lynn Baker in late 1994. The transmitter was in operation by early 1997. Figure 11.25 shows the dome probably just after the transmitter was installed but before it was operational.

With klystrons being about 50% efficient, a 1 MW transmitter power required approximately 2.2 MW of power to operate. Power for the previous transmitter was provided by two 750 kW turbine generators installed in 1974 and clearly beyond their use date. It was decided to procure a single large turbine generator with both a new and used one as options due to possible budget issues. José Maldonado and John Kiefer developed a request for proposals, which was released in October 1993 with proposals due 16 December. Four proposals were received. Two proposals were selected in mid-February for in person talks, which were held on 5 April. This resulted in the decision to give a contract to Solar Turbines for a refurbished Centaur 40 2.5 MW turbine generator at a cost of $1.544M. The purchase order was issued in May 1995 with delivery expected in December or January. The two existing 750 kW generators and their control cabinets were sold to Solar. Since the turbine generator required electrical power to start it was decided to replace the old site emergency backup generator with a new 750 kW diesel one rather than buy one just to start the turbine generator. The Solar turbine was delivered on 2 February 1996 and installed over the following weeks in time for the testing of the transmitter.

Fig. 11.25 The Dome in August 1996. The heat exchanges for the new transmitter are on the roof of the dome. Access doors to the receiver and transmitter rooms are on the left side of the dome. Heavy equipment was lifted from the center of the primary reflector and then lowered onto the relevant platform outside the doors after moving the dome into place. Credit: Cornell University/ NSF

11.9 Impact of the Gregorian Upgrade

The Gregorian upgrade was likely the most radical and comprehensive modification made to any large astronomical telescope. With the exception of the 430 MHz transmitter and its 96 ft (30 m) line feed, every major system on the telescope was changed or modified as well as the telescope structure itself. This included the Gregorian dome with its horn-based receivers, the new S-band transmitter, a dual transmitting capability for the 430 MHz incoherent scatter radar and new telescope drive systems. What has not been discussed in this chapter was the development of a new telescope pointing and user interface, CIMA, new local oscillator systems and the development of new correlators for spectral line observations. All telescope and instrument control and data lines were converted to fiber optic ones, which involved designing and building a multi-fiber cable wrap system at the azimuth rotary joint designed by José Maldonado. As discussed in Chap. 12, the "new" telescope when augmented with new instrumentation opened up many opportunities for new science, the objective of all the time, effort and funding that it took to bring the project to a successful conclusion.

11.10 The Angel Ramos Foundation Visitor Center

Tourists and groups of school children had been visiting the observatory since its completion in late 1963. While the view of the telescope from the viewing area adjacent to the control room was clearly impressive, an audio recording and a few photos conveyed little real information about the telescope and its astronomical and ionospheric programs. Some groups were allowed into the control room area but this could be disruptive. There was strong need for something better.

As an adjunct to the Gregorian project, in the early 1990s the then new observatory director, Daniel Altschuler, initiated a design, fund raising and construction project to build a modern visitor center with the goal to both enhance tourism and promote STEM (science, technology, engineering and mathematics) education in Puerto Rico. A number of local Puerto Rican architectural firms were invited to submit ideas with no constraints on the location. Several chose to place the Center on the hill on which tower 12 (T12) was located while at least one placed it where the cable car house was located just below the operations building and put the cable car house on top. The firm of Méndez, Brunner Architects and Engineers was selected based on a large open flat roofed structure with enclosed areas underneath just north of T12, and a path to an overlook of the telescope. In 1985 Luis Badillo, a graduate of the University of Puerto Rico's School of Architecture, joined the firm rising to Partner in 1991 when the firm's name was changed to Méndez, Brunner, Badillo and Associates. Badillo took another look at the design of the Visitor Center and radically changed it bringing the entire building to a relatively small but adequate flat area at the base of T12 that had a direct overlook of the telescope.

In conjunction with the design work, a funding campaign was started as it was clear that the NSF would not fund the construction of the Visitor Center. Construction costs were estimated to be about $3M and a local Puerto Rican foundation, the Angel Ramos Foundation, agreed to provide a matching grant for up to one half of the cost if the other half could be raised. Angel Ramos was the owner and publisher of the El Mundo newspaper in Puerto Rico starting in 1944 and, later, founded radio and TV stations. The Foundation was established by his wife Argentina after he died in 1960.

After a successful fund raising campaign including a donation from Cornell University, construction started in 1995 and was completed in 1997 (Fig. 11.26) with the inauguration taking place on 1 March (Fig. 11.27). A large central area housed the displays, there was a large auditorium and a viewing platform gave an excellent view of the telescope's reflector and suspended platform. There was also, of course, a shop, important for the financial viability of the project. This, plus an entry fee covered the operating costs as NSF funds could not be used. A successful proposal to the NSF funded the displays.

With the auditorium often in use for the visiting public, another venue was needed for the educational mission of the Visitor Center. After a fund raising campaign again supported by the Angel Ramos Foundation, a new building, the Learning Center, was constructed close to the Visitor Center. It was completed in 2001 and was essentially a large classroom with associated service facilities. It was in constant use for educational programs, small science conferences and NAIC advisory committees.

Fig. 11.26 The newly constructed Angel Ramos Foundation Visitor Center in 1997. Tower T12 is in the background and the telescope can just be seen to the left. Credit: Cornell University

Fig. 11.27 The Visitor Center inauguration event under a tent over the volley ball court; Daniel Altschuler, the Observatory Director, is at the podium. First from the right, Yervant Terzian, Chair of the Cornell Department of Astronomy; second, Frank H. T. Rhodes, President of Cornell University; fourth Charlie Rodriguez, President of the Senate of Puerto Rico; sixth, Angel Roman, Mayor of Arecibo. Credit: Cornell University

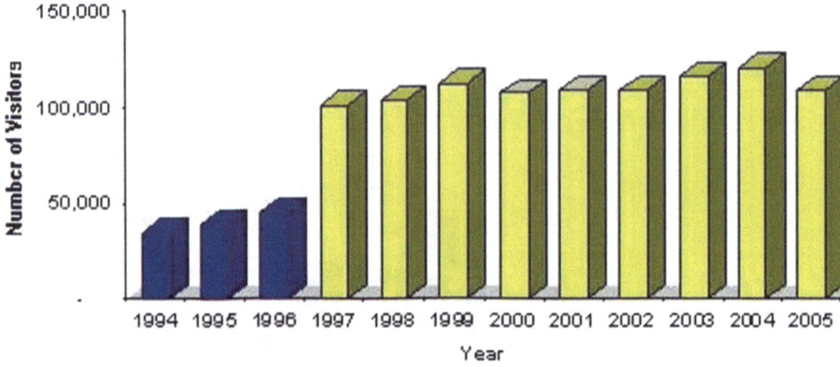

Fig. 11.28 The number of visitors including tourists and special groups, primarily school children, increased substantially once the angel Ramos Visitor Center was opened. From the 2005 NAIC/Arecibo Observatory Progress Report and 2006 Program Plan. Cornell University archives under Arecibo Ionospheric Observatory

The Angel Ramos Visitor Center was a great success, a tribute to Daniel Altschuler's foresight in initiating its construction, and to its first director, José Alonso, who remained in the position until 2008. The center was the forerunner for similar visitor centers at the other US national astronomy centers. The number of visitors per year increased from about 40,000 just prior to its opening to over 100,000 (Fig. 11.28) including an average of approximately 33,000 in group visits, primarily school groups.

11.11 At the Movies

Parts of two movies were shot at the observatory during the Gregorian upgrade years. The 1995 James Bond movie GoldenEye, made by Eon Productions, featured the Arecibo telescope rising from a lake supposedly in Cuba. A model of the telescope was used in many scenes but there was extensive footage of the actual telescope: if it looked a bit rusty in places, it was the real telescope. Along with the production staff, Pierce Brosnan, playing James Bond, spent time at the observatory and interacted with the observatory staff, signing photos and creating considerable excitement.

The movie Contact released in 1997 was based on the book of the same name by Carl Sagan. It featured Jody Foster as Dr. Ellie Arroway, whose character was inspired by Jill Tarter of the SETI Institute and Project Phoenix. Early scenes in the movie were taken at the Arecibo Observatory, where it rained incessantly. The production soon moved to the NRAO's VLA site.

References

Garber, S.J. 1999. Searching for Good Science - The Cancellation of NASA's SETI Program. *Journal of the British Interplanetary Society* 52 (1): 3–12.

Kildal, Per-Simon, and Svein Andreas Skyttemyr. 1989. Diffraction Analysis of a Proposed Dual-Reflector Feed for the Spherical Reflector Antenna of the Arecibo Observatory. *Radio Science* 24 (5): 601–617.

Chapter 12
The Post Gregorian Upgrade Era

The Gregorian upgrade, which provided a true focus for incoming radio waves plus increased sensitivity and frequency coverage, opened up significant new opportunities in both radio and radar astronomy. In particular, a multi-beam system for, mainly, surveys of atomic hydrogen in the Milky Way and external galaxies were now a possibility. For the newly-named Space and Atmospheric Sciences (SAS), the dual beam capability would allow improved wind and plasma line measurements. Before these improvements could be utilized, the main reflector needed to be repaired and reset, new horns and low noise receivers built and multiple back-end processors designed and constructed. All this was achieved over the next few years with the 7-beam Arecibo L-Band Feed Array (ALFA) being installed in 2004. This was followed by the initiation of a number of extensive surveys. VLBI flourished and several SETI programs were conducted with, sadly, no positive detection. Radar astronomy saw a change in emphasis from the study of planets and satellites to astrometric and imaging observations of asteroids and, when possible, comets. For SAS the LIDAR program flourished with the acquisition of new instruments, and a new HF facility utilizing the 305 m primary reflector was designed and constructed after the HF transmitting facility at Islote on the coast east of Arecibo was destroyed during Hurricane Georges in 1999.

12.1 Post-Gregorian Upgrade Events

With the completion of the Gregorian upgrade in 1997, the planetary radar program was well positioned to take advantage of the very significant increases in sensitivity that the upgrading provided. For the radio astronomy program, the sensitivity improvements were significant and there was also a considerable reduction in the beam sidelobe levels at all frequencies. The upgrading combined with the resetting of the surface of the primary reflector also allowed much higher sensitivity at frequencies above 3 GHz and the possibility of a multi beam system at 21 cm

D. Campbell, *The Arecibo Observatory: A History of Innovation and Discovery*, Historical & Cultural Astronomy, https://doi.org/10.1007/978-3-031-75225-4_12

wavelength, the hydrogen line frequency. The Australia Telescope National Facility (ATNF) of the CSIRO had built a 13-beam dual polarization system at 21 cm for the Parkes telescope that had enabled a very productive HI sky survey, HIPASS, and an equally productive search for new pulsars. For space and atmosphere sciences, the dual beam capability allowed much more accurate wind measurements, and the increased bandwidth of the Gregorian horn system over that of the 430 MHz line feed allowed plasma line measurements over this larger bandwidth.

In order to take advantage of the telescope's new capabilities especially for the radio astronomy program, improvements to the telescope and its instrumentation were necessary. This involved another major effort to bring the surface accuracy of the telescope as close to a rms of 2 mm as possible and a push to construct the needed new receiver systems including a 7-beam receiver system at the 21 cm HI line frequency, which would enable large scale HI surveys and pulsar searches. These surveys and searches also required the construction or acquisition of banks of spectrometers to cover the seven dual polarization channels.

Internet at last: Communication to and from the observatory with the outside world during most of the 1960s was severely limited. There was a microwave link to the Ramey Air Force base at the NW corner of the Island and the possibility a connection to Patrick Air Force base in Florida allowing telephone calls to be made, but the system rarely worked. The alternative was a microwave link via a Collins radio in a booth in the office the administrative assistant (secretary in those days) for the scientific staff, that connected to a similar booth in the Space Sciences building at Cornell University, normally answered by Jane Lee. You could then talk to someone in the building or be patched through to an outside telephone number. Everyone became used to saying "over". Getting in touch with observatory scientists or maintenance staff living in the Arecibo region also required radio sets. All the relevant people had one, and a call sign, radar 1, radar 2, etc. All conversations were public but the system was essential in case of maintenance problems or, if telescope time suddenly became available at night or the weekend, to allow a scientist to take the time.

In the late 1960s, Rolf Dyce organized a commercial microwave link to Arecibo via about 6 ft (2 m) antennas mounted on the top of Tower 12 and on the roof of a commercial building close to the Arecibo campus of the University of Puerto Rico that had direct line-of-sight to T12. From there a link to the Puerto Rican Telephone Company (PRTC) was made allowing regular telephone calls. A normal telephone switchboard was installed at the observatory allowing local and external calls—a tremendous improvement. At some later time, when is not clear, copper lines were installed from Barrio Esperanza, about 3 miles (5 km) from the observatory, that directly connected the observatory to the PRTC network. NASA's High Resolution Microwave Survey (HRMS, a SETI search, see later in this chapter) in preparation for their 1992 observations arranged for additional capacity enabling internet service although limited to 56 kbits/s. The data rate was not enough to transmit significant amounts of data, so data tapes were still routinely posted.

In August 1999 an agreement was signed with the PRTC, in which the PRTC agreed to provide full internet services to the observatory by installing 3 miles (5 km)

Fig. 12.1 Gene Bartell on the occasion of his retirement receiving the gift of a painting of the Puerto Rican countryside from Daniel Altschuler, the Observatory Director. Credit: Cornell University

of fiber optic cable from Barrio Esperanza to the observatory. The bandwidth would be 1.544 Mbits/s, about 30 times the previous bandwidth, and it included a voice link utilizing a new switchboard also provided by PRTC. True internet at last! The agreement was due to the initiative of Arun Venkataraman (head of the computer department), Eddie Castro (head of electronics) and Rey Medina (head of human resources).

In about 2005, an agreement with the University of Puerto Rico's main campus in Rio Piedras (a suburb of San Juan) led to a further increase in the internet bandwidth. This eventually included an as-needed special late night only bandwidth of 512 Mbits/s enabling real-time VLBI observations (E-VLBI) with the European VLBI network.

Gene Bartell: In the summer of 2001, Gene Bartell,[1] who had been the administrative director for NAIC for 30 years, retired (Fig. 12.1). Gene was the epitome of a self-effacing but very effective administrative director, developing the yearly budgets, dealing with Cornell's business and human resources departments, assisting with the preparation of the yearly Progress Reports and Program Plans and looking after myriad details of the management of NAIC and the observatory.

---

[1] An interview with Gene Bartell can be found on the Cornell University library's e-commons site under Arecibo Observatory.

Fig. 12.2 A view of the completed North VSQ. It has 12 double occupancy rooms and was a much needed expansion of the observatory's ability to house visitors. Credit: Cornell University

VSQ expansion: In late 2002, a second addition to the Visiting Science Quarters (VSQ) was completed. It was situated above the entry road to the observatory just before the main gate and had 12 double occupancy rooms and a small kitchen building and generally referred to as the North VSQ (Fig. 12.2). The building was designed by José Maldonado and constructed by observatory maintenance staff led by Victor Santiago with some help from temporary workers. With this new building there was accommodation for over 40 people at the observatory, significantly improving the observatory's ability to host small workshops, teacher training, advisory panels, the yearly Research Experience for Undergraduates (REU) participants and, of course, visiting observers. The construction in 2001 of the Learning Center near the Angel Ramos Visitor Center made an expansion of the VSQ a necessity if the Learning Center was to be fully utilized.

Hurricane Georges: Hurricane Georges was the first hurricane to seriously affect the observatory. It struck on the evening of 21 September 1998. The predicted path had it passing directly over the observatory, but at the last moment it turned slightly south. However, with winds of 130 mph (210 kph) it was a major storm. Fortunately, there was no major damage to the telescope. Six sections of the catwalk walkway grating fell into the reflector damaging a number of panels and multiple trees were blown over. As discussed in Chap. 10, the major damage was to the heating facility built in a marshland on the coast just east of Arecibo. Essentially all the antennas were destroyed and multiple supporting towers collapsed. The damage was so

Fig. 12.3 Past and current NAIC and Observatory site directors attending the 2003 40th anniversary celebration. From left to right; William Gordon (Telescope Builder and Observatory), Donald Campbell (Observatory), Frank Drake (Observatory and NAIC), Paul Goldsmith (NAIC), Michael Davis (Observatory), Daniel Altschuler (Observatory), Riccardo Giovanelli (Observatory), Robert Brown (NAIC), Sixto Gonzalez (Observatory), Tor Hagfors (Observatory and NAIC), Gordon Pettengill (Observatory), where NAIC and Observatory indicate NAIC and Observatory directors. Missing are John Findlay (Observatory, passed away in 1994), Harold Craft (Observatory) and Timothy Hankins (Observatory). Credit: Cornell University

extensive that repair was not possible, especially as there was a strong environmental movement pushing to return the marshlands to their original state.

40th Anniversary celebration: This was an occasion to celebrate 40 years of amazing discoveries in all three research areas and the people who made it all possible. There was, of course, the usual celebratory dinner, but most important were the series of talks recounting the major events and those discoveries over the 40 years. Tor Hagfors gave the first Gordon Lecture, a yearly event that continues to this day at Cornell University funded by a bequest from Tom and Betty Talpey, and Bill Gordon recounted the history of the design and construction of the telescope. This was followed with talks by Herb Carlson, Marshall Cohen, Gordon Pettengill, Richard Behnke, Joanna Rankin, Don Campbell and many others.[2] With so many of the past NAIC and Arecibo Observatory directors present it was a great opportunity for a group photo (Fig. 12.3).

[2]A full description of the 40th event is in the April 2004 #37 NAIC/Arecibo Observatory newsletter.

12.2 Resetting the Primary Reflector

One of the main objectives of the Gregorian upgrade was to allow observations at frequencies up to, hopefully, 10 GHz. Observations in 1991 with the mini-Gregorian at 10.67 GHz indicated that the primary reflector surface had a rms accuracy of about 3.5 mm at least when measured over the relatively small areas illuminated by the mini-Gregorian. This was acceptable, but higher accuracy would be required after the upgrade. A combination of items dropped during the upgrade from the suspended structure including a cable that destroyed almost 100 panels, and further settling of a fill area in the south east portion the bowl left a large number of panels either damaged and/or misaligned. A program that lasted for over 2 years to repair and realign the reflector surface was started in November 1999.

The positional errors of the panels were not the only problem that needed to be fixed. Over the 27 years since the panels and their height adjusters were installed, significant corrosion had occurred including to many of the galvanized collection plates and threaded adjusters for the 38,778 panels (Fig. 12.4). Sections of a number of the light E-W cables to which the panel collection plates were attached had also corroded and had to be replaced. José Maldonado organized the repair effort, which was carried out by observatory personnel. It was important that every one of the threaded adjusters worked and, eventually, this required repairing or replacing about 6000 of the collection plates and adjusters. An immense task that was finally completed in early 2001.

Fig. 12.4 Two edge panels of the reflector attached to a collection plate that normally holds the corners of four panels. It is attached to the light east-west cables, one of which can just be seen below the edges of the two panels, via an adjustable threaded stud. The height adjustment is made by turning the stud at the top or bottom—normally the bottom. Credit: Cornell University/NSF

Lynn Baker, who had built the mini-Gregorian, took on the task of organizing the surface measurements working with John Brown, president of Geodetic Systems Inc. José Maldonado working with Lynn organized the actual adjustment of the surface panels, which was done by observatory personnel. Film photogrammetry was the preferred measurement method since film had at that time a much higher density of pixels than digital video photogrammetry. This was important as the preferred location for imaging the reflector surface were the tops of the three towers. Geodetic Services Inc., who had developed the film photogrammetry system including the measurement system and the V-Stars software, refurbished one of its systems that had been in use at the Boeing Aircraft Co. and provided it to NAIC. Most users of these systems had switched to video photogrammetry.

In late 2000, targets were placed on about 1900 locations on the reflector surface primarily where the tie backs to the concrete anchors were located (see Fig. 9.13) but also on some small patches in order to examine the panel-to-panel setting errors. Measurements showed that the rms for the tie back locations had degraded to 15 mm. Based on these measurements, the tie back locations were adjusted by changing the lengths of the tie back cables. A new survey in January 2001 gave a rms for these locations of 5 mm resulting in a significant improvement in telescope performance (Edmundson and Baker 2001). A third measurement and adjustment of the tie back locations in March 2001 resulted in a rms of 2.9 mm.

With the repair of the surface completed and the tie back locations adjusted, 7.5 cm (3 inch) diameter targets were placed close to where the rest of the 38,778 panels were connected to the collection plates. A full survey, 12 photos from the top of Towers 4 and 8 and 24 from the top of Tower 12 due to the obstruction caused by the catwalk, was carried out in May of 2001. The measurement accuracy in all three directions was close to 0.7 mm. Figure 12.5 shows the resulting distribution of errors for the 36,699 measured locations. The rms was 5.3 mm and there was a mean offset from the ideal spherical shape set by the measurements of the tie down locations of 1.7 mm. This was due to the scalloping of the E-W light cables to which the collection plates are connected over the 25 ft spacing between the N-S main reflector cables. All the photos were taken after sunset as the scalloping increases with the daytime rise in temperature.

The errors from the ideal sphere at each collection plate were translated into turns of the threaded stud and the adjustments were made over the succeeding months. In October/November of 2001 another set of photos was taken with the results as shown in Fig. 12.5. The rms of the 37,889 locations on the reflector was 1.6 mm, and there was no significant mean, a very substantial improvement over the first set of measurements. 11,234 points had errors greater than 1.5 mm. These were finally adjusted starting in November 2004 based on the measurements, but no follow on survey was made to check the results. It is estimated that the final rms was very close to 1.5 mm. Since the 1×2 m (3×6 ft) panels had an intrinsic rms error of about 1 mm combining this in quadrature with the 1.5 mm rms for the target locations, the final error for the primary reflector surface would have been close to 1.8 mm.

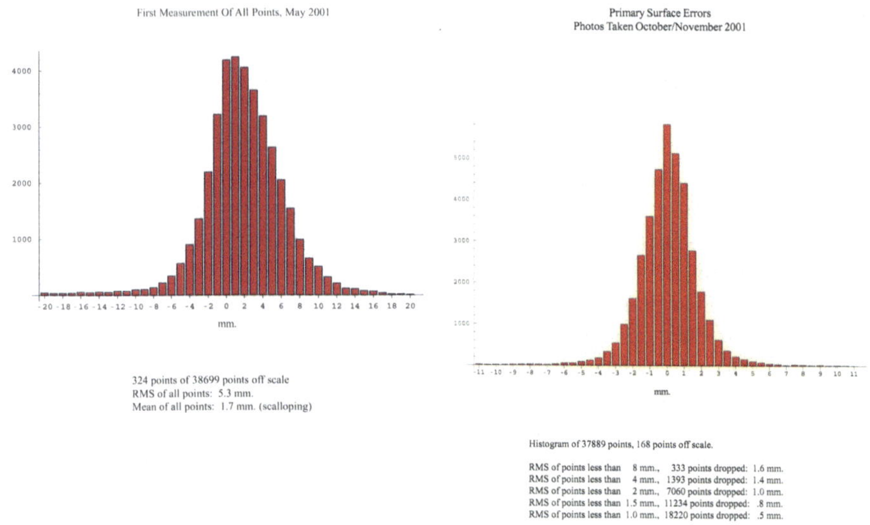

Fig. 12.5 Left: The results from the measurements of all the panels after the May 2001 survey. The tie back points had been corrected prior to this survey by adjusting length of the tie back cables based on a January 2001 survey of the tie back locations. Right: The results of adjusting the studs for all the panels based on the May 2001 survey. Credit: Lynn Baker

Video photogrammetry was used between November 2003 and May/June 2004 to survey the 415 panels of the secondary reflector and adjust the panels. The final rms was 0.6 mm. In October 2004, the tertiary was also measured and adjusted resulting in a 0.3 mm rms. Combining the values for the three reflectors the overall rms was close to 1.9 mm. Considerable effort also went into aligning the horns and the three reflectors with respect to each other. However, radio frequency measurements indicated that the rms was close to 2.3 mm. This was thought to be related to misalignments in the system such as the position of the dome. However, 2.3 mm would give very good performance at 5 GHz and acceptable sensitivity at 8 GHz.

12.3 The Arecibo L-Band Feed Array

One important byproduct of the 1990s upgrade was that the Gregorian optics allowed multi-beam systems for the first time since the late 1960s innovative 10 short line feeds at 600 MHz (see Chap. 9). While such a system was theoretically possible using line feeds before the Gregorian upgrade, the need for high sensitivity would have required 40 ft line feeds, which have limited bandwidth, and multiple cryogenically cooled receivers. In the early 1990s the Australian Telescope National Facility (ATNF) of the Australian CSIRO started development of a cryogenically cooled L-band 13-beam dual linear polarization receiver for the Parkes 64 m telescope (Staveley-Smith et al. 1996). It was installed on the telescope in January 1997.

Over the next 4 years the system was used to do the first very large scale survey of neutral hydrogen in the local universe, the HIPASS survey. Approximately 5000 galaxies were detected in HI. Over a similar period more than 700 new pulsars were discovered. The first Fast Radio Burst (FRB)[3] was detected in 2007 in Parkes survey data taken in 2001.

The success of the Parkes multibeam system spurred Riccardo Giovanelli to think of a similar system for Arecibo. Because of its complex Gregorian optics, Arecibo's field of view was smaller than Parkes and only seven horns would have reasonable performance. Since the ATNF had built a very capable system, the simplest and quickest way to obtain one for Arecibo was to have the ATNF build it. Discussion with the ATNF resulted in an agreement signed in March 2002 that they would design, construct and participate in the antenna tests of a 7-beam dual polarization system with a 300 MHz bandwidth between 1.225 and 1.525 GHz. German Cortes of NAIC and Trevor Bird of the Australian CSIRO settled on a stepped circular horn design very similar to the one used for the Parkes 13 horn system. In September 2002, Graeme Carrad, head of ATNF's electronics division, accompanied by Graham Moores and Pat Sykes visited Arecibo to review the draft system design and discuss the physical and electronic interfaces with the observatory's systems. Designs were approved in late 2002 allowing construction to proceed.

The International Astronomical Union (IAU) meeting in Sydney in July 2003 was attended by NAIC Director Paul Goldsmith and Don Campbell allowing a review of the multibeam receiver work at the ATNF. By this time the horn feed array and testing had been completed and the receiver was being manufactured. By early 2004, the entire system was completed and shipped to the Arecibo Observatory. On 1 April, ALFA was installed on the telescope with the involvement of ATNF personnel and commissioning started (Fig. 12.6).

12.4 Instrumentation Development

To utilize ALFA two major instrument developments were needed, a 14 channel IF and Local Oscillator system and a 14 channel backend spectrometer and pulsar processor. Work on the design of the LO system started in late 2002 led by Eddie Castro, the electronics department head, and his team. Construction including the installation of the needed fiber optic lines was completed in late 2003 prior to the arrival of the ALFA frontend.

In the late 1990s Andrew Dowd, William Zisk and Jeffrey Hagen of the observatory's electronics department started to develop a Wideband Arecibo Pulsar

[3]FRBs are single short bursts of radio emission that originate mainly in external galaxies. Most of them do not repeat but a few do repeat at random intervals. A great many of them have been detected mainly by large field-of-view radio telescopes like CHIME in Canada, ASKAP in Australia and, recently, the 500 m FAST telescope in China that went into full operation in 2020.

Fig. 12.6 Left: The 7-horn ALFA system installed on the rotary floor of the Gregorian Dome. The high-powered water-cooled S-band transmitting horn is to its left. Right: ALFA installed on the rotary floor showing the rotation rack gear at the bottom and the cable wrap at the top used to keep the 7-beams aligned on the sky while tracking. The system was built by the Australia Telescope National Facility of the Australian Commonwealth Scientific and Industrial Research Organization (CSIRO). Credit: Cornell University/NSF

Processor (WAPP) (Dowd et al. 2000). It came into operation in early 2000. This was an autocorrelator based design that utilized an updated version of the Canaris/Jon Hagen chip used in the Interim Correlator with two IF inputs allowing two polarizations, bandwidth options of 100 MHz or 50 MHz and, for pulsar observations, an output dump time as short as 16 microsec. Four were initially built and in anticipation of the arrival of the ALFA system, an additional four were constructed allowing 7 input channels and a spare (Fig. 12.7).

For mapping HI in our own galaxy, spectrometers covering a relatively narrow bandwidth at high spectral resolution were required. In late 2003, the SETI Berkeley group under Dan Wertheimer initiated a design and construction effort to produce seven Fast Fourier Transform (FFT) dual polarization spectrometers covering a bandwidth of 6.7 MHz with a spectral resolution equivalent to a velocity resolution of 0.184 km/sec at 1.4 GHz. It would also provide simultaneously low resolution spectra covering 100 MHz to allow accurate baseline calibration. The main designer of the GALSPEC system was Jeffrey Mock, a University of California at Berkeley graduate, who was a brilliant San Francisco based independent electronics expert. The system was installed at Arecibo in late 2004 (Fig. 12.8).

The limited bandwidth of the WAPP spectrometers meant that the 300 MHz bandwidth of the ALFA system could not be fully utilized especially for pulsar searches. Jeff Mock was contracted to design and build Fast Fourier Transform (FFT) spectrometers covering the full bandwidth. Each unit had two input channels, each with 172 MHz bandwidths allowing coverage of the full 300 MHz ALFA bandwidth. Each channel could accept two polarizations and for the output the user could select the total power spectrum, spectra for the individual polarizations or

Fig. 12.7 Six of the WAPP spectrometers, two more are in the rack to the far right. Despite the PP standing for Pulsar Processor the WAPPs were also used extensively for observations of external galaxies. Credit: Cornell University/NSF

Fig. 12.8 The GALSPEC seven dual polarization channel spectrometer used for galactic surveys. Credit: Cornell University/NSF

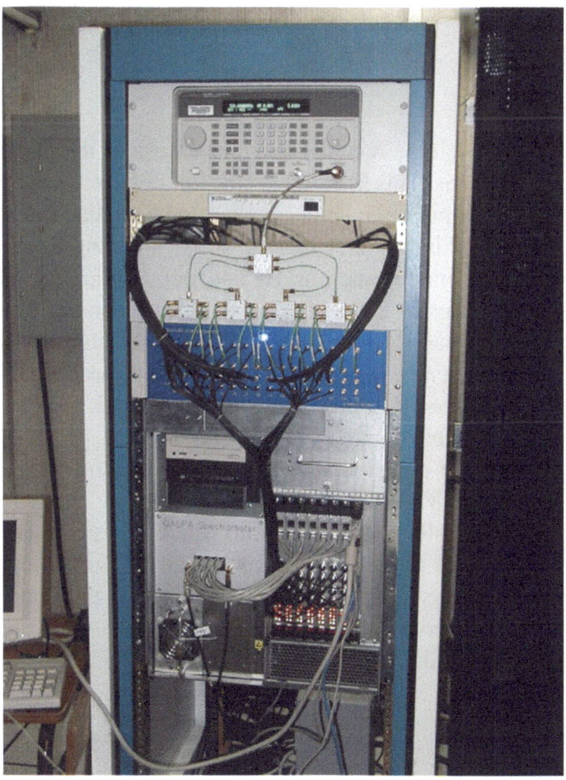

Fig. 12.9 The three racks holding 14 Mock spectrometers in the equipment room of the Observatory. Credit: Cornell University/NSF

these as well the real and imaginary parts of the complex product of the two input polarizations. This last allowed the full polarization properties of the received signal, needed for pulsar and other studies, to be obtained. A total of 14 units were installed at Arecibo in the first half of 2008, allowing seven each for two simultaneous observing modes such as HI and pulsar searches (Fig. 12.9). Sadly, Jeff Mock passed away in 2009 after a battle with cancer. Several additional spectrometer units and parts were acquired from his estate providing spares that proved very valuable over the succeeding years. The spectrometers were initially called the pDev spectrometers, this was changed to the Mock spectrometers in recognition of Jeff Mock's contributions.

With the completion of the Mock spectrometers, Arecibo had the 100 MHz bandwidth WAPP spectrometers, the narrow bandwidth GALSPEC ones and two sets of the 300 MHz bandwidth Mock spectrometers. This meant that within the limitation of the telescope pointing at the same location in the sky, two or three independent surveys could be conducted at the same time in what is called commensal observations, greatly increasing the efficiency with which the telescope could be used.

12.5 Some Post-Gregorian Radio Astronomy Programs

The ALFA Surveys: The installation of the ALFA system had a profound affect on the nature of the astronomy program at the observatory. Because ALFA enabled efficient large-scale surveys of the sky, there was a transition from mostly small investigator programs with one or a few participants plus a very small number of large-scale surveys such as pulsar searches to many more large-scale surveys, each with many participants while still maintaining some small investigator programs. Clearly, NAIC had to play a role in organizing these large-scale surveys. In late March of 2003, two workshops took place at Arecibo, one to organize potential surveys and form consortia for extragalactic HI (EALFA) surveys and the second to do the same for galactic (GALFA) surveys. The pulsar search community was already well organized and proposed a survey, PALFA (Pulsar ALFA) with James Cordes, Cornell University as the PI. There was interest in all varieties of pulsars but especially millisecond pulsars, which have the very stable rotation periods needed for the detection of nano-Hz gravitational waves. Most of them are in binary orbits with either another neutron star or a white dwarf star, allowing the masses of both the pulsar and the companion to be determined. After the first discovery of a Fast Radio Burst (FRB) in 2007, PALFA included searching for these enigmatic sources.

The outcome of the meetings was multiple extra-galactic (EALFA) and galactic (GALFA) surveys. As can be seen from the list below, most but not all of these involved the detection of atomic hydrogen (HI) in our galaxy and its vicinity or in external galaxies:

ELFA:

1. ALFALFA (Arecibo Legacy Fast ALFA Survey) a low sensitivity survey of about 7000 sq. deg., roughly 50% of the Arecibo observable sky, mapping the distribution of galaxies with HI content in the local universe out to redshifts of about 0.06. The PIs were Riccardo Giovanelli and Martha Haynes of Cornell University. This survey, which utilized about 4400 h of observing time, began in 2005 and was largely completed by 2011. It mapped the 3-D distribution of 31,500 galaxies (Haynes et al. 2018). ALFALFA involved a very large number of researchers and graduate students from many institutions. An undergraduate program funded by the National Science Foundation, PI Rebecca Koopman of Union College, involved a large number of students in ALFALFA and included yearly meetings at the Arecibo Observatory.
2. AGES (Arecibo Galactic Environments Survey) an intermediate sensitivity survey covering about 200 sq. deg. of the sky. PIs John Davies of Cardiff University and Robert Minchin of the Arecibo observatory.
3. AUDS (Arecibo Ultra Deep survey) a very deep survey of approximately 1 sq. deg. of the sky. PIs Wolfram Freudling of the European Southern Observatory and Lister Staveley-Smith of the International Center for Radio Astronomy Research in Australia.
4. ZOA (Zone of Avoidance) a search for HI rich galaxies in the area obscured by our own galaxy, the Zone of Avoidance. PI Patricia Henning of the University of

New Mexico. This survey ran commensally with Galactic HI and the PALFA surveys.

5. GASS (GALEX Arecibo SDSS Survey), was a directed survey aimed at examining the HI distribution in 1000 massive galaxies selected from the SDSS optical survey to obtain an "unbiased view of the processes responsible for the transition between blue, star-forming spirals and red, passively-evolving ellipticals". This program used the 21 cm line feed not the ALFA array. PI David Schiminovich of Columbia University.

GALFA:

1. I-GALFA, a survey of HI in the galaxy covering $\pm 10^{\circ}$ of galactic latitude and the longitudes visible to Arecibo with GALSPEC's very high frequency resolution corresponding to184 m/s. The PIs were Steve Gibson of Western Kentucky University and Bon-Chul Koo of Seoul National University. It had been intended that this 2 year survey, which had to be done between May and September when the galaxy was observable with Arecibo at night, would begin in 2006, but delays in approval of the proposal prevented this and the 2007 season was cancelled due to the lead paint remediation and re-painting of the telescope structure. The survey finally got underway in 2008 and was completed in 2009 (Peek et al. 2010).
2. TOGS (Turn On GALFA Survey) was a high spectral resolution HI survey with the same frequency resolution as I-GALFA that ran commensally primarily with the ALFALFA and AGES surveys so was limited to the part of the sky included in these surveys. This did not include the galactic plane. The survey leaders were Mary Putman now of Columbia University and Snezana Stanimirovic now at the University of Wisconsin, Madison. The survey was aimed at mapping galactic HI clouds and high velocity clouds. It began in 2005 and covered over 7000 sq. deg. of Arecibo's observable sky. A brief description of results and some images can be found at Gibson et al. (2008).
3. TOGS2 was very similar to TOGS but ran commensally with GALFACTS (see below), an all Arecibo sky survey so that TOGS2 covered the galactic plane. TOGS2 began in 2008 at the same time as GALFACTS.
4. GALFACTS (Galactic ALFA Continuum Transit Survey). This survey's primary objective was "sensitive imaging of polarized radiation and Faraday Rotation Measure from both diffuse emission and against a high-density grid of extragalactic sources". Since it utilized the full 300 MHz ALFA bandwidth, it required the use of the Mock spectrometers so did not start until 2008. The PI was A. Russell Taylor of the University of Calgary.
5. GALFA-RRL (GALFA-Radio Recombination Lines) aimed at mapping the distribution of radio recombination lines. It was led by Yervant Terzian of Cornell University.

Arecibo's tremendous sensitivity and its excellent "back-end" equipment combined with the ALFA array resulted in surveys that were a very significant advance over previous ones with similar objectives.

VLBI: The post-Gregorian period was a transformative time for VLBI at Arecibo.[4] A Maser frequency standard was obtained from the Jet Propulsion Laboratory followed by a S2 VLBI recorder from the Institute of Space & Terrestrial Sciences (ISTS) in Toronto, Canada. This was in preparation for observations in support of the Japanese HALCA mission, an orbiting 8 m antenna reaching a maximum distance from Earth of 21,000 km. The first observation in support of HALCA were made on 28 July 1997 and this was also the first post-Gregorian upgrade use of the telescope for radio astronomy observations. Sensitivity was significantly lower than what would be achieved later due to the platform not yet being at its design height, but fringes were detected. Multiple observations in support of HALCA followed with about 40 radio telescopes world-wide participating. Arecibo's tremendous sensitivity made it an important contributor.

A 1998 proposal to the NSF for funding to purchase a recording system compatible with the NRAO's Very Long Baseline Array (VLBA) at a cost of $502,000 was successful. Cornell University covered the required cost share of 30%. The system was received and installed in early 1999. This allowed Arecibo to participate in a wide range of VLBI observations. The first observation in conjunction with the VLBA and the VLA was of a nearby luminous infrared galaxy NGC7674 (Fig. 12.10).

Arecibo's tremendous sensitivity was utilized with the formation of the High Sensitivity Array (HSA), a combination of the VLBA, Arecibo, the Green Bank 100 m telescope (GBT), NRAO's Very Large Array in New Mexico and the 100 m Effelsberg telescope in Germany. The addition of these telescopes to the VLBA array gave an increase in sensitivity of more than a factor of ten. The first call for proposals was made in early 2004 with proposals due 1 June. Of course, Arecibo could only participate in observations of sources that fell within its declination coverage. Arecibo also participated in observations with the European VLBI Array.

With the launch by Russia in July 2011 of the RadioAstron VLBI mission in an orbit that took it almost to lunar distances from Earth, a new era in high resolution VLBI began. RadioAstron was equipped to observe in four frequency bands, 0.327, 1.665, 4.830 and 18,392–25.112 GHz. Multiple Earth based telescopes were used in conjunction with RadioAstron including the high sensitivity Arecibo, GBT and VLA instruments. Arecibo could observe at the three lowest bands. Commissioning took the first 6 months and an early science phase extended from February 2012 to June 2013. It ceased operations in 2019.

12.6 The Search for Extraterrestrial Intelligence (SETI)

The first serious search at radio wavelengths for intelligent life in the Universe, Project OZMA, was conducted by Frank Drake at NRAO's Green Bank observatory in 1960. He used a 26 m antenna at the 21 cm wavelength of the hydrogen line

[4] A brief 2014 discussion of VLBI at Arecibo can be found at: https://areciboscience.org/White_Papers/vlbi_wp_2014_complete.pdf

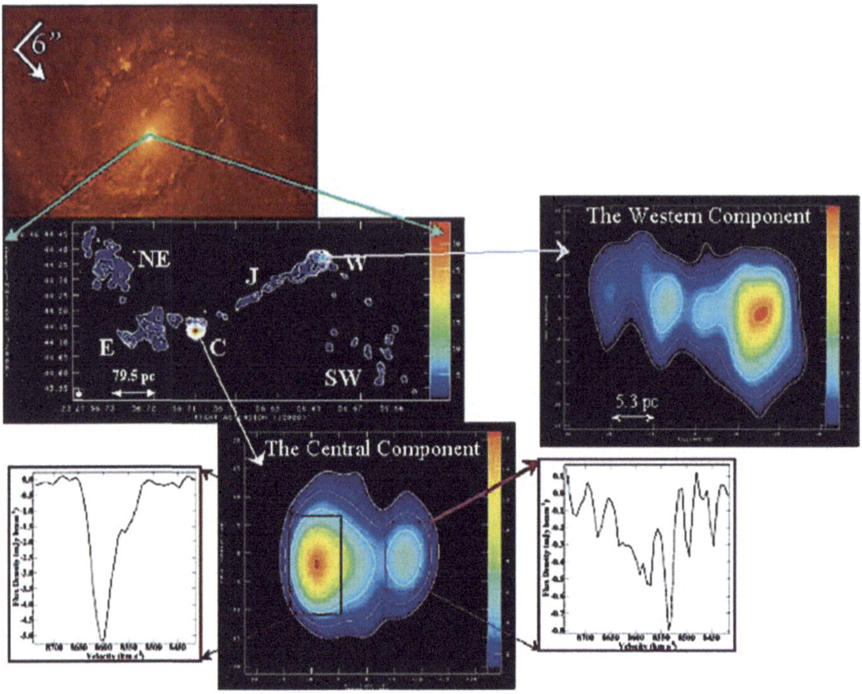

Fig. 12.10 Upper left: An optical image of NGC 7674 with the HST (Malkan et al. 1998). Center left: The 1380-MHz VLBI continuum image of the galaxy nucleus at a resolution of 20 mas (10.6 pc). Six structures are marked in this image, namely the western (W), the central (C), the eastern (E), the jet (J) connecting the C and W components, and the diffuse north-eastern (NE), and south-western (SW) components. The contour levels are at −5, 5, 10, 20, 40, 640 × the rms noise level of 34 μJy beam-1. Center right and bottom center: Full-resolution continuum images of the C and W components, respectively. The restoring beam size is 11 × 5 mas in position angle 3°. The rms noise in both images is 67 μJy beam-1. Bottom left and right: HI absorption spectra obtained against the eastern and western peaks of the central (C) component. The effective velocity resolution of the HI spectra is 14 km/s. Figure assembled, and caption written, by C. Salter and E. Momjian. It is in the 2004 NAIC Arecibo Progress Report and 2005 Program Plan; Cornell University archives under the Arecibo Ionospheric Observatory. See Momjian et al. (2003) for the details of the observations

looking from signals from two nearby solar type stars, Tau Ceti and Epsilon Eridani. After 150 h of observations over 4 months, no signals were detected. Drake's interest in SETI carried over to his time at Cornell, where, as discussed in Chap. 9, he initiated the transmission of the Arecibo Message during the dedication event in November 1974 for the first upgrading of the telescope. In the late 1970s, Drake and Carl Sagan carried out a symbolic search with Arecibo for intelligent signals from a nearby galaxy based on the premise that a huge number of stars would be encompassed in a single telescope beamwidth and that an associated planet of one or more of them may be transmitting an extremely strong signal.

The Berkeley SETI program: In 1979 the University of California Berkeley SETI group initiated the SERENDIP program. The limitation at that time was real time processing power. SERENDIP I, which used Berkeley's Hat Creek radio telescope, had 100 approximately 1 Hz channels centered on the 21 cm line. It was succeeded in 1986 by SERENDIP II which had 65,000 real time channels using Fast Fourier Transforms (FFT) techniques. In 1992 the group brought SERENDIP III to Arecibo. Arecibo's spherical reflector meant that all the line feeds on both carriage houses were always focusing incoming RF radiation but from directions dictated by where the feed that was being used for observations was pointed, the locations of the carriage houses and the feeds on the carriage houses. For a random search using one of the feeds, this was fine as long as it was known where it was pointed as a function of time. Unusually, SERENDIP III observed at 430 MHz using the 96 ft long line feed. It used a 4 million channel FFT spectrometer to analyse a 2.5 MHz bandwidth giving a spectral resolution of 0.6 Hz. The telescope's pointing information was recorded every 5 or 10 s.

In 1997 SERENDIP IV was initiated utilizing a specially built 16 ft line feed at the 21 cm wavelength hydrogen line mounted uphill from the 430 MHz line feed on carriage house 1. The bandwidth used was 96 MHz centered on the hydrogen line, and a 168 million FFT spectrum analyzer was used to again give a 0.6 Hz resolution. This program ended in 2007. SERENDIP V was initiated in 2009 utilizing the 7-Beam ALFA system. It went through a number of iterations with the final system utilizing 200 MHz centered on the hydrogen line and two polarizations for a total of 14 channels. Frequency switching allowed the full 300 MHz bandwidth of ALFA to be covered. While it could only be used when the ALFA system was at the focal location, this was not a significant problem due to the large amount of observing time devoted to the ALFA surveys, and the availability of seven beams more than compensated. No clear detections of signals from extraterrestrials were made and the program came to an end with the collapse of the Arecibo telescope in late 2020.[5]

SETI@home: Realizing that the millions and millions of computers in the world spent much of their time idle, the Berkeley group came up with the very innovative idea of utilizing those unused CPU cycles to search for intelligent life. They initiated the SETI@home project in 1999 using a 16 ft line feed on Arecibo and utilizing 2.5 MHz centered on the hydrogen line. The data was then broken into small packages and distributed for processing to the computers of everyone who had signed up to participate. It was an amazing organizational achievement. The program continued to use the 16 ft line feed until the 7-beam ALFA system was available in 2005. Again, while ALFA could only be used when it was in the focal location this was not a significant problem given the very large amount of time that the L-band ALFA surveys used and any lost time was more than compensated by having seven beams. Like SERENDIP, SETI@home came to an end with the collapse of the

[5]A brief discussion of the SERENDIP program through 2009 can be found at https://www.planetary.org/articles/setiathome_20090526

Fig. 12.11 The Project Phoenix equipment trailer being installed in front of the transmitter bay. Credit: Cornell University/NSF

Arecibo telescope in late 2020. The status of Berkeley's SETI program as of 2011 can be found in Korpela et al. (2011).

Project Phoenix: As discussed in Chap. 10, the $100M NASA High Resolution Microwave Survey (HRMS) SETI program had started a targeted search with Arecibo in 1992 but was cancelled by the US Congress in 1993. The cancellation was after NASA had committed $2.3M to the Gregorian upgrading program in anticipation of NAIC providing 2500 h of Arecibo telescope time to HRMS. Given this, NAIC did commit to awarding the 2500 h of telescope time to Project Phoenix, the privately funded successor to HRMS based at the SETI institute close to the NASA Ames Research Center in Mountain View, California (Fig. 12.11). Jill Tarter was the lead scientist for Project Phoenix. Observations with Arecibo started in late 1998 utilizing evening and nighttime hours to limit the effects of radio frequency interference. Due to the high demand for use of the telescope, only two 3 to 4 week sessions per year were scheduled meaning several years were needed to complete the program which finished in 2004. The search covered the frequency range of 1.2 GHz to 3.0 GHz with a spectral resolution of 1 Hz. Over 800 Sun-like stars within 200 light years of Earth were observed, unfortunately with no clear detections of signals from an intelligent source.[6]

---

[6]A brief description of Project Phoenix can be found at https://www.seti.org/seti-institute/project/details/project-phoenix

12.7 Planetary Radar Program and Some Results

With the completion of the Gregorian upgrade of the telescope, two factors opened up new opportunities for new or improved investigations of solar system bodies; the large increase in the sensitivity of the 2.38 GHz planetary radar system (Fig. 12.12) especially at high telescope zenith angles, and the concurrent rapid increase in the discovery rate for asteroids, especially near-Earth asteroids (NEAs). The very large increase in sensitivity at high zenith angles opened the possibility for studying the radar scattering properties of the larger satellites of Saturn, especially Titan, the largest and only planetary satellite with a dense atmosphere. Given Arecibo's maximum tracking time of close to 2 h and 40 min, at about 2.25 h the round-trip-light time to the Saturn system means that transmitting to the Saturn system must take place at high zenith angles during the 25 min just after the planet is first observable and the echo is then received in the last 25 min of tracking time, also at high zenith angles.

In 1998 the US Congress passed a bill that directed NASA to find within 10 years 90% of Near Earth Objects (NEOs), primarily NEAs but possibly a few comets, with diameters greater than 1 km that may threaten Earth. The search would be done by ground-based optical telescopes owned by the NSF or other institutions with some funding from NASA. As can be seen from Fig. 12.13, the search started slowly but by the early 2000s the rate started to increase. The initial objective of finding 90% of NEOs larger than 1 km was met in 2011. However, in 2005 Congressman Dana Rohrabacher sponsored the George E. Brown Jr. Near Earth Object Survey Act, which would expand NASA's program by requiring 90% of NEOs 140 m and larger to be cataloged within 15 years. This would, hopefully, provide long warning times

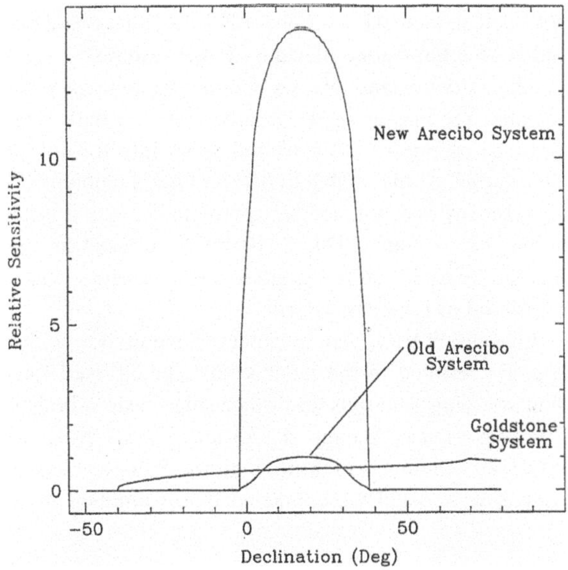

Fig. 12.12 The relative sensitivities of the Arecibo planetary radar system before and after the Gregorian Upgrade. The relative sensitivity of NASA's Goldstone 70 m DSS14 antenna's 8 GHz radar system is also shown. Its large declination coverage compensates for its lower sensitivity

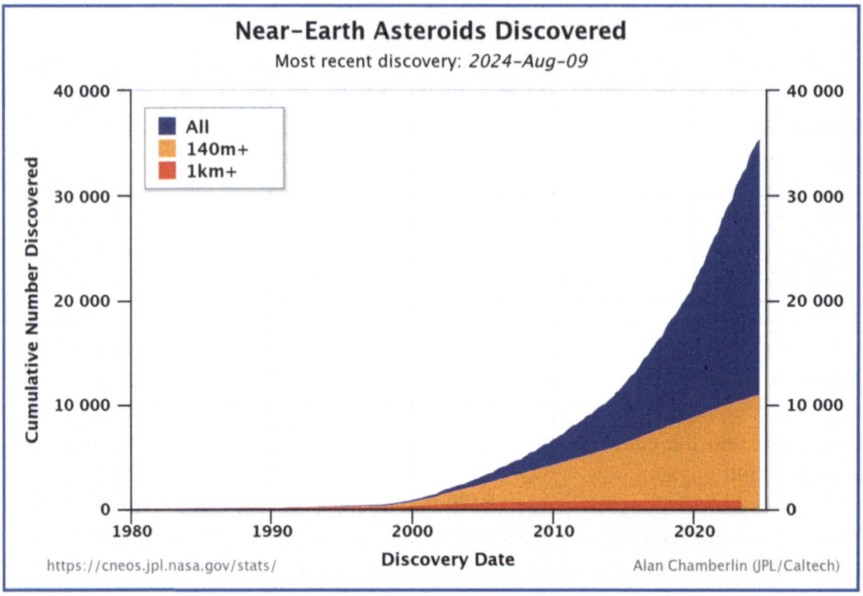

Fig. 12.13 The cumulative number of NEAs discovered through August 2024 and the number of with sizes greater than 1 km and greater than 140 m. Figure courtesy of the Center for Near Earth Object Studies at NASA's Jet Propulsion Laboratories (A large amount of information about NEOs including search statistics and deflection plans can be found at http://neo.jpl.nasa.gov/stats/)

for potential city destroying impacts allowing the possibility of the asteroid to be deflected into a safe orbit. One side effect of the searches for NEOs was that it provided an increasing number of objects that could be investigated with the Arecibo and NASA Goldstone planetary radar systems.

The Arecibo and NASA Goldstone planetary radars' ability to measure the distance and line-of-sight velocity of NEAs with very high precision allowed their orbits to be predicted up to 400 years into the future based on a single epoch of measurements, invaluable in determining the potential hazard due to Earth impact. If the future orbit is not precisely predictable, then it is not possible to determine which of the, as of June 2024, 2340 NEAs classed as potentially hazardous asteroids (PHAs) as their orbits come close to, or cross, that of Earth's have a significant likelihood of impacting Earth.

Finding SOHO: An unexpected event not long after the completion of the upgrade resulted in the radar system being used for an entirely different purpose than studying asteroids and other solar system bodies. On 24 June 1998, the joint European Space Agency (ESA) and NASA Solar and Heliospheric Observatory (SOHO), which was in orbit about the Sun, developed problems and on the 25th all contact was lost. ESA contacted Don Campbell to see if the Arecibo radar could

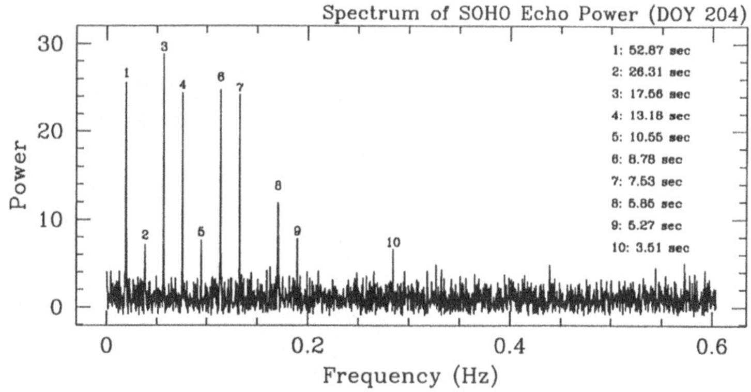

Fig. 12.14 The frequency spectrum of the radar echo power returned from the SOHO solar orbiter. The fundamental rotation period (1) was 52.87 s, about the expected period so good news. The remaining peaks are harmonics originating from the non-spherical shape of the spacecraft or from attachments to it. The main participants in the observations and data reduction were Gregory Black and Don Campbell (Cornell University) and Marty Slade (JPL)

verify whether SOHO was in its expected location and whether it was rotating with its nominal period of about 1 min. After discussions with JPL's Marty Slade, it was decided to attempt the detection by transmitting from Arecibo and use NASA's 70 m DSN Goldstone antenna to receive the echo. This worked well, an echo was obtained from SOHO indicating that it was at its nominal position and that it was rotating with a period of 52.87 s, very close to what was expected (Fig. 12.14). This was the most distant radar detection ever of a man-made object. ESA did eventually make contact with SOHO and slowly brought it back into full operation. Despite the loss of its gyroscopes as of this writing SOHO is still operational.

The asteroid program: From 2000 to 2010, the Arecibo planetary radar system was used to observe about 20 asteroids per year. Numerous people were involved with the observations at Arecibo and NASA's Goldstone facility with many, but not all, associated with the Arecibo Observatory, Cornell University and the Jet Propulsion Laboratories (JPL). One person who played a very important role for both his contributions to, and public advocacy of, the program was Steven Ostro of JPL (Fig. 12.15). Ostro obtained his PhD at MIT under the guidance of Gordon Pettengill and Irwin Shapiro based on Arecibo radar studies of Saturn's rings and the icy Galilean satellites of Jupiter. After a stint as an Assistant Professor at Cornell, he moved to JPL as a Senior Scientist, where he became deeply involved with radar studies of asteroids. For his work he was twice awarded NASA's Medal for Exceptional Scientific Achievement. Sadly, Steve Ostro died in December 2008 of complications related to cancer. He was 62.

A few of the major achievements of the radar program between 1998 and 2011were:

1992 UY4: Only high powered radar systems have the capability of imaging NEAs. A prime example—there are many others—is the images of 1992 UY4

Fig. 12.15 Steven Ostro in 1979 in the receiver room of the Arecibo observatory during an observation. The large IBM printer sheet to his right lists for every minute of the observation the predicted position in Right Ascension and Declination of the object being observed as well as the telescope's zenith and azimuth angles plus the round-trip-light time and Doppler shift for the object being observed. Credit: The author

(Fig. 12.16) a 1 km diameter NEA that came within 0.03 AU from the Earth in 2005. It was imaged with 10 m resolution with the Arecibo radar showing its surface in considerable detail highlighting the ability of radar to characterize these objects. At a distance of 0.03 AU, the entire asteroid subtends an angle of about 0.04 arcsec, close to the limit of the resolution of very large optical telescopes. No current or planned optical telescope can come close to 10 m resolution at the then distance of UY4.

Binary NEAs: During the close approach of the NEA 2000 DP107 in late September and early October 2000, it was observed by both the Goldstone and Arecibo radar systems. The Goldstone observations on September 22 and 23 indicated that DP107 may be a binary system. Arecibo observations between September 30 and October 7 clearly showed that DP 107 is a binary system, the first known binary asteroid system. The approximate sizes of the two bodies were 800 m and 300 m and the orbital period was 1.77 days (Margot et al. 2001).

In 2001 the NEA 1999 KW4 made a close approach to Earth at a distance of 0.032 AU, a perfect target for the Arecibo radar system. It, too, turned out to be a binary system with interesting dynamical properties (Fig. 12.17) (Ostro et al. 2006, the Science issue had an image of the shape model on its cover). A large amount of information about the system was derived from the observations, the size, shape, density and rotation period of each of the components of the binary system as well as the orbital period. Many binary systems were observed over the following years;

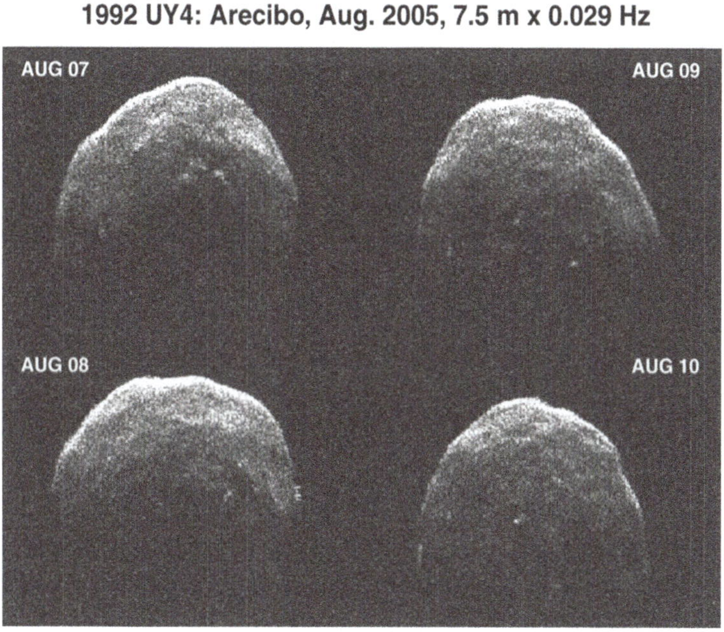

Fig. 12.16 Delay-doppler mages of 1992 UY4 which came within 0.03 AU of Earth in August 2005. Time delay is from the top down and doppler spectral power is left to right. Courtesy of Lance Benner (JPL) and colleagues

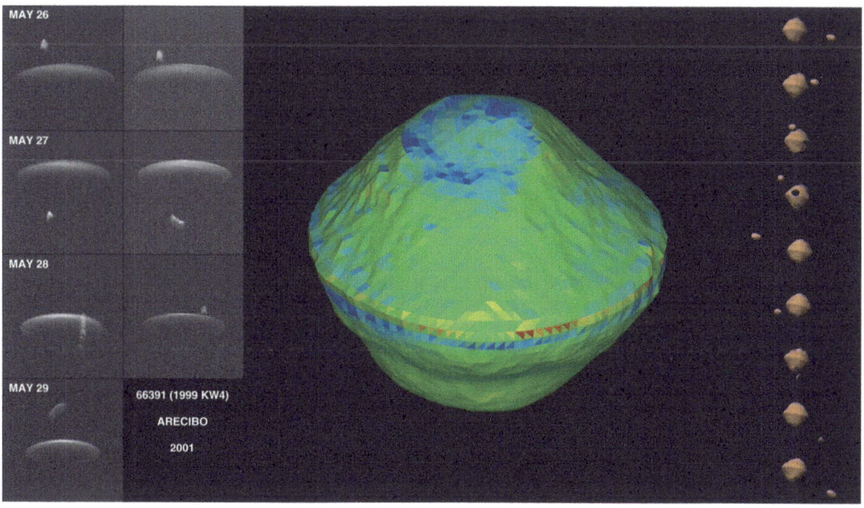

Fig. 12.17 Left: A series of delay-Doppler images of the binary NEA 1999 KW4 made in 2001with the Arecibo planetary radar system. Time is vertical starting at the top and rotational Doppler shift is along the horizontal axis. The broad feature is the radar echo from the 1.5 km main body of the asteroid while the small features are the echoes from the companion as it orbits the main body (From Ostro et al. 2006). Center: The shape model derived for KW4 based on the delay-Doppler images. Right: A modeled sequence showing the small companion orbiting the main body of KW4. Images courtesy of Steven Ostro

about 16% of NEAs were found to be binary systems. As of 2023, three triple systems have been discovered, with the first being 2001 SN263, which was shown to be a triple system from Arecibo observations in 2008.

The Yarkovsky effect: Just as on Earth where the hottest part of a fine day is usually in the mid-afternoon, asteroid surfaces heated by the Sun tend to re-radiate in the infrared most strongly not directly back to the Sun but in a direction offset from this due to the thermal inertia of the surface material. This means that there is a non-Sun directed very small recoil force on the asteroid that can, over a long time, modify its orbit. Observations of the 0.53 km NEA 6489 Golevka in 1991, 1995 and 1999 with the Arecibo and Goldstone radar systems allowed its orbit to be very precisely predicted for the next observations by Arecibo in May 2003. Making this prediction with and without taking the Yarkovsky effect into account and comparing the results with the actual measurements in 2003 very clearly showed that the Yarkovsky effect could modify NEA orbits (Chesley et al. 2003). Measuring the Yarkovsky effect for small non-binary asteroids is the only way to obtain their mass short of a space mission, and the effect is also part of the process that moves small bodies in the main belt of asteroids into the inner solar system.

Titan: Photochemically produced haze layers in Titan's atmosphere make it difficult to observe its surface at optical and IR wavelengths. The Saturn system became observable with Arecibo in 2001, and radar observations were made on 16 successive days corresponding to one rotation period of Titan. Nine additional observations were made in 2002, receiving the echo with both Arecibo and NRAO's 100 m Green Bank telescope. Strong specular returns—mirror like reflections from surfaces smooth at the scale of the 13 cm wavelength of the radar—were detected at a number of longitudes indicating the possible presence of liquid methane lakes (Campbell et al. 2003). Subsequent analysis based on the later radar imaging of the surface of Titan from the Cassini Saturn orbiter, indicated that the Arecibo specular returns likely came from smooth dry lake beds as the Cassini radar did detect liquid methane lakes at high latitudes but not in the equatorial regions corresponding to the Arecibo observations (Hofgartner et al. 2020).

12.8 Space and Atmospheric Science

In mid-1999 there was a reorganization of the ionospheric program and a name change to the Department of Space and Atmospheric Science (SAS) along with a new job title, Assistant Director of NAIC for SAS. Donald Farley, a long term Cornell faculty member in the School of Electrical Engineering and former director the Jicarmarca Radio Observatory in Peru, was appointed as the first assistant director. Michael Sulzer, who had been in charge of the ionospheric modification program, was appointed head of the radar sciences department and Craig Tepley was appointed head of the optical sciences department. The intention was to give the Arecibo SAS program a more defined role since it was funded through the Upper

Atmosphere Facilities section of the NSF while the astronomy program and most observatory operations were funded through the Division of Astronomical Sciences.

The incoherent scatter radar (ISR) continued on a limited basis during the Gregorian upgrade work as the upgrade did not affect Carriage House 1 and the 430 MHz line feed. With the completion of the upgrade, the ISR program had access to the new dual beam capability allowing much more precise wind measurements in the upper atmosphere, and the larger bandwidth of the transmitting horn in the Gregorian dome compared to the line feed allowed plasma line measurements over a ±15 MHz band. During the very early telescope upgrade period, the HF facility was also upgraded by replacing the wire-based pseudo coaxial lines (see Chap. 10) with solid coaxial lines significantly increasing the radiated power and also reliability. The upgrade pushed the effective radiated power (ERP) above 80 MW, the transition zone that allowed artificial auroras (i.e. airglow) to form in sporadic E-layers, wind induced layers of high electron and ion densities in the 90 km to 120 km height range (the upper mesosphere and lower thermosphere). CCD (charge-coupled-devices) imagers at the observatory could then be used to study the variety of structures in the sporadic E-layers (Bernhardt et al. 2003). As recounted in Chap. 10, hurricane George in September 1998 severely damaged the HF facility at its coastal site and this, combined with pressures from the environmental movement to restore the coastal marshes, resulted in it being dismantled.

Observations in support of the international World Day programs continued to dominate the 430 MHz incoherent scatter radar (ISR) telescope time. However, a very active meteor observational program that utilized both the ISR radar and passive and active optical instruments continued.

LIDAR program[7]: As recounted in Chap. 10, investigations of the neutral atmosphere using lidars began in 1989 when Chet Gardner of the University of Illinois brought his resonance scattering lidar, which was sensitive to neutral sodium atoms in the mesosphere, to participate in the AIDA campaign (see Chap. 10). These metals are deposited through meteoric input, and a permanent layer exists in the mesosphere and the lower thermosphere region, maintained by a balance between photochemistry, production and loss processes. Unlike passive optical systems, resonance lidars provide both the height and time variation of metal ions in the mesosphere and lower thermosphere and are excellent tools for inferring neutral ion concentrations, temperatures and winds in the Mesosphere and lower thermosphere (MLT) regions.

The AIDA campaign was followed by the installation of a Doppler Rayleigh lidar system in 1990. In 1995, an Alexandrite laser was purchased to enhance the lidar program. These lidar acquisitions led to the construction of a separate LIDAR laboratory on the top of the hill just above the photometer laboratory. It was completed in August 1997 (Fig. 12.18). During the Coquí II NASA sponsored

[7]This section was written with extensive assistance by Dr. Shikha Raizada, a long time member of the Space and Atmospheric Sciences group at the observatory specializing in LIDAR studies.

Fig. 12.18 The new LIDAR laboratory was finished in August 1997. This image is from 2023. The laboratory is located on the top of the hill above the west side of the operations building. Credit: Arecibo Observatory/NSF; photo by Phillip Perillat

rocket campaign in 1998[8] simultaneous measurements using the Na lidar and Incoherent Scatter Radar were made to study sudden sodium layers and their coupling with sporadic E (Friedman et al. 2000). Funding from the NSF and via a NASA EPSCOR grant through the University of Puerto Rico (UPR) led to the development of a potassium lidar system, which became operational in July 1999. Seasonal variations of potassium concentrations were the made by Friedman et al. (2002).

Soon after, a further NASA EPSCOR grant through UPR allowed modifications to the sodium resonance lidar that allowed the first measurements of neutral iron in the mesosphere and to infer temperatures (Raizada and Tepley 2002). Several new and innovative lidar related activities occurred in conjunction with the ISR until the telescope's collapse in 2020. The majority of the developmental and lidar upgrades were funded by NSF awards.

A new HF system: With the decommissioning of the Islote HF facility after the damage caused by Hurricane Georges and the push from environmental groups to return the site to marshland, there was a strong desire within the community engaged in HF heating investigations to restore a HF transmitting capability at Arecibo. Arecibo was the only facility that could combine HF heating with a highly sensitive

[8]See the 15 February 2000 special issue of the Geophysical Research Letters on the Coqui II rocket campaign results.

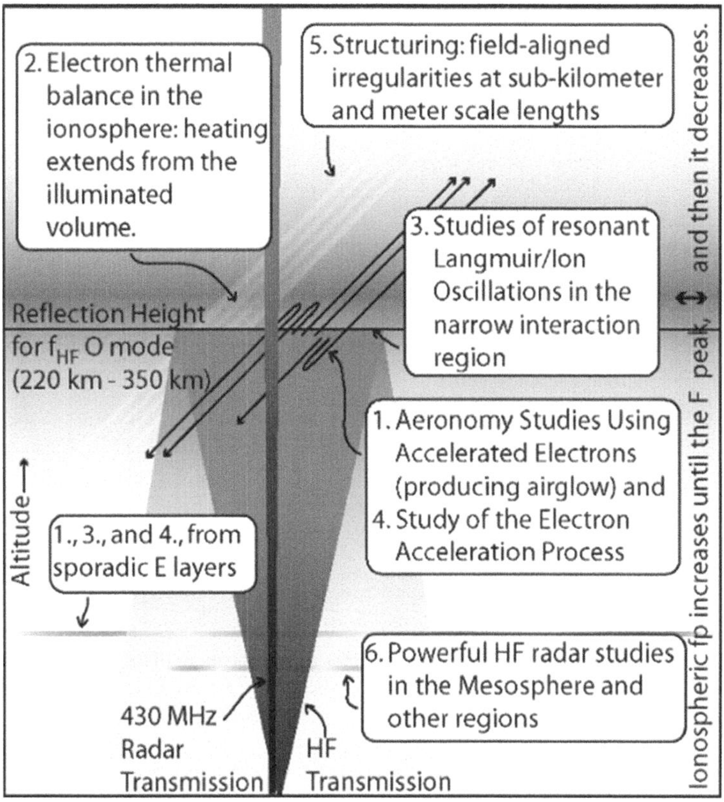

Fig. 12.19 Six of the scientific drivers for the construction of a new HF heating capability on the Arecibo telescope. Figure C.1-1 from the 2007 funding proposal to the National Science Foundation, PI Michael Sulzer

incoherent scatter radar and an extensive suite of optical instrumentation for diagnostic purposes. Six of the primary scientific drivers for a new HF system are illustrated in Fig. 12.19.

A group led by Mike Sulzer, a longtime member of the observatory's Space and Atmospheric Sciences Department and Head of its Radar Sciences group, investigated options for installing a HF transmitter on the 305 m telescope. The first idea was to hang a dipole or Yagi antenna from the telescope's suspended structure and feed it from a high-powered commercial transmitter on the ground, very similar to the first HF transmitting feed in the early 1970s (see Chap. 9). The design finally settled on (Fig. 12.20) consisted of: (1) A sub-reflector suspended under the telescope's platform by cables from the towers; (2) Three radiating dual linear polarization dipole feeds at each of two frequencies, 5.1 MHz and 8.175 MHz, located concentric with the center of the bowl above the reflector; (3) Seven 100 kW transmitters (six to feed the three dual polarization dipoles plus one spare) plus long copper coaxial feeder cables obtained from the decommissioning of the DOD's Over

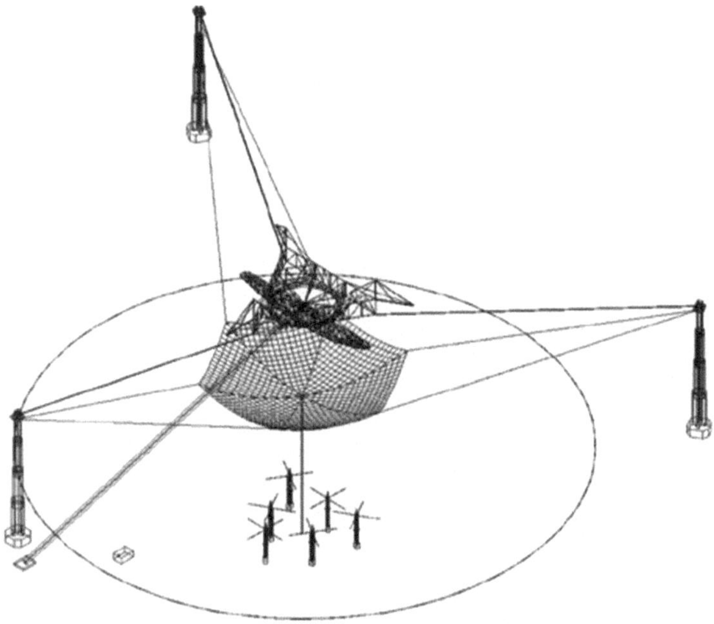

Fig. 12.20 A schematic showing the HF sub-reflector and the locations of the six cross polarization transmitting dipoles at 5.1 and 8.175 MHz. Credit: Michael Sulzer and Cornell University

the Horizon radar system in Maine; (4) A new prime power generating system consisting of four 1-MW diesel generators which was shared with the planetary radar program. The generators were eventually procured using part of the $3M that the Puerto Rican government contributed to Arecibo infrastructure projects in 2009.

The ground work for the funding for the new HF system was done by Herb Carlson, who was a strong advocate for HF heating investigations since his time on Arecibo's staff in the late 1960s and early 1970s. He organized a collaborative funding program between the NSF Division of Space and Atmospheric Sciences, the Air Force Office of Scientific Research (AFOSR) and the Naval Research Laboratories (NRL). This was followed by funded proposals with Mike Sulzer as the PI, to the three institutions with the NSF providing $1.22M, AFOSR $942K and NRL via the Office of Naval Research $442K. The estimated yearly operating costs were close to $300K. As part of Carlson's negotiations with AFOSR and NRL, the two institutions also contributed $250K each per year to the HF operations and other costs.

The electrical design of the sub-reflector and radiating dipoles was done by Prof. James Breakhall of Penn State University (PSU) under a sub-contract with Cornell. A small engineering firm located near PSU, STARR-H, did the detailed mechanical design. Critical aspects of this design, especially ones related to telescope and staff safety, were reviewed by NAIC's consulting engineering firm, Ammann & Whitney, resulting in some modifications of the design. In early 2011 STARR-H was awarded

Fig. 12.21 The completed transmitter building with the transmitters installed. The lines to the heat exchanges still need to be connected. Credit: Cornell University/NSF

a $369,478 contract to produce all the components of the sub-reflector and radiating dipoles. Work at STARR-H on fabricating the components extended beyond the end of Cornell's management of NAIC on 30 September 2011. However, Cornell continued to monitor STARR-H's work and final delivery of all the components to the Arecibo Observatory was made in late June, 2012. The towers to hold the dipoles, basically cell towers, were procured separately under design specifications by STARR-H.

The six HF transmitters were located in a large pre-fab building constructed at the end of the observatory's maintenance complex closest to the telescope's reflector rim and assembled by observatory maintenance staff. The transmitters were installed in the building in 2009 (Fig. 12.21) and subsequent work involved construction of a control room, installation of cooling systems and the building of a sub-station and switches to provide the high voltage prime power to the transmitters (Fig. 12.22). Actual power depended on the installation of the four 1 MW diesel power generators and this was accomplished in October, 2010.

Bringing the system into operation took longer than expected. The six dipoles needed some remedial work, installing the long coaxial feeder lines to the dipoles from the transmitters was an arduous task and getting the transmitters fully operational with limited manpower was difficult. However, it was finally done (Fig. 12.23) and the first regular observations took place in June 2015.

Fig. 12.22 The building constructed to house the HF transmitters. The transmitter cooling system heat exchanges can be seen across the road to the left of the building and the transmitter power supply sub-station is next to the building. The large building in the foreground houses the observatory maintenance shops. Credit: Arecibo Observatory/NSF

Fig. 12.23 The completed HF heating facility showing the mesh sub-reflector and the six transmitting dipoles above the center of the primary reflector. Credit: Michael Sulzer and Arecibo Observatory/NSF

References

Bernhardt, P.A., et al. 2003. Using Radio-Induced Aurora to Measure the Horizontal Structure of Ion Layers in the Lower Thermosphere. *Journal of Geophysical Research* 108: 1-1-1-11.

Campbell, Donald B., Gregory J. Black, Lynn M. Carter, and Steven J. Ostro. 2003. Radar Evidence for Liquid Surfaces on Titan. *Science* 304: 431–434.

Chesley, Steven R., Steven A. Ostro, et al. 2003. Direct Detection of the Yarkovsky Effect by Radar Ranging to Asteroid 6489 Golevka. *Science* 302: 1739–1742.

Dowd, A., W. Zisk, and J. Hagen. 2000. WAPP - Wideband Pulsar Processor, Pulsar Astronomy 2,000 and Beyond. In *ASP Conference Series*, ed. M. Kramer, N. Wex, and R. Wielebinski, 202. ASP.

Edmundson, K and L. Baker. 2001. Photogrammetric Measurement of the Arecibo Primary Reflector Surface. In *17th Annual Coordinate Measurement Systems Committee Conference*, August 13–17.

Friedman, J.S., et al. 2000. Simultaneous Atomic and Ion Layer Enhancements Observed in the Mesopause Region Over Arecibo During the Coqui II Sounding Rocket Campaign. *Geophysical Research Letters* 27: 449–452.

Friedman, Jonathan S., Stephen C. Collins, Ruben Delgado, and Paul A. Castleberg. 2002. Mesospheric Potassium Layer over the Arecibo Observatory, 18.3°N 66.75°W. *Geophysical Research Letters* 29 (5): 15-1-15-4.

Gibson, S.J., K.A. Douglas, C. Heiles, E.J. Korpela, J.E.G. Peek, M.E. Putman, and S. Stanimirovic. 2008. Mapping Hydrogen in the Galaxy, Galactic Halo, and Local Group with ALFA: The GALFA-H I Survey Starting with TOGS. In *The Evolution of Galaxies Through the Neutral Hydrogen Window*, vol. 1035, 249–251. AIP Conference Proceedings.

Haynes, M.P., R. Giovanelli, et al. 2018. The Arecibo Legacy Fast ALFA Survey: The ALFALFA Extragalactic HI Source Catalog. *Astrophysical Journal* 861: 1–19.

Hofgartner, Jason D., et al. 2020. The Root of Anomalously Specular Reflections from Solid Surfaces on Saturn's Moon Titan. *Nature Communications* 11: 2829.

Korpela, E.J., et al. 2011. Status of the UC-Berkeley SETI Efforts. *Proceedings of the SPIE* 8152.

Malkan, Matthew A., Varoujan Gorjian, and Raymond Tam. 1998. A Hubble Space Telescope Imaging Survey of Nearby Active Galactic Nuclei. *Astrophysical Journal Supplement Series* 117 (1): 25–88.

Margot, J.-L., et al. 2001. Radar Observation of Binary Asteroid 2000 DP107, 1754, Abstract. In *Lunar and Planetary Science Conference XXXII*. NTRS.

Momjian, Emmanuel, Jonathan D. Romney, Christopher L. Carilli, and Thomas H. Troland. 2003. Sensitive VLBI Continuum and H I Absorption Observations of NGC 7674: First Scientific Observations with the Combined Array VLBA, VLA, and Arecibo. *Astrophysical Journal* 597 (2): 809–822.

Ostro, Steven J., et al. 2006. Radar Imaging of Binary Near Earth Asteroid (66361) 1999 KW4. *Science* 314: 1276–1280.

Peek, J.E.G., A. Begum, K.A. Douglas, S.J. Gibson, J. Grcevich, C. Heiles, E.J. Korpela, M. Lee, M. Putman, D.R. Saul, and S. Stanimirovic. 2010. The GALFA-HI Survey. In *The Dynamic Interstellar Medium: A Celebration of the Canadian Galactic Plane Survey, ASP Conference Series*, ed. R. Kothes, T.L. Landecker, and A.G. Willis, vol. 438. Astronomical Society of the Pacific.

Raizada, Shikha, and Craig Tepley. 2002. Iron Boltzmann Lidar Temperature and Density Observations from Arecibo - An Initial Comparison with Other Techniques. *Geophysical Research Letters* 29 (12): 1-1-1-4.

Staveley-Smith, L., W.E. Wilson, T.S. Bird, M.J. Disney, R.D. Ekers, and K.C. Freeman. 1996. The Parkes 21 cm Multibeam Receiver. *Astronomical Society of Australia* 13 (3): 243–248.

Chapter 13
Difficult Times

Despite the research opportunities opened up by the Gregorian upgrade as discussed in Chap. 12, after a few short years of relative calm, NAIC entered into an extended period of management and budgetary turbulence. New directors for both NAIC and the Arecibo Observatory were appointed. NASA cut off funding for the planetary radar program. The NSF's 2006 Senior Review of NSF supported facilities recommended that NAIC's budget be significantly reduced, resulting in the closure of the Ithaca based instrumentation laboratory. This loss of funding threatened the planetary radar program leading to considerable "push back" from the planetary research community and from Congress although without immediate effect. In 2009 there was some good news, a total of $3M in infrastructure funds was allocated by the Puerto Rican government, and the NSF provided $3.1M under the American Recovery and Reinvestment Act. At about the same time, NASA restored funding for the planetary radar program at a level of $2M per year. Concurrently, the NSF carried out a competition for the management of the observatory and eventually appointed SRI International as the new management institution. After almost 50 years, Cornell University's management ended on 30 September 2011.

13.1 Management Changes

The years 2003 to 2008 were difficult ones for NAIC and the Arecibo observatory due to a series of management issues. In early 2002 Paul Goldsmith stepped down as director of NAIC and a search for a new director was initiated. Don Campbell stepped in as interim director until a new appointment was made. The search committee recommended that Robert Brown, a radio astronomer with the National Radio Astronomy Observatory (NRAO) since 1969 and its then deputy director, be appointed to the position. Brown accepted and took up the position in early May 2003. This must have been a difficult transition for Brown as he had little experience of academic life after 34 years with NRAO, and he was the first NAIC director not to

D. Campbell, *The Arecibo Observatory: A History of Innovation and Discovery*, Historical & Cultural Astronomy, https://doi.org/10.1007/978-3-031-75225-4_13

be a member of the permanent Cornell faculty. One of Brown's very early actions was a request to Daniel Altschuler that he step down as director of the Arecibo Observatory after over 11 years in the position. Altschuler remained on the observatory's staff until 2007, heading up a new observatory office devoted to the Public Understanding of Science (OPUS). Sixto Gonzalez, the assistant director for Space and Atmospheric Sciences (SAS), was appointed director, the first native born Puerto Rican to hold the position. In 2004 Eddie Castro, the head of the Electronics Department, resigned and in 2006 José Maldonado, the Assistant Director for Facilities, also resigned, very significant losses to the observatory. Ganesh Rajagopalan replaced Castro as head of electronics and Felipe Soberal, an engineer who was a longtime assistant to Maldonado, was appointed as Assistant Director of Facilities.

Sixto Gonzales's 3 year appointment as director was a difficult period for him due to management issues. His stepped down after the 3 years reverting to being Assistant Director for SAS, a position he had maintained while serving as the Director. On 26 September 2006, Timothy Hankins, a former Arecibo staff member and Emeritus Professor at the New Mexico Institute of Technology, was appointed as interim director while a search for a permanent one was made. Robert Kerr, an atmospheric scientist who had done his PhD thesis research at Arecibo and was then on the staff of the NSF, was appointed director as of 26 February 2007.

Robert Brown's appointment as NAIC Director ended in May 2008 and he left Cornell. Don Campbell was appointed to the position as of 1 June 2008. This did not sit well with Robert Kerr as he felt strongly that the observatory should be entirely managed from Arecibo with Cornell just providing the needed support services such as payroll. In other words, the Arecibo director should also be the Principal Investigator on the NSF Cooperative Agreement for the operation of NAIC, the role of the director of NAIC. One problem was that it would effectively cut ties with the Cornell Arecibo related faculty, almost all of whom had been on the observatory's staff or had done their PhD thesis research at Arecibo. Most of them were supervising graduate students working with Arecibo data. It would also likely deprive Arecibo of the expertise provided by the engineering faculty, notably Anthony Ingraffea and William McGuire who, although retired, still provided valuable advice on engineering issues. Robert Kerr submitted a letter of resignation in July 2008 and departed on 1 September 2008. Michael Nolan, the head of both the planetary radar group and technical services at the observatory, was appointed to succeed Kerr, initially on an interim basis and then permanently. Nolan remained director until 30 September 2011, the end of Cornell's management of NAIC. At Cornell, David Howe, the Administrative Director of NAIC, resigned to take another position with the university. After a search, James Blair, the former Commander of Cornell's Air Force Reserve Officer Training Course (ROTC) program and Professor of Aerospace Science, was appointed to the position. Over the next 3 years, Jim Blair made significant contributions to the management of NAIC and the Observatory.

13.2 A Tremendously Sad Event

On 23 June 2003, three students went on a hike down to the Rio Tanamá, a relatively small and mostly shallow river except after heavy rains, which flows in a deep valley immediately south of the telescope. One section of the river flows through a cave under some high ground, so its southern side, which is relatively flat, can be accessed via a path over this high ground that leaves from the observatory grounds (see Fig. 9.12 of Chap. 9). Apparently, two of the students, Kristopher Reilly from New College, Florida, who was spending the summer at the observatory under the NSF sponsored Research Experience for Undergraduates (REU) program, and Colin (Mike) Ewers from Carlton College, who was visiting the observatory along with another student and a Carlton faculty member, attempted to cross the river just above a waterfall. In unknown circumstances, both students disappeared over the waterfall. The other student from Carlton College was a short distance up river and noticed that they had disappeared. After a fruitless search for them, she made her way to the observatory and raised the alarm. A search by local rescue teams over the next 2 days eventually recovered the bodies of both students.

This was the most traumatic event in the observatory's history up to that time. Harold Craft, former observatory director and Cornell Vice President for Administration and Chief Financial Officer, Robert Brown, counselors from the University's health service and others travelled to Arecibo to help as much as they could. Counseling was provided for the other summer students who were at the observatory, and Robert Brown organized a memorial service at the Visitor Center. However, nothing could soften the reality of the loss of the two students.

13.3 Budgetary Issues and the Senior Review

Despite the impressive results related to near-Earth asteroids that the new planetary radar system on the Arecibo telescope was obtaining, in early December 2001 Don Campbell, then Associate Director of NAIC and head of the planetary radar program at Arecibo, received a letter from the NASA program manager canceling NASA's $500,000 per year support for the Arecibo planetary radar program as of 1 January 2002, only 1 month later. The root cause of the problem seemed to be the transfer of Arecibo's NASA funding from NASA's Planetary Astronomy Program to the then newly funded Near-Earth Object Observations Program, which had been created after the US Congress in 1998 directed NASA to find at least 90% of all near-Earth asteroids larger than 1 km by 2008. Asteroid searches were done by institutionally owned optical telescopes not NASA ones and it took some time for the program to ramp up and utilize all of the approximately $3M per year that NASA had budgeted. It is likely that the Arecibo program was transferred to make use of these funds. Given the congressional mandate, NASA regarded the search programs as more important than the characterization observations done with Arecibo. Once the search programs ramped up, they needed the entire program budget leading to the cancellation of the Arecibo funding. Officially, the cancelation was related to NASA's support of ground-based facilities not being part of its mission.

Once the news of the cancellation of the Arecibo funding was public, there was a significant reaction by the planetary community especially the Division of Planetary Sciences of the American Astronomical Society (DPS) and the Planetary Society.[1] The DPS statement was signed by its Chair, Wesley Huntress, a former NASA Associate Administrator for Space Science. This bought the matter to the attention of Colleen Hartman, the head of NASA's Solar System Exploration Division, who contacted Campbell and told him that NASA would restore $400,000 in funding for the 9 month period of 1 January to 30 September 2002 to change from calendar year funding to the Federal fiscal year one starting on 1 October. NASA would then provide $500k for FY2003 and phase out its funding over the following 2 years. The NSF agreed that NAIC could continue to support the planetary radar program using NSF funds provided that it could be done within the existing budget.

NASA's cancellation of its support for the Arecibo planetary radar program was a prelude to more serious long term funding problems. The U.S. National Academy of Sciences' 2001 Decadal Report "Astronomy and Astrophysics in the New Millennium" recommended that, about every 5 years, the National Science Foundation astronomy facilities be competitively reviewed and prioritized based on past performance and future expectations. A single committee should conduct this review across all subfields of astronomy. New facilities should first be competitively evaluated between 5 and 10 years after they become operational and on the normal cycle thereafter. The committee went on to suggest that all NSF astronomy facilities be subjected to a competitive review, with prioritization and privatization or closure as possible options.

Acting on this recommendation, in 2005/2006 the NSF conducted a "Senior Review" of all of its facilities, primarily those managed by the National Radio Astronomy Observatory (NRAO), the National Optical Astronomy Observatory (NOAO), NAIC and the National Solar Observatory (NSO) but also including NSF supported university based facilities. As stated on the NSF Division of Astronomical Sciences website:

> This review examined the balance of our investments in the various facilities that we support. The primary goal of the review and the adjustment of balance that will result is to enable progress on the recommendations of the Decade Survey and other priorities. At the same time, we must preserve, indeed grow, a healthy core program of astronomical research. We regard this as essential to support the scientific programs that will be undertaken with the new facilities, to seed the next generation of capability, and to attract, train, and retain the next generation of astronomical researchers.

---

[1] A discussion of these events can be found at Space.com's WEB site. https://web.archive.org/web/20051128090915/. http://www.space.com/scienceastronomy/astronomy/arecibo_cuts_011220.html

The immediate mandate of the review committee was to recommend actions that would reduce the cost of the operations of the four National Centers by $30M per year by the 2011 fiscal year. This would free up funds to support new initiatives including the operating costs of the North American component of the Atacama Large Millimeter Array (ALMA). The committee took this charge seriously and graded the major instruments in optical/infrared, radio/millimeter/sub-millimeter and solar astronomy as either important to the long term future of each field and ones that could be phased out over the following years.

The report, "From the Ground Up: Balancing the NSF Astronomy Program" dated 22 October, 2006,[2] was not good news for NAIC. It stated that only ALMA, the 100 m Green Bank Telescope in West Virginia and the Extended Very Large Array in New Mexico plus the university-based programs and technology support should comprise the Radio/Millimeter/Submillimeter base program. While recognizing Arecibo Observatory's past and current contributions to the field of astronomy and astrophysics and being very supportive of the potential of the ALFA surveys, the committee was not persuaded of the primacy of the science program beyond the end of the decade and found that the case for long term support at the present level was not as strong as that for other facilities. As a result, the report recommended that Arecibo's budget from the NSF division of astronomical sciences be reduced over the following 3 years from the then current $10.4M to $8M, the minimum needed to operate the program. The $1.9M provided by the NSF division of atmospheric sciences in FY2007 would be in addition to this. The report also recommended that 80% of the astronomy program's telescope time be reserved for finishing the ALFA surveys and 20% for non-survey individual programs in order to preserve some "discovery potential." After the 3 years, the report stated that the observatory should either be closed or operated on a greatly reduced NSF budget if alternative sources of funding could be found. There was some concern about the cost of closing the observatory leading to a recommendation that the NSF initiate an independent study of the close down costs.

Robert Brown, the NAIC Director since May 2003, responded very quickly to the Senior Review recommendations. In a 13 November 2006 "Notice" to NAIC staff he proposed a series of actions:

1. We will eliminate all AST-supported NAIC program elements that are not directly related to operation of the NAIC astronomy research program. Closure of the NAIC Ithaca shop is one example of this.
2. We will reduce the number of telescope observing hours for astronomy to approximately 3800 h/year. (Now it is 4800 h/year).
3. Nearly all astronomy observations on the telescope will be scheduled at night, leaving the days free for maintenance activities.
4. We will do telescope and instrument maintenance 8-h/day, 5-days/week (Monday-Friday).

---

[2]The Senior Review report can be found at https://www.nsf.gov/mps/ast/seniorreview/sr_report_mpsac_updated_12-1-06.pdf

5. Because 80% of the astronomy observations will be for survey observations, we will reduce the number of receivers available for astronomy observations from 13 to 6. This will make it easier to maintain the remaining receivers.
6. The survey observations require less local support. Accordingly, scientific and support services in many areas will be reduced.
7. We will encourage remote observations to reduce the number of on-site visitors requiring Observatory assistance. This will enable us to reduce the cafeteria hours, and trips to the airport, for example. Student visits for training still will be encouraged.
8. We will continue to operate the Arecibo Planetary Radar in FY2007 as we seek additional support for it from NSF and other sources.

These were, in general, sensible responses to the report. However, the closure of the "NAIC Ithaca Shop", the NAIC Ithaca Instrumentation Laboratory that had supported Arecibo for almost 40 years providing cryogenically cooled receivers, line feeds, the mini-Gregorian, etc. would be a major loss depriving the observatory of the ability to initiate similar large instrumentation upgrades in house.

Bob Brown's desire to move rapidly in response to the Senior Review report was driven partly by the need to paint the suspended structure of the telescope. The last full strip down and painting had been in the early 1970's prior to the banning of lead based paint. This meant that all the paint had to be stripped off the structure using enclosures and special procedures, a complex and expensive task that was estimated to cost about $2M. By closing the instrumentation laboratory and reducing staffing quickly, funds would be freed up to enable the painting project to proceed. This was in a sense looking to the future, the painting project was necessary for the observatory's long term future.

The other issue referred to in Bob Brown's "Notice" was the planetary radar program. The Senior Review report made no reference to this program in its recommendations. How was it to be paid for or would it have to be terminated? This caught the attention of the planetary community and on 13 November 2006 Guy Consolmago, the Chair of the Division of Planetary Science (DPS) of the American Astronomical Society (AAS), wrote a letter to Wayne van Citters, the head of the NSF Division of Astronomical Sciences, expressing concern and pointing out:

> There is in fact only one reference to radar in the entire 78 page document, and no mention at all of asteroids. But the Arecibo radar results are key to understanding near earth object sizes, shapes, and dynamics. Besides having a central scientific importance, both of themselves and as samples derived from the main asteroid belt, near earth asteroids may represent a significant hazard to Earth and also a potential source of future resources. To decommission one of our primary tools for studying them would deal a serious blow to both our science and our safety. The letter was incorporated in a 16 November 2006 DPS press release.[3]

On 14 November Bob Brown issued a statement[4] highlighting the importance of Arecibo's Planetary radar program by referencing a recent Nature paper on Arecibo's search for water ice on the Moon and a Science paper on the observations

[3]http://www.spaceref.com/news/viewpr.html?pid=21300

[4]http://www.naic.edu/~astro/NSFSR/Arecibo_Planetary_Radar_Statement.pdf

of the near-Earth binary asteroid KW4. He pointed out that the absence of any significant reference to Arecibo's planetary radar program in the report implied the absence of any future funding for the program. He also stated that the program would continue through 2007, but it was not clear what would happen after that. The DPS placed Brown's statement on its WEB site.

During 2007 the issue of the future of Arecibo including its planetary radar program drew significant attention. One of the strongest congressional supporters of action to protect Earth from future impacts of near-Earth objects (NEOs) was Representative Dana Rohrabacher. In a 25 June 2007 press release and an accompanying letter to Congressman Mark Udall, Chairman of the sub-committee on Space and Aeronautics of the Committee on Science and Technology, Rohrabacher requested a hearing "to Assess NEOs Threat".[5] He pointed out that while NASA was pursuing the discovery of Near-Earth Objects with sizes greater than 1 km, it was doing little to meet the objectives of the 2005 "George E. Brown Jr Near-Earth Object Survey Act" which required that 90% of NEOs 140 m and larger be catalogued within 15 years. He went on to say:

Taking the aforementioned factors into consideration, I respectfully request the Space and Aeronautics Subcommittee convene a hearing for the following purposes:

- To review the status of the Spaceguard program in terms of its ability to meet the goals of the Survey Act,
- To explore the impact the closing of Arecibo will have on current and future NEO programs,
- To ascertain what other resources may be brought to bear to achieve the object of the Survey Act,
- To determine future actions necessary to implement the Survey Act.

This brought the issue of the impact of the potential closing of the Arecibo to the attention of the Congress.

In an 18 September 2007 letter to Arden Bement, the Director of the National Science Foundation, Congressman José E. Serrano, whose district was in the Bronx, NY and who was born in Mayaguez, Puerto Rico, pointed out the importance of the Arecibo observatory for both its research programs and its public outreach and educational work including 25,000 school children visits per year. He ended the letter saying: It is my hope that you will reconsider this decision. However, please be assured that I will do everything in my power to ensure that federal funding does not fall below the critical level that would force the Arecibo Observatory to close. As a member of the House of Representatives Appropriations Committee, Serrano had considerable budgetary influence, something that he later exercised to restart NASA funding for Arecibo.

In a letter dated 2 October, the Governor of Puerto Rico, Aníbal Acevedo Vilá, wrote to the NSF Director, Arden Bement, expressing strong concerns about the impact of the Senior Review recommendations on the Arecibo observatory's

[5] http://www.spaceref.com/news/viewpr.html?pid=22937

research programs and pointing out that the NSF's budget was likely to significantly increase under the Obama administration's America Competes act rather than the premise of the Senior Review of a flat or decreasing budget. A similar letter dated 3 October 2007 was sent to Bement by Congressional Representatives Luis Fortuño, Dana Rohrabacher and Nydia Velázquez pointing out that by being on the forefront of basic research involving Near-Earth Objects, Space Weather and Global Climate Change, the Arecibo Observatory is an outstanding resource to Congress and to the American people. Neither of these letters seemed to have the desired effect.[6]

A hearing on NEOs was held in October 2007[7] and included testimony by Representative Luis Fortuño, the Resident Commissioner for Puerto Rico in the US House of Representatives and a future governor of Puerto Rico, Donald Yeomans, the manager of the Near-Earth Object program office at the Jet Propulsion Laboratory, and Don Campbell, representing NAIC and Cornell University. All three of these presentations directly addressed the importance of the Arecibo radar system for obtaining precise orbits of NEOs and characterizing their properties such as size, rotation state, mass, density and binary configuration. Representative Rohrabacher also gave a strongly worded statement on the importance of Arecibo. In his presentation Representative Fortuño referred to a Bill, HR 3737, that he had submitted on 3 October to the House of Representatives, that directs the National Science Foundation and NASA to work together to ensure continued full funding of the Arecibo Observatory and in particular, the radar. Unfortunately, the hearing did not immediately lead to renewal of NASA's funding for Arecibo's planetary radar program. It would be 2 years before NASA funding became a reality, just in time to mitigate the large reduction in NSF funding based on the Senior Review report.

From 13 to 15 March, 2007 the NSF held a management review of NAIC at the Arecibo Observatory. This review covered all aspects of NAIC activities, management, science and public outreach with the primary question being whether the NSF should re-compete the management of NAIC when the current Cooperative Agreement expires. During his exit review presentation on 15 March, Richard Green, the Chair of the committee, gave his overall impressions as recorded by Professor Martha Haynes of Cornell:

(a) AO is unique, highly productive, enabling forefront science, truly meritorious of support.
(b) Staff is dedicated, capable, and hardworking. Impressed that staff remain motivated, even under recent very stressful conditions.
(c) Decision to invest in ALFA surveys is really paying off. Committee agrees with Riccardo Giovanelli's point that metric of success, even for a management review, should be scientific productivity, which in the case of Arecibo is at an all-time high.

[6]Both letters are in the Cornell University archives under Arecibo Ionospheric Observatory.

[7]All the testimony at the hearing can be found at: https://www.govinfo.gov/content/pkg/CHRG-110hhrg38057/html/CHRG-110hhrg38057.htmt

(d) Planetary radar program is unique for both surface topography and positional data.
(e) SAS is making a unique contribution by doing extremely important work very well.

On the issue of public outreach, the committee was extremely complimentary "The Visitor Center and EPO programs are the model and envy of other national centers."

These were much more complimentary statements than those contained in the Senior Review report and addressed the contributions of the Planetary Radar and Space and Atmospheric Sciences programs.

On the issue of re-competition of the management of NAIC, the committee was clear that the panel will recommend to the NSF that Cornell be offered the opportunity for an uncompleted renewal provided that by the end of CY 2007 Cornell delivers to NSF/AST:

I. A plan for how CU would operate AO/NAIC in FY2011 given a series of a projected base budgets, ranging from realistically low to considerably higher.
II. A more detailed and specific strategic plan for pursuing additional support beyond the NSF base support.

Due to a widely held belief by the observatory staff and throughout Puerto Rico that the observatory might be shut down in the near future plus strongly felt concerns at the NSF that Cornell was not actively pursuing alternative sources of funding, Wayne van Citters, the director of the NSF Division of Astronomical Sciences, made a 4 day visit to Puerto Rico from 25 to 29 July 2007. He was accompanied by several NSF staff including Dana Lehr, NAIC's program officer, and by several Cornell NAIC related persons including Joseph Burns, the Vice Provost for Research for Science and Engineering, a well known solar system dynamics expert, and Robert Brown, the NAIC director. As he stated in a memorandum dated 17 July 2007, van Citters organized the visit in order to (1) to provide the public and AO staff with accurate information regarding SR recommendations and NSF planning for AO, including assurance that closure would not occur prior to 2011, and (2) to meet with officials from the PR government, universities and private foundations to brainstorm about potential new partnerships that could lead to shared funding of AO operations in the post-2011 timeframe.[8]

On 26 June, a Forum was held in Arecibo attended by about 200 people including the mayor of Arecibo, the president of the Puerto Rican Senate, Kenneth McClintock, and representatives of the University of Puerto Rico. McClintock offered the possibility of including 3 million dollars for the Observatory in a bond issue; this did happen in 2008. The forum was followed by a meeting with the staff of the Arecibo observatory to clarify the process and timeline for the implementation of the Senior Review recommendations. On 27 June there was a lunch in San Juan with Cornell University alumni in Puerto Rico followed by perhaps the most important meeting of

[8] Wayne van Citters Memorandum dated 17 July 2007 and addressed to the NSF Assistant Director for Mathematical and Physical Sciences (MPS). Cornell University Archives.

the visit, which included the President of the PR Senate, the Executive Director of the Angel Ramos Foundation (which provided funding for the Visitor Center and Learning Center at AO), and representatives from the PR Governor's office, the PR Federal Affairs Administration, the PR Secretary of Labor, the Office of the Resident Commissioner, the Mayor of Arecibo and local business interests, UPR, and the PR Tourism Company. After much discussion, the possibilities for raising funds included Federal programs for infrastructure and education, such as the U.-S. President's High Job Growth Training Initiative, contributions or matching funds from private industry to support research projects with attribution, partnerships with the growing aerospace industry in north-central PR, water and power assistance to reduce operations costs, and a rebalancing of AO activities towards engineering and atmospheric sciences.

The following day there was a meeting of the NSF attendees with Manuel Gomez, Professor of Physics and Director of the Resource Center for Science and Engineering at the University of Puerto Rico's Rio Piedras (San Juan) campus. As stated in van Citters memorandum, Gomez indicated that he had wanted stronger collaborative ties with the observatory in the area of astronomy but had not been successful. He also noted that if UPR was part of the observatory's management organization, this would give the possibility of university and government financial support.

In order to build on the momentum of the NSF and Cornell meeting in Puerto Rico, van Citters authorized NAIC to set up an office at the observatory and hire two new staff members to pursue possible funding opportunities from Island institutions including the local government. However, van Citters' idea of raising operating funds within Puerto Rico ignored the realities of the Island's situation. The Island was in a serious financial situation resulting from a loss in manufacturing jobs and the migration of professional and young people to the United States. The government was borrowing money via bond issues to cover its operating costs, a situation that imploded a few years later with the Puerto Rico government and its independent agencies such as the Electric Power Authority in debt by about 70 billion dollars, equal to its gross national product, and no clear way to pay it. The NSF's budget was about two thirds that of the Puerto Rican government, which had the responsibility of looking after the Island's over 3 million people. Given this, it was clear that the PR government was not going to commit to operating funds for Arecibo. It did make a gesture via the $3M one off grant for infrastructure improvements obtained as part of a bond issue thereby increasing the Island's debt. Similarly, the University of Puerto Rico, which is almost entirely funded by the PR government, was in no position to contribute significantly to the observatory's budget. It was more likely that they would have been hoping that some NSF funding would come their way as a result of being part of the observatory's management.

This reality was made clear in a 2 October letter from the Governor of Puerto Rico, Aníbal Acevedo Vilá, to the NSF Director, Arden Bement, in which he pointed out that the NSF's budget is being augmented by the Congress and that, despite Puerto Rico's fiscal problems, the government is providing $3M for infrastructure improvements at the Observatory. He went on to say that "the NSF bears the responsibility for managing and funding facilities such as the Arecibo

Observatory—not state or local governments." In other words, Puerto Rico would not be contributing to the Observatory's operational costs.[9]

On 13 and 14 September 2007, Bob Brown and others organized a meeting "Frontiers of Astronomy with the World's Largest Radio Telescope" also known as the "Arecibo Futures Meeting" in Washington D.C. close to the Capital with the concentration being on plans for future research opportunities in radio astronomy, planetary studies and ionospheric physics. Invitations to the session on Near-Earth Objects (NEAs) were sent to Congressmen Mark Udall and Dana Rohrabacher, chair and member of the House Subcommittee on Space and Aeronautics, respectively, and to Richard Obermann, Staff Director of the sub-committee plus other staffers. The speakers for this session were Donald Yeomans and Steven Ostro of JPL and William Bottke of the South West Research Institute, all experts on NEAs with Steven Ostro being the leading expert on radar studies of NEA's. There were over 70 attendees discussing future research plans and possibilities for Arecibo in astronomy over the 2 days of the meeting, which ended on an optimistic note despite the funding outlook.

Andrew Lawler wrote an article in Science[10] on the meeting with emphasis on the potential loss of Arecibo's radar studies of NEAs. He quoted Congressman José Serrano as saying, "I will be monitoring its (Arecibo's) funding situation and making sure that the correct decisions are being made." Given the location of the Futures meeting, Wayne van Citters was clearly concerned that the Cornell administration was looking for congressional support. Lawler quoted him as saying, "To involve Congress in one aspect of it (the recommendations of the Senior Review) is not a productive way to go."

The outlook for Arecibo's future was given a significant boost in the report of the Radio, Millimeter and Sub-Millimeter panel of the Astronomy 2010 Decadal Survey, New Worlds and New Horizons. Quoting the report,[11] "Since the time of the Senior Review, developments in the areas of transient astronomy and astrometry have dramatically changed the context in which these instruments (Arecibo and the Very Long Baseline Array) are viewed. Because of the importance of Arecibo for pulsar timing and galaxy evolution, the panel recommends restoring $2 million per year to its baseline budget." It is not clear if this affected the NSF AST budget for 2011. The Senior Review had recommended a drop to $4M and the actual AST funding was $6M, so perhaps it did.

Ten years later, the view of the importance of Arecibo by the Radio, Millimeter and Sub-millimeter panel of the Astronomy 2020 Decadal Survey, Pathways to Discovery in Astronomy and Astrophysics in the 2020s, was stronger than that of

[9]The letter from the Governor to the NSF director is in the Cornell University Archives under Arecibo Ionospheric Observatory.

[10]Science Vol. 317 21 September 2007, page 1663.

[11]The Astro 2010 report of the Radio, Millimeter and Sub-Millimeter panel can be found at: https://nap.nationalacademies.org/read/12982/chapter/13

the 2010 panel,[12] "The RMS panel views continued operational support for Arecibo and the GBT over the next decade as essential." The report was issued in July 2020, just months before the collapse of the telescope.

13.4 Budgets and a Reprieve

The Senior Review recommended that NAIC's NSF AST budget starting with FY2008 be successively reduced to $8M in 2010 and then to $4M in 2011. In practice, from FY2006 to FY2008, NAIC's budget from the NSF was relatively stable with $10.4M from NSF AST for each of these years plus about $1.9M from NSF ATM. In 2009, AST dropped to $9.6M and ATM increased to $2.2M. In 2010 AST dropped again to $8.4M and ATM increased to $2.35M. The promised drop in AST funding to $4M in 2011 did not happen, but it did drop to $6M. ATM funding increased to $3M for a total of $9M.

Despite a significant reduction in staff and the closing of the Ithaca based instrumentation laboratory, it was not possible to operate the observatory and support all three of the research fields on $9M a year. Just in time, Representative José Serrano came to the rescue by inserting into NASA's FY2010 budget a requirement that $2M be allocated to support the planetary radar program at the Arecibo Observatory. The then director of NASA's Planetary Science Division, James Green, and the head of the NEO Observation Office, Lindley Johnson, had been supportive of providing funding for Arecibo for some time, but the NASA administration was reluctant to provide operational funding for an NSF facility. With the congressional directive overriding this, it was suggested that Cornell submit a 5-year proposal for about $2M per year. A proposal was submitted to NASA in March of 2010 requesting approximately $2M per year for 5 years. It was favorably reviewed and the requested amount was awarded for calendar year 2010 with similar amounts for the succeeding 4 years. These funds could be used not just for the operation of the S-band planetary radar system but for general observatory operations that supported the planetary radar program thus relieving the NSF provided budget of some of these costs. In later years NASA greatly increased its funding support.

Additional good budgetary news was the allocation of $3.1M to NAIC under the Obama administration's "American Recovery and Reinvestment Act (ARRA)," a 787 billion dollar stimulus in response to the 2008/2009 economic crisis, and the allocation of the promised $3M in Puerto Rican bond funds to the Observatory for infrastructure improvements. This $6.1M allowed badly needed infrastructure improvements which were not feasible under the reduced NSF operations budget

[12] The Astro 2020 report of the Radio, Millimeter and Sub-Millimeter panel can be found at: https://nap.nationalacademies.org/read/26141/chapter/23

and also relieved the operations budget of some costs for critically needed infrastructure items such as new klystrons for the planetary radar transmitter.

The ARRA was passed by Congress and signed by President Obama on 17 February 2009. These funds were for infrastructure improvements and could not be used for operations but the operations budget included critically needed improvements to the observatory's infrastructure and these costs could be transferred to the stimulus budget. As required, a proposal was submitted to the NSF in June 2009 resulting in the $3.1M award in August 2009. The proposal requested funding for environmental and energy saving projects such as new air conditioning systems for the operations and office buildings, a telescope structural condition survey by Ammann & Whitney, the acquisition of a 12 m phase referencing antenna for VLBI observations and three new klystrons for the 2.38 GHz planetary radar transmitter.

The bond issue that contained the $3M for infrastructure improvements at the Arecibo Observatory was approved by the Puerto Rican legislature in September 2008 and an assistance agreement was signed by Cornell University with the Puerto Rican Infrastructure Financing Authority (PRIFA) in September 2009. It laid out an initial set of projects for which the bond funds would be used. Included were the purchase of four 1 MW diesel generators to power the S-band transmitter and the new HF transmitting facility replacing the aging turbine generator installed in the mid-1990s plus the construction of a building to house the generators; the painting of the telescope's cable system and towers; the purchase of a 12 m antenna primarily for use as a phase reference for VLBI observations, making VLBI observations much more efficient; the construction of an engineering office building if sufficient funds were available. Since the 12 m antenna did not use any resources in Puerto Rico, its purchase was later transferred to the ARRA funds and substituted with the completion of the road to the top of the hill above the observatory's entrance gate on which the 12 m telescope was installed plus the design work and some of the costs for a major expansion of the Visitor Center.

In other good news, the Puerto Rican Department of Education and the observatory's Visitor Center initiated an "Inspiration to Science" program in which over the next few years thousands of school children came to the Visitor Center on organized tours and class room activities in the Learning Center. The program's $2,352,800 budget was provided by the Education Department.

All of the infrastructure projects financed by the ARRA and Bond funds were completed with the exception of the actual construction work for the planned Visitor Center expansion and the painting of the cables and towers. Both of these ran up against the 30 September 2011 deadline, the date on which Cornell University ceased to manage the Arecibo Observatory. The cable painting was delayed due to a structural issue with the telescope that took considerable time to resolve (see below). The 12 m antenna was installed (Fig. 13.1) and its access road completed. Very importantly, the four 1 MW diesel generators were acquired and installed in a new building providing considerably improved reliability for the planetary radar observations.

Fig. 13.1 The 12 m antenna erected on a hill above the entrance gate to the Observatory. Credit: Cornell University/NSF

13.5 A Welcome Distraction

On 27 May 2008, the observatory was caught up in US politics; former President Bill Clinton came to visit (Fig. 13.2). Hillary Clinton was running for president and was a candidate in the Democratic primary in Puerto Rico to select the delegation of the Democratic party in Puerto Rico at the upcoming Democratic party convention where the democratic party candidate for president would be selected. Bill Clinton was in Puerto Rico to campaign for Hillary Clinton. US citizens resident in Puerto Rico, which includes almost all Puerto Ricans and any other US citizen living in Puerto Rico, cannot vote in US presidential elections. However, Puerto Rico does send delegations to the National Conventions of both parties where they can vote for their preferred candidate.

13.6 Infrastructure Projects

Visitor Center expansion: Cornell engaged Luis Badillo, who did the designs of the original Visitor Center and Learning Center, of the architectural firm Mendez, Brunner and Badillo to design an expansion of the Visitor Center that included a planetarium and a conference room or learning area for school children on a lower floor. The final design (Fig. 13.3) was completed in August 2010 with an estimated design, construction and outfitting cost of about $2.8M. Using Puerto Rican Bond funds and accumulated Visitor Center funds, since the Visitor Center was owned by Cornell University its funding was independent of any Government funding,

Fig. 13.2 Welcoming former President Bill Clinton to the observatory. The observatory staff are assembled on the steps of the viewing platform in front of the Angel Ramos Visitor Center. Credit: Cornell University

approximately 50% of the needed funding was immediately available. It was hoped that the Angel Ramos Foundation would contribute significantly to the cost of the expansion of the Angel Ramos Visitor Center. Unfortunately, the competition for the management of NAIC/Arecibo Observatory and the consequent end of Cornell's management on 30 September 2011 ended Cornell's plans for the expansion. The succeeding management organization did not pursue the project.

Telescope structural failure: In February 2010, one side of a diagonal truss member in the triangular platform failed. Ammann and Whitney (A&W), NAIC's consulting engineering company since 1972, was immediately contacted and a temporary fix put in place under their instructions until a permanent one could be designed and installed. There are six identical trusses symmetrically placed relative to the three apexes of the triangular platform. None of the other five showed signs of problems, leading to the possibility that there could have been a small defect in the member. In May 2010, A&W reported on a finite element analysis of the structure which indicated that these six members could be over stressed and that two nearby members and one connection could be slightly over stressed. A program to reinforce these members was carried out over the following months.

Fig. 13.3 The final design of the Visitor Center expansion showing the planetarium "dome,' which had seating for 80 people, and the expanded museum and store space. The bottom floor had a conference room that could seat 205 people or 100 in a workshop arrangement plus a gathering area for social events. The tilted to the south top of the planetarium is covered in solar panels. The existing building is in grey. Credit: Architect Luis Badillo and Cornell University

Telescope structural condition survey: NAIC's consulting engineering company, A&W, carried out a detailed inspection of the suspended structure of the telescope including the cable system on approximately 5 year intervals. The May 2003 inspection and report indicated no significant problems with the structure or the cable system including the cable sockets. The following inspection was scheduled for 2009 but was delayed until December 2010 extending into early 2011 by the installation of a new fully approved fall safety system—the old safety lines had been removed during the painting project—and the reinforcement work resulting from the structural failure in early 2010. A&W's report was delivered in March 2011. Again, no major deficiencies were found and no changes since the 2003 inspection were observed with the main and auxiliary cables and their sockets.

13.7 The Management Re-Competition and Its Outcome

Despite the recommendations of the March 2007 NSF sponsored Management Review of NAIC, the NSF decided to recompete the management of NAIC. This was likely driven by Wayne van Citters concern that Cornell University was not being active enough in trying to raise additional funding for Arecibo operations and the general feeling in the National Science Board that the management of the national astronomy centers should be recompeted periodically. There had never been a change in the management of a national center. As by far the smallest of the centers, the only one managed by a single university and its future NSF funding very uncertain based on the recommendations of the Senior Review, NAIC was an obvious candidate for a change in management if the NSF was intent on showing that it was willing to do at least one change. Having a single university, Cornell, manage an NSF funded National Center had been a somewhat contentious issue dating from the formation of NAIC in the 1971. However, one big advantage was that it brought scientific and engineering support for the Arecibo observatory at minimal cost. There were many faculty members, research associates and graduate students who were active users of Arecibo in both astronomy and atmospheric sciences with the faculty members in particular contributing significantly to the planning of telescope and instrumentation improvements. On the engineering side, faculty members such as William McGuire over several decades and, in later years, Anthony Ingraffea were always available with much needed advice about telescope structural issues.

The current Cooperative Agreement, which began on 1 October, 2005, had an expiration date of 31 March 2010. In January 2010 this was extended to 30 September 2010 and, later, to 30 September 2011. A Request for Proposals for the management and operation of NAIC was issued on 30 April 2010. Letters of intent were due 2 August 2010 and the full proposals on 2 October 2010. The anticipated NSF combined astronomy and atmospheric sciences budget for the 5 years of the new cooperative agreement started at $8.7M in FY 2012, declined to $8M in FY2013 and 2014 followed by a small increase to $8.2M in FY2015. NASA's $2M per year would augment this for a yearly budget close to $10M, still significantly below the $12.5M in 2008 especially when including inflation.

Cornell submitted its proposal by the 2 October deadline. A major change that it proposed for the management of NAIC was the creation of a Management Board to replace the Visiting Committee. The Board, which, like the Visiting Committee, would report to the President of Cornell, would have representatives from four US mainland universities, the University of Puerto Rico, the Albert Einstein Institute of the Max Planck Institute for Gravitational Physics in Germany and a Consortium of Canadian Universities. It had members with expertise in both astronomy and atmospheric sciences. Cornell as the PI institution would retain overall responsibility for the management of NAIC. Apart from their oversight responsibilities, all of the partner institutions had a direct interest in some aspect of the research programs at the observatory and, potentially, would station scientific staff at the observatory.

On Wednesday 11 May 2011 Dana Lehr, the NSF's Program Manager for NAIC, informed Don Campbell that Cornell University's proposal to manage NAIC and the Arecibo Observatory was not successful. This was not totally a surprise but nonetheless a considerable shock. On Thursday, there was a telephone meeting to formally convey the decision between Cornell representatives and NSF representatives including the NSF Assistant Director for Mathematics and Physical Sciences, the Assistant Director for Geosciences and Dana Lehr. That same day NAIC's staff in Arecibo and Ithaca were informed of the NSF's decision and that the new NAIC management organization would be California based SRI International. The changeover date would be 1 October 2011. SRI operated several NSF supported incoherent scatter radars and so had a group with research objectives similar to some aspects of the atmospheric sciences program at Arecibo.

On the Friday, SRI transmitted a message requesting that it be distributed to the Arecibo staff. In it they stated that SRI would be managing the observatory in partnership with the Universities Space Research Association (USRA) and Universidad Metropolitana (UMet, a private university in Puerto Rico now the Ana G. Mendez University). They also made it clear that Robert Kerr would be returning to the observatory as the director and that "all Arecibo Observatory management would reside on site." This was a distinct change from the Cornell model of the NAIC director as the Principal Investigator under the Cooperative Agreement with the NSF working from the Cornell campus and an Arecibo director reporting to the NAIC director.

This was the beginning of a very intense period for the NAIC and Cornell management dealing with all the issues that a change of NAIC/Arecibo management would entail. Numerous trips were made to Arecibo by Cornell and NAIC Ithaca based personnel to talk with the Arecibo staff, interact with NSF and the SRI partnership representatives and facilitate the Partnership's interactions with the staff regarding their possible future employment.

A major concern was the Arecibo and Ithaca NAIC staff, all of whom were Cornell employees. Would their jobs, especially for the staff in Arecibo, be maintained by the new partnership? There was no requirement in the NSF's Proposal Solicitation for the management of NAIC that all or some of the current staff be retained by a new management organization. Each of the partners, SRI, USRA and UMet would employ some of the staff with the possibility that staff benefits would vary depending on the employer. This was a significant change from everyone working for Cornell University. For Cornell, the immediate question was whether the management change was a transfer or, for Cornell, a closure. The decision determined whether or not the employees who were rehired by the Partnership were entitled to severance payments. It was finally decided, and the NSF agreed, that it was officially a closure as far as Cornell was concerned. This meant that, with the exception of those with academic appointments, all the staff would receive severance payments, a not inconsiderable sum as very many of them had been working for Cornell for several decades. In addition, the staff would also receive payments for any unused vacation time. Needless-to-say this was very much welcomed by the staff.

Another major issue was the Angel Ramos Foundation Visitor Center and Learning Center. They had been built with private funds raised by Cornell but were on U.S. government owned land and its staff was paid out of Cornell funds, primarily from the shop income and entrance fees. It was clear that Cornell could not continue to run them even if had wished to. The solution was to transfer them to the NSF and, after 30 September, they were incorporated into general observatory operations and managed by UMet under its partnership agreement with SRI. The

Fig. 13.4 Observatory staff, retirees and their families at the 16 September dinner under a tent next to the swimming pool. Credit: Cornell University; photo by the author

transfer was conditional on maintaining the visitor center's name, the Angel Ramos Foundation Visitor Center.

Three events were the culmination of the transfer of the management of Arecibo: A Remembering Bill Gordon event on 15 September, a staff appreciation dinner on 16 September and the final lowering of the Cornell and other flags, and departure of the Cornell based management on the afternoon of 30 September. Speakers at the Bill Gordon event included Miguel Feyjoo, the first engineer hired by Bill Gordon after his arrival in Arecibo in the summer of 1960; Tom Talpey, who came to Arecibo with Bill Gordon that summer as a receiver engineer; Herb Carlson, Bill Gordon's graduate student resident in Arecibo in the early 1960s and, later, a long time Arecibo staff member; Don Farley, Carl Heiles from UC Berkeley and Don Campbell. The staff appreciation dinner, which included the then current observatory staff and any retirees that could attend plus families (Fig. 13.4), took place under a large tent erected over the volley ball court next to the swimming pool. It was also attended by Robert Buhrman, Cornell's vice-provost for research, Harold Craft, Jim Blair, the NAIC administrative director, Don Campbell, and a number of other Cornell based NAIC staff and faculty. It was a happy but also sad affair with the uncertainty of the future hanging over everyone.

The last moments of Cornell's management on 30 September were certainly not happy ones. At about 4:30p Harold Craft, Jim Blair and Don Campbell plus a sizable number of the observatory's staff witnessed the lowering of the Cornell, Puerto Rican and US flags, which were taken back to Cornell. After handshakes and hugs, the Cornell University based people got into a car and drove out the gate.

Chapter 14
Epilogue

On 1 October 2011, SRI International took over the management of the observatory and Robert Kerr became the Observatory Director again as well as the Director of NAIC. However, in mid-2015, SRI relieved Kerr of his position as director of NAIC resulting in his resignation as observatory director and departure from Puerto Rico. Francisco Cordova was appointed as the new observatory director. In early 2017, the NSF issued a solicitation for the management of the observatory. This resulted in the University of Central Florida (UCF) being selected to manage Arecibo as of 1 April 2018. In 2013, the 50th anniversary of the observatory's inauguration on 1 November 1963, there was a celebration of its contributions to the fields of radio astronomy, planetary studies and ionospheric physics. Two natural events, an earthquake in 2014 and hurricane Maria in 2017 resulted in some damage to the telescope including the loss of the 430 MHz line feed. Despite a strong endorsement of the importance of Arecibo for radio astronomy and planetary studies by the 2012 NSF sponsored Portfolio Review, and NASA significantly increasing its support for the observatory's planetary program, the NSF continued to reduce Arecibo's budget, predicting that it would be only $2M per year at the end of UCF's five year management term. On 1 December 2020 the suspended structure of the telescope collapsed after progressive failures of the cables from tower T4 to the suspended structure.

Supplementary Information The online version contains supplementary material available at https://doi.org/10.1007/978-3-031-75225-4_14.

D. Campbell, *The Arecibo Observatory: A History of Innovation and Discovery*, Historical & Cultural Astronomy, https://doi.org/10.1007/978-3-031-75225-4_14

14.1 Observatory Management

The transition of the management of NAIC/Arecibo Observatory to SRI and its partners went relatively smoothly and most of the observatory staff were re-hired by SRI or one of its two partners. Robert Kerr replaced Mike Nolan as observatory operations director and was also the principal investigator (PI) on the NSF Cooperative Agreement (i.e. the NAIC Director). Zaven Arzoumanian, a former Cornell University graduate student who worked for USRA and was not resident in Arecibo, was appointed as head of the astronomy group. He was followed in March 2012 by Fernando Camilo, who was resident in Arecibo. Camilo left in June 2014 and was eventually followed in June 2015 by Joan Schmelz, who had done her PhD thesis research at Arecibo. Mike Nolan filled in during the intervening year. Sixto Gonzales remained as the head of the Space and Atmospheric Sciences (SAS) group until 2015 when he left the observatory and was replaced by Christiano Blum.

In mid-2015 Yuri Milner, a tech billionaire, initiated the Breakthrough Listen project funded with $100M to support the search for extraterrestrial intelligence, SETI. Included in the funding was the purchase of significant amounts of observing time on three large radio telescopes, the NSF's 100 m Green Bank telescope in West Virginia, the 64 m Parkes telescope in Australia and the Automated Planet Finder telescope in California. Arecibo, an obvious choice given its long history of SETI searches, was not included due, as reported in an article in Scientific American[1] (Billings 2015), to a disagreement with the NSF as to how much the Foundation would reduce Arecibo's funding if they received funds from the Breakthrough Listen project. Kerr was extensively quoted or paraphrased in the article. Having the PI on the Arecibo Cooperative Agreement in a public disagreement with the NSF apparently did not sit well with SRI, and Kerr was replaced as PI on the Cooperative Agreement by John Kelley, a longtime atmospheric scientist at SRI. Shortly after, Kerr resigned as Arecibo Operations Director. A brief description of the events can be found in a November 2015 Nature article (Watson 2015) and in a Nature editorial (2015). Francisco Cordova, a civil engineer who had been working with the Boeing Company, was appointed as the new operations director. Cordova was the second Puerto Rican to fill this position.

As discussed in detail below, on 25 January 2017 the NSF issued a solicitation for the management and operation of the Arecibo Observatory with a proposal due date of 25 April. The University of Central Florida (UCF) was awarded the new Cooperative Agreement and took over the management of Arecibo on 1 April, 2018. Ramon (Ray) Lugo, the head of the Florida Space Institute at UCF was the PI on the Cooperative Agreement and Francisco Cordova was retained as the Arecibo based Director of Operations. UCF operated the observatory in partnership with Yang Enterprises, a NASA contractor which employed all of the observatory's technical staff, and the Ana G. Méndez University, which continued its role in managing the

[1] https://www.scientificamerican.com/article/search-for-alien-life-ignites-battle-over-giant-telescope/

Fig. 14.1 Attendees at the 50th anniversary of the 1 November 1963 inauguration of the Arecibo telescope. Frank Drake is standing at the center with Herb Carlson to his left. Mike Nolan, the last Cornell director, is sitting in front. Don Campbell is fourth to the left of Frank Drake. Credit: Arecibo Observatory

Visitor Center, cafeteria and onsite housing. The reported budget from the NSF was \$20.15M over the five years of the cooperative agreement, an average of only \$4M per year. It was higher than the average in the early years and predicted to taper to ~\$2M in the last year.

14.2 A Celebration

1 November 2013 was the 50th anniversary of the dedication of the Arecibo telescope, an event to be celebrated. Several of the still living previous NAIC and Observatory directors and many others who had connections with Arecibo over the years attended (Fig. 14.1) plus, of course, the current staff of the observatory. A series of talks (Fig. 14.2) over two days in the Visitor Center auditorium recounted the history of the telescope including its upgrades and many of the scientific results obtained over the years in astronomy, planetary science, and atmospheric and space physics. The first day was capped with a dinner under a large tent next to the swimming pool, a longstanding tradition at the observatory.

14.3 Major Natural Events and Impacts on the Telescope

On 13 January 2014 at 00:01:05, a 6.4 magnitude earthquake about 80 km (50 miles) NNW of Arecibo caused substantial vertical motions as recorded by the Puerto Rican seismic network accelerometer located at the observatory. There was no damage to the telescope except for 11 cable wire strands pulling loose from the zinc splice block of one of the Tower 8 main cables. This splice block was installed at the time of the telescope's construction because one of the delivered main cables was about 9 ft short (see Chap. 5). To avoid waiting several months for a replacement cable, the short cable was spliced to a 9 ft section of cable and the splice

Monday [Oct28]	
Speaker	Topic
Breakfast	
Welcome + Key Note 1 - Rich Behnke	
Key Note 2 - Don Campbell	
Key Note 3 - John Matthews	50 Years of Radio Sciences at Arecibo: A Brief Overview
Break	
Min-Chang Lee	
Joanna Rankin	
Leonid Gurvits	Arecibo Observatory and space Radio Astronomy
Xinzhao Chu	
Key Note 4 - Alex Wolszczan	
Dan Wertheimer	
Frank Djuth	
Lunch	
Key Note 5 - Carl Heiles	
Mike Sulzer	
Allison Smith	
John Meriwether	
Jay Lockman	
Paulo Freire	
Paul Bernhardt	
Break	
Key Note 6 - Herb Carlson	
Ganesh	
Wolfram Freudling	The ALFA Ultra Deep Survey
Jim Breakall	
Jim Gould	Ammann & Whitney's History with the Telescope
A. J. Ferraro	Early Ionospheric Experiments at the Arecibo Observatory
Loris Magnani	The Life and Times of the Mini-Gregorian
Dinner	
Key Note after dinner - Sheldon Reynolds	

Tuesday [Oct 29]	
Speaker	Topic
Breakfast	
Key Note 7 - Frank Drake	
Asti Bhatt	
Marshall Cohen	
Rick Jenet	The Arecibo Remote Command Center: Changing lives, building leaders, and altering the socio-economic climate of South Texas
Dave Jauncey	
Abel Mendez	
Snezana Stanimirovic	GoldenEye's view of the ISM in the Milky Way
Break	
Key Note 8 - Lynn Carter	
Gordon Pettengill	Early Planetary Radar
Lance Benner	
Mike Nolan	
Jonathan Lunine	Reflections on Arecibo by a (Non-Radio) Astronomer
Amy Lovell	
Lunch	
Key Note 9 - Joe Taylor	
Anthea Coster	Measuring the rise and fall times of HF-induced irregularities at Arecibo in the 1980s
Robert Minchin	The Arecibo Galactic Environments Survey
John Noto	Innovating Airglow - Facilitating the Aeronomy Ecosystem
Tim Hankins	
Phil Kronberg	
Brett Isham	Plans for an HF imaging Array in Aguadilla, Puerto Rico
Break	
Key Note 10 - Karen O'Neil	
Carl Melis	
Dave Hysell	
Travis McIntyre	The ALFA Zone of Avoidance Survey
John Foster	
Andrea Lommen	Gravitational Waves at Arecibo
Yervant Terzian	

Fig. 14.2 The speakers at the 50th anniversary of the 1 November 1963 inauguration of the Arecibo telescope held in the Visitor Center at the observatory

encased in a zinc block. The strands that pulled out from the block were on the tower side of the block. Ammann & Whitney recommended that two steel beams be placed on the far side of the block and tied back to the steel tower saddle with long threaded rods. For unknown reasons, this temporary fix remained in place for a number of years, UCF ordered a new cable shortly before the collapse of the telescope.

On 20 September 2017 Hurricane Maria passed directly over Puerto Rico with the eye of the hurricane passing over Arecibo. Maximum average wind speeds as recorded by the telescope's platform weather station were close to 80 mph (~130 kph) with gusts up to 100 mph (160 kph). As the eye passed over the observatory, the wind speed dropped to about 10 mph (16 kph). These were by far the strongest hurricane winds to ever strike the observatory. Damage to the telescope was surprisingly small with the only major loss being the 96 ft (30 m) long 430 MHz line feed, which broke about 25% of its length below the carriage house with the pieces falling into the primary reflector. There was some relatively minor damage to the catwalk. The dome survived intact. However, it is possible that the wind induced extra load on one or more of the auxiliary cables may have initiated cable slippage in their sockets (see the section of the collapse of the telescope in 2020). No serious effort was made to rebuild the line feed despite many of the pieces still largely intact and the existence of Allan Love's original "cook book" giving all the information needed to build a new one. This limited incoherent scatter radar observations of the ionosphere to the use of the 430 MHz horn in the dome.

The hurricane caused serious damage to the electric power infrastructure and many buildings in Puerto Rico. This included the homes of many of the observatory's staff and a GoFundMe site was instituted by Nicholas White, USRA's Vice President for Science, to provide some assistance to those staff most affected. Many Arecibo related people sent food and other essential items for distribution. Much of this activity was coordinated through Arecibo staff member Angel Vasquez, an experienced amateur (Ham) radio operator, via his contacts with other Hams in the United States. Telephone and the internet were seriously disrupted especially in the countryside so for some time Angel provided a much needed means of communication.

Between 6 and 11 January 2020 there were seven earthquakes with magnitudes ranging from 5.0 to 6.4 at distances of 36 km to 54 km from the observatory on the south side of Puerto Rico. This unnerving sequence of quakes caused significant damage along the south coast. The telescope was spared any damage but, again, it is possible that the quakes exacerbated the issues with the auxiliary cable sockets (see discussion related to the collapse of the telescope).

14.4 National Science Foundation and Arecibo

In 2012, the NSF instituted another Portfolio Review as a follow on from the 2010 Decadal Survey, New Worlds New Horizons. Reviewing the NSF's RMS (Radio, Millimeter and Sub-Millimeter) portfolio, the Review report[2] recommended that the NSF continue support for the ALMA mm/sub-mm array in Chile and for the recently upgraded Very Large Array in New Mexico. For the remaining NSF RMS facilities, Arecibo, the GBT and the VLBA, it concluded that Arecibo was the most important for continued funding followed by the VLBA and then the GBT. The Executive Summary for the Report specifically recommended that just Arecibo be included in the NSF portfolio of funded telescopes as of 2017 although with the caveat that this be re-evaluated again at the end of the decade. Unlike the 2006 Portfolio Review report, the 2012 Report paid attention to studies of our solar system referencing the 2011 National Academies Planetary Decadal Survey, Visions and Voyages, which

[2] The Portfolio Review 2012 report can be found at: https://www.nsf.gov/mps/ast/portfolioreview/reports/ast_portfolio_review_report.pdf

included a strong endorsement of Arecibo's contribution to the study of asteroids and comets; "The Arecibo radio telescope is essential for detailed characterization of the shape, size, morphology, and spin dynamics of NEOs that make close approaches to Earth. These radar observations also provide highly accurate determinations of orbital parameters for primitive bodies critical to modeling and planning future exploration."[3]

Despite the Portfolio Report, on 26 October 2015 the NSF issued a "Dear Colleague letter: Concepts for future operation of the Arecibo Observatory"[4] which requested "community responses on viable concepts for the future continued operation of the Arecibo Observatory, specifically including strategies and goals for continued operations that involve a substantially reduced funding commitment from NSF." The options included the "possible transfer of title of the Observatory to another party". Responses were requested by 15 January 2016. It had been over four years since SRI was awarded the Cooperative Agreement for the operation of NAIC/ Arecibo Observatory and the NSF was clearly preparing for a new management competition. However, the Letter made it clear that any management agreement after the current one would come with a significantly reduced budget.

In another ominous move, in June 2016, the NSF initiated an Environmental Impact Assessment for Arecibo. This may sound benign, but such an assessment is a requirement in preparation for withdrawing funding and, perhaps, closing a federal government owned facility. However, shutting down the telescope would, inevitably, require that it be dismantled and, likely, returning the site to its natural state. The costs of doing this were in the range of $100M so keeping the telescope operational may have, at least in the short run, seemed the cheaper option given its NSF operating budget of about $8M per year.

On 18 April, James Green, the director of NASA's Planetary Science Division, wrote to James Ulvestad, the director of the NSF Division of Astronomical Sciences, pointing out the importance of Arecibo's planetary radar program to the study of near-Earth asteroids and committing NASA to continuing to fund the radar program at a level of about $3.6M per year. Adding to the pressure to keep Arecibo funded, on 21 June Manuel Fernós, the President of Inter American University (IAU), a multi-campus institution in Puerto Rico, sent a letter to France Cordova, the director of the NSF, pointing out the importance of Arecibo to STEM education in Puerto Rico and that IAU was involved with two projects, a cube sat and a radio array, both of which would be used to complement Arecibo's ISR radar and/or Arecibo's new HF modification facility.

[3]Chapter 4 of "Visions and Voyages for Planetary Science in the Decade 2013 to 2022 (2011)". The full report can be found at: https://nap.nationalacademies.org/catalog/13117/vision-and-voyages-for-planetary-science-in-the-decade-2013-2022

[4]The letter can be found at: https://www.nsf.gov/pubs/2016/nsf16005/nsf16005.jsp?WT.mc_id=USNSF_25&WT.mc

On 27 July 2017, the NSF released the 302 page Final Environmental Impact Study (FEIS).[5] As listed in the abstract, six alternatives for the future of Arecibo were considered and their environmental impact assessed. The alternatives were: "(1) collaboration with interested parties for continued science-focused operations (the Agency-preferred Alternative); (2) collaboration with interested parties for transition to education-focused operations; (3) mothballing of facilities; (4) partial demolition and site restoration; (5) complete demolition and site restoration; and (6) the No-Action Alternative. The environmental resources and concerns considered in the FEIS were biological resources, cultural resources, geology and soils, groundwater, hazardous materials, solid waste, health and safety, noise, socioeconomics, traffic and transportation, and visual resources." The NSF preferred alternative (1) of continued science-focused operations was not as benign as it sounds. It would involve "collaborations with new stakeholder(s) who would use and maintain Arecibo Observatory for continued science-focused operations. NSF would reduce its funding of the observatory and the new stakeholder(s) would be responsible for future upgrades." Up to 26 structures would be demolished including the administration (office) building, the new HF transmitter building and, puzzling, the S-band transmitter power supply building eliminating all planetary radar observations and, hence, any NASA funding. The other options involved even more radical changes to the observatory. In the end, it was option 6, the no action alternative, that prevailed at least until the collapse of the telescope on 1 December 2020.

On 30 September 2016, the NSF issued a Notice of Intent to solicit proposals for the Continued Future Operation of the Arecibo Observatory. On 25 January 2017, the solicitation was issued with a proposal due date of 25 April. It referred to a draft of the Environmental Impact Study, which included the six alternatives for the future of Arecibo with Alternative 1 the NSF's preferred alternative and stating that the "solicitation supports this Preferred Alternative." The NSF budget for the then current year was \$8.3M, \$4.2M from Astronomy and \$4.1M from Geosciences. NASA was contributing \$3.6M per year to support the planetary radar program. Anticipating a 1 April 2018 start for the new Cooperative Agreement, the projected NSF budget for the first year was \$7.150M reducing to \$2M in the fifth year for a total over the five years of \$20.150M. Based on James Green's letter of 2016, it must have been assumed that NASA would continue its support at about \$3.6M per year provided that the planetary radar program continued at an acceptable level. Even if the NASA funding continued, it was clear, as stated in the solicitation, that partners that would provide additional funds would be needed.

It is not clear how many proposals were received by the NSF. SRI quite possibly did not propose and USRA likely did independently. The award was made to the University of Central Florida (UCF), a state university, with a start date in April 2018. There was probably a hope within NSF that the state of Florida would eventually take over the title to Arecibo. As discussed earlier, UCF had partners, Yang Enterprises, a NASA contractor, would be in charge of maintenance and

[5] The FEIS can be found at: https://www.nsf.gov/mps/ast/env_impact_reviews/arecibo/eis/FEIS.pdf

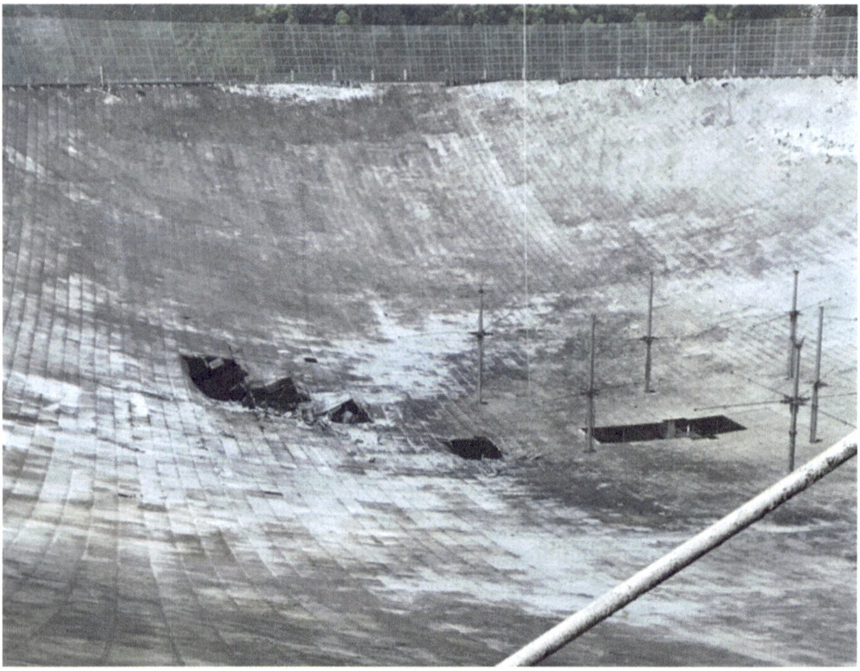

Fig. 14.3 The damage to the reflector from the fallen cable, which can be seen hanging from the platform as the vertical white line. About 250 panels were destroyed. The HF radiating dipole elements can be seen at the right. Credit: Arecibo Observatory

operations for the observatory and UMet (later the Ana G. Méndez University) would continue to operate the Visitor Center with it educational programs plus the observatory's cafeteria and housing. Almost all the staff with the exception of the scientific staff, would be employed by either Yang Enterprises or the Ana G. Méndez University. Management at UCF would be through its Florida Space Institute and its director, Ramon (Ray) Lugo, a former NASA Center director, was the PI on the Cooperative Agreement. Francisco Cordova was retained as Arecibo site director.

14.5 The Telescope's Collapse and Its Aftermath

At 02:36 in the morning of 10 August 2020 the north auxiliary main cable that was installed during the 1990s upgrade pulled out of its socket at Tower 4. It dropped on to the primary reflector causing considerable damage (Fig. 14.3) and ended up draped over the side of the dome damaging some of its panels. Despite some problems, it was possible to move the azimuth to get the dome away from the hanging cable. Based on advice from former staff of Cornell's engineering consulting firm Ammann & Whitney – A&W was now part of WSP, a Montreal, Canada based engineering firm - the dome was placed in the hurricane survival position and the azimuth rotated to minimize the load on the T4 cables.

Two other engineering firms, Thornton Tomasetti and WJE (Wiss, Janney and Elstner Associates Inc) were engaged to assist with the planning for initial remediation work and to assess safety issues. Thornton Tomasetti was the lead engineering firm. Despite the loss of the cable, it was thought that the telescope was reasonably safe as the factor of safety for the original four main cables going from the apex of the suspended structure to tower T4 was still 1.66. Two temporary cables were ordered with delivery in months.

On 6 November at 19:39 one of the Tower 4 original main cables, the furthest south cable, failed due again to a socket failure. This failure was a significant surprise given the nominal safety factor of 1.66. It clearly put in serious doubt the ability of the remaining main cables to support the now higher loads on them. There was an immediate ban of anyone except essential people being at the observatory and no one was to be in the vicinity of the towers and prime reflector. The three engineering firms made separate recommendations; two, Thornton Tomasetti and WSP recommended decommissioning of the telescope (i.e. a controlled dismantling) while WJE suggested a number of steps that could "save" the telescope by lowering the load on the remaining cables to T4. Included were lowering the tension in the backstays of T4 allowing the tower to tilt inwards thus lowering the load on the main cables, and removing the approximately 20 tons of lead ballast at the end of the feed arm opposite to the dome's location, a non-trivial task if no one could be on the platform. In the end, the NSF decided that decommissioning was the only viable option.[6]

On the morning of 1 December, a second T4 main cable failed leading to rapid failure of all the remaining T4 main cables and the collapse of the entire suspended structure of the telescope, an incredibly sad and shocking event for the staff and everyone with a long involvement with the telescope (Fig. 14.4).

The first task was to assess the damage. The top section of the towers T4 and T12, and two sections of T8 broke off due to the drag of the cables. The collapsed backstay cables from T12 damaged the roof of the Visitor Learning Center and did some minor damage to the Visitor Center. Neither the operations building nor any other significant structure apart from the Visitor Center complex was damaged.

In the months that followed, Thornton Tomasetti was contracted to clear and dispose of all of the pieces of the platform, azimuth arm and dome, which had ended up on the west side of the bowl, and the large number of cables. It was also contracted to conduct a forensic study of the root causes of the collapse; many of the cable related pieces such as the cable sockets were collected as part of this study. The cleanup operation was large, approximately 900 tons of, mainly, steel from the suspended structure had to be cut up and disposed of. This required removal and

[6]NSF's announcement of the decommissioning plan including links to the recommendations of the three engineering firms can be found at: https://beta.nsf.gov/news/nsf-begins-planning-decommissioning-arecibo

Fig. 14.4 The moment of the collapse of the telescope, a shocking event for everyone that was, or had been, associated with the Arecibo observatory. Credit: Arecibo Observatory, photo by Carlos Perez

disposal, except for the good panels, of about 25% of the primary reflector surface and its supporting cables. In addition, there were the 18 main cables, 21 backstay cables and the collapsed sections of the towers that needed to be removed. The cleanup was completed in December 2021.

In parallel with the cleanup work, the NSF chartered a salvage committee, the Arecibo Observatory Survey Salvage Committee (AOSSC), "to work with the Observatory management and Thornton Tomasetti to identify possible items of historic significance that could be found within the wreckage, to suggest which large sections of the structural framework to retain, and to make initial recommendations about conservation, display, and educational outreach aspects of the retrieved material."[7] A committee of twelve was formed chaired by Carlos Padin of the Ana G. Méndez University with membership distributed over Arecibo management, technical and scientific staff, an NSF representative and three Arecibo users. A total of 114 items including the ALFA receiver and the two klystrons from the 2.38 GHz transmitter, were collected and stored. Thornton Tomasetti's cleanup work was documented using photos and drone footage as an historical record and, if possible, to show the condition of the wreckage around the recovered items.

After over a year's work, Thornton Tomasetti delivered their report, The Arecibo Telescope Collapse: A Forensic Investigation, on 25 July 2022.[8] It described a very detailed investigation of the root causes of the collapse with emphasis on the cable support system, especially the socketing of the ends of the cables. The socket is a

[7]The report of the AOSSC can be found at: https://www.naic.edu/ao/blog/arecibo-observatory-survey-salvage-committee-report

[8]https://www.nsf.gov/news/special_reports/arecibo/Arecibo-Telescope-Collapse-Forensic-Investigation-508c.pdf

metal cone with an opening at the narrow end just wide enough for the cable to pass through. The individual wires that comprise the cable, 126 in the case of the 3.25 inch diameter auxiliary main cables, are then splayed out in a process called brooming so that the wires fill the cone. Molten zinc is then poured into the cone adhering to the wires and forming a solid plug that can resist the load on the cables. As stated in the report, the sockets are expected to have at least the full breaking strength of the cables but this is dependent on the individual wires being properly splayed out in the cone. The factor of safety, the ratio of the minimum cable breaking strength to the tension in the cable, for the auxiliary cables installed in the 1990s as tabulated in the report varied from 2.0 to 2.2 with one having a higher value. These safety margins were similar to the ones for the original main and backstay cables.

The Executive Summary in the Thornton Tomasetti report listed five factors that contributed to the failure of the telescope's cable system; "(1) the manual and inconsistent splay of the wires during cable socketing, (2) the design of the cable system with relatively low safety factors, (3) the occurrence of extreme environmental events such as hurricanes and earthquakes, (4) the non-replacement, repair or bypass of the sockets where large cable slips were observed and (5) the addition of auxiliary cables as isolated cables." As the report details, a significant slip of one of the Tower 12 auxiliary backstay cables out of its anchor socket was measured after hurricane Maria struck the telescope on 20 September 2017 and a similar slightly smaller slip was measured in early 2019 for the northern Tower 4 auxiliary main cable at the tower socket, the one that failed on 10 August 2020. In its Executive Summary the Thornton Tomasetti report went on to state that "to our knowledge, they (the slips) were not identified as an immediate structural integrity issue."

At the same time as Thornton Tomasetti was doing its forensic study, the NSF requested that the National Academies for Science, Engineering and Medicine (NASEM) conduct an "Analysis of Cause(s) of Failure and Collapse of the 305-Meter Telescope at the Arecibo Observatory in Puerto Rico". A committee of eight experts was formed chaired by Roger L. McCarthy and a series of open hearings including public presentations took place starting on 24 and 25 January 2022 with the last one on 16 and 17 October 2022. There was a site visit to the Arecibo Observatory on 19 and 20 May 2022.[9] As of late August 2024 the report has not been released. The findings of the National Academies report on the Arecibo telescope collapse will be discussed in the Electronic Supplemental Material, ESM 14.1.

The report with the extensive appendices is also at: https://www.thorntontomasetti.com/sites/default/files/2022-08/TT-Arecibo-Forensic-Investigation-Report.pdf?utm_source=social&utm_medium=share&utm_campaign=AreciboReport

[9]The detailed charge to the committee and links to videos of some of the public talks can be found at: https://www.nationalacademies.org/our-work/analysis-of-causes-of-failure-and-collapse-of-the-305-meter-telescope-at-the-arecibo-observatory

14.6 The Future of the Arecibo Observatory

Future possibilities: With the telescope's collapse the major issue, apart from the cleanup of all the debris, was the feasibility of building another telescope in the sinkhole. Ideas ranged from rebuilding a modified version of the original telescope to building an interferometer system with many small dishes mounted on one or more tiltable flat plates. There did not appear to be strong support for the first option although it clearly would be an acceptable option for solar system radar and ionospheric studies. A considerable effort by a group under the leadership of Anish Roshi went into developing the interferometer concept (Roshi et al. 2021). As of July 2024, the concept remains just that and there has been no progress in replacing the 305 m telescope and radar system.

The shutdown: As of 14 August 2023 the NSF had the University of Central Florida terminate all activity at the observatory site and dismiss all the staff. Virtually no one was allowed entry to the site. The NSF contracted a maintenance company to look after the site infrastructure; they did employ a number of the original staff. All the observing equipment was left in place for now except for some of the upper atmosphere optical instruments which were sent to the Island of Culebra just off the east coast of Puerto Rico. During the management of SRI International, a small laboratory was established on the island equipped with some optical observing instruments since the sky over Culebra has many more clear nights than that over the Arecibo observatory. UCF continues to manage this facility.

Current plans for the observatory site: In October 2022 the NSF issued a solicitation for proposals to "re-imagine the existing Arecibo Observatory site (AO) as the Arecibo Center for STEM Education and Research (ACSER)".[10] This effectively announced that it would not be building a new telescope at the Arecibo site but that it would be setting up a STEM (Science, Technology, Engineering and Mathematics) center (Nature News 2022). Needless-to-say, this was a tremendous disappointment to the relevant science communities and to the people of Puerto Rico. In September 2023, the NSF elaborated on the plan to have the Angel Ramos Foundation Visitor and Learning Center be the home for a STEM Center.[11] The Arecibo Center for Culturally Relevant and Inclusive Science Education, Computational Skills, and Community Engagement (Arecibo C3) would be operated by Cold Spring Harbor Laboratory, The University of Puerto Rico Rio Piedras, Universidad del Sagrado Corazón in San Juan, and the University of Maryland Baltimore County under an over $5M five year award. On 15 May 2024, the NSF announced that the

[10]The solicitation can be found at: https://new.nsf.gov/funding/opportunities/arecibo-center-stem-education-research-acser/nsf23-505/solicitation

[11]The NSF announcement can be found at: https://new.nsf.gov/news/nsf-announces-over-5m-funding-create-new-stem

Arecibo C3 center will likely open in November 2024 with a pilot project starting in the summer of 2024.[12]

14.7 What Has Been Lost?

A small sampling of the discovery opportunities that have been lost due to the telescope's collapse:

- The North American Nanohertz Observatory for Gravitational Waves (NANOGrav) project aimed at detecting gravity waves at nano-Hertz frequencies (10^{-9} Hz) by precise measurements over many years of the arrival times of pulses from a carefully selected set of millisecond pulsars distributed over the sky. Agazie et al. (2023) reported the detection of a stochastic background of gravity waves based on measurements over 15 years of the arrival times of pulses from 68 ms pulsars. Due to its tremendous sensitivity, Arecibo played a significant role in this detection with the other contributing telescope being the 100 m Green Bank telescope (GBT) in West Virginia. With the loss of Arecibo, the continuation of this project in North America will be solely dependent on measurements of pulsar arrival times with the GBT. The number of pulsars that can be timed will be reduced due to limits on the observing time available on the GBT and because some pulsars are too faint to be timed by the GBT. The new FAST telescope in China is also being used to time millisecond pulsars to detect gravity waves at nano-Hertz frequencies and this program may become the dominant one.[13]
- At the time of the Arecibo telescope's collapse, a 40-beam L-band phased array feed system funded by the NSF was being developed at Brigham Young University and Cornell University as a successor to the 7-beam ALFA system discussed in Chap. 12. Its main objectives were the discovery of more millisecond pulsars for inclusion in the NANOGrav's detection of gravity waves, and deep surveys looking for small low hydrogen mass galaxies. As of mid-2024, the construction of the instrument was almost finished, and it will be mounted on the GBT where it will make significant contributions to galactic and extra-galactic astronomy.
- Radar astrometry and imaging observations of Near Earth Asteroids (NEAs) will be severely curtailed by the loss of Arecibo. Precise measurements of the distance and line-of-sight velocity of NEAs, especially when combined with optical

[12] The announcement can be found at: https://new.nsf.gov/news/nsf-announces-pilot-phase-anticipated-opening-date-of-arecibo

[13] Until not long before its collapse in December 2020, the 305 m Arecibo telescope was by far the largest and most sensitive single dish radio telescope in the world. In 2020 the 500 m diameter FAST telescope in China came into full operation after several years of commissioning tests (Jiang et al. 2020). The effective aperture of FAST is 300 m compared with Arecibo's 225 m. FAST is now very much the dominant single dish radio telescope at frequencies up to its maximum operating frequency of 3 GHz, especially as it is equipped with a 19-beam system covering the 1420 MHz band of atomic hydrogen.

position measurements, allows their orbits to be precisely known far into the future. This allows the determination of which ones pose a serious threat to Earth at some time in the future, and the necessary planning to mitigate the threat. Arecibo's ability to also measure the size, shape, rotation state and possible binary configuration of NEAs was also important for mitigation planning and for the general study of these small bodies. The only other radar system capable of studying NEAs is NASA's Goldstone's 64 m antenna equipped with an 8 GHz transmitter. However, this radar is not always available due to the 64 m antenna's primary task of supporting spacecraft missions and its radar system is more than a factor of 10 less sensitive that the Arecibo system. As of 2024 there is a design study underway for a transmitting capability on the GBT with another antenna system to be used to receive the echo. However, it will be a number of years before such a system could be operational.
- The loss of Arecibo, the world's most sensitive incoherent scatter radar system, will seriously impact future studies of the Earth's upper atmosphere especially in the D-region at 60 km to 90 km and above the F-region out to several thousand km, regions where high sensitivity is required.
- The loss of Arecibo leaves a large gap in the magnetic latitude coverage of the chain of incoherent scatter radars that monitors the properties of the ionosphere at a regular period, the World Day program. One objective of the program is the long term monitoring of the temperature of the Earth's thermosphere (see Chap. 1) to validate models of atmospheric evolution that may be affected by climate change. Arecibo provided the most accurate ion temperature measurements.
- The loss of the telescope and incoherent scatter radar system also meant the loss of the HF transmitting system used for ionospheric modification (heating) experiments. As the only HF transmitting facility in the United States coupled with an incoherent scatter radar, this ended studies of artificial airglow, field-aligned irregularity generation, and strong Langmuir turbulence.[14]

What has Puerto Rico lost: The Arecibo observatory was the most prominent scientific research installation in Puerto Rico and, of course, the city of Arecibo was widely known around the world. Over 100,000 people including about 30,000 school students, came to the Angel Ramos Foundation Visitor Center every year. Many of the visitors were from outside Puerto Rico so there was a local decline in tourism with the loss of the telescope. The school children received an introduction to astronomy and atmospheric sciences which, hopefully, opened the minds of many of them as to their future career possibilities. At the more advanced level, the yearly NSF sponsored Research Experience for Undergraduates summer program recruited a mix of students from Puerto Rico and the US mainland, providing them with direct experience with scientific research. Finally, the observatory's closure had a

[14] Input for the discussion of the impact of the collapse on the atmospheric sciences program was provided by Brett Isham and David Hysell.

significant impact in the local area and the city of Arecibo, where most of the approximately 120 scientists, engineers, technicians, maintenance and support staff lived.

A final reflection:

Over its 57 years, Arecibo made many significant discoveries in our quest for knowledge of the Earth's upper atmosphere, the solar system, our galaxy and the local universe. It was a key contributor to pulsar science and its ramifications, to our knowledge of the structure of the local Universe, to revealing the surface of Venus under its cloud cover, to determining the properties and orbits of asteroids, and to the structure, composition and dynamics of the upper atmosphere, the ionosphere. But more than that, Bill Gordon's Big Dish represented the vision, initiative and drive needed to build large scientific instruments and develop innovative techniques to use them to reach the original research objectives and the many more that arise from the initial work or fortuitously as was the case with pulsars for Arecibo. Arecibo's size and iconic design inspired both scientists and the public for decades.

References

Agazie, G., et al. 2023. The NanoGrav 15 Year Data Set: Evidence for a Gravitational-Wave Background. *Astrophysical Journal Letters* 951: L8.

Billings, L. 29 July 2015. Search for Alien Life Ignites Battle Over Giant Telescope. *Scientific American* 313 (1).

Jiang, Peng, et al. 2020. The Fundamental Performance of Fast with 19-Beam Receiver at L-Band. *Research in Astronomy and Astrophysics* 20 (5): 64.

Nature Editorial. November 2015. *Radio Interference: Conflict at the Arecibo Observatory Highlights the Need for Funders to Become More Flexible* 527: 134

Nature News. 2022. *Renowned Arecibo Telescope Won't Be Rebuilt – and Astronomers Are Heartbroken*, 14 October 2022.

Roshi, A., et al. 2021. *The Future of the Arecibo Observatory: The Next Generation Arecibo Telescope*. arXiv:2103.01367.

Watson, T. 2015. Arecibo Observatory Director Quits After Funding Row. *Nature* 527: 142–143.

Appendices

Appendix A

Arecibo Observatory and NAIC Directors*

Arecibo Observatory Directors

	From	To
William E. Gordon	1963 November	1965 September (operations years)
John Findley	1965 September	1966 August
Frank D. Drake	1966 September	1968 August
Gordon H. Pettengill	1968 November	1970 December
Rolf Dyce	1970 December	1971 August Acting Director
Tor Hagfors	1971 September	1973 July
Harold D. Craft Jr	1973 August	1981 July
Donald Campbell	1981 July	1982 July Acting Director
Donald Campbell	1982 July	1987 November
Riccardo Giovanelli	1987 December	1988 December Interim Director
Michael Davis	1988 December	1992 March
Daniel Altschuler	1992 April	2003 August
Sixto Gonzalez	2003 September	2006 September
Timothy Hankins	2006 September	2007 February
Robert Kerr	2007 March	2008 September
Michael Nolan	2008 September	2011 September
Robert Kerr	2011 October	2015 October
Anthony van Eyken	2015 November	2016 May Acting Director
Francisco Cordova	2016 June	2022 July
Olga Figueroa Miranda	2022 September	2023 August

D. Campbell, *The Arecibo Observatory: A History of Innovation and Discovery*, Historical & Cultural Astronomy, https://doi.org/10.1007/978-3-031-75225-4

CRSR and NAIC Directors

	From	To
Thomas Gold	1959 November	1969 October CRSR Director
Thomas Gold	1969 October	1971 May Ass. Vice Pres. Res.
Frank D. Drake	1971 July	1981 June 1st NAIC Director
Harold D. Craft	1981 July	1982 July Interim Director
Tor Hagfors	1982 July	1992 July
Donald Campbell	1992 July	1993 April Interim Director
Paul Goldsmith	1993 April	2003 April
Robert Brown	2003 May	2008 May
Donald Campbell	2008 June	2011 September
Robert Kerr	2011 October	2015 October
John Kelly	2015 November	2018 March
Ramon Lugo	2018 April	2021 October
Julie Brisset	2021 November	2023 August

*The months are approximate

Appendix B

Acronyms

AAG Arecibo Advisory Group
ARRA American Recovery and Reinvestment Act
AFCCDD Air Force Command and Control development Division
AFCRC Air Force Cambridge Research Center
AFCRL Air Force's Cambridge Research Laboratories
AFOAR Air Force Office of Aerospace Research
AFOSR Air Force Office of Scientific Research
AFSC Air Force Systems Command
AGES Arecibo Galactic Environments Survey
AIDA Arecibo Initiative in the Dynamics of the Atmosphere
AIO Arecibo Ionospheric Observatory
ALFA Arecibo L-Band Feed Array
ALFALFA Arecibo Legacy Fast ALFA Survey
AMC Air Material Command
A&W Ammann and Whitney
ARDC Air Research and Development Command
ARL Army Research Laboratory
ARPA Advanced Research Projects Agency
ALMA Atacama Large Millimeter Array
ATNF Australia Telescope National Facility
AUDS Arecibo Ultra Deep Survey

AUI	Associated Universities, Inc.
AUSAC	Arecibo Users and Scientific Advisory Committee
BMD	Ballistic Missile Defense Group
BMEWS	Ballistic Missile Early Warning System
CCI	Contemporary Constructors Inc.
CALTECH	California Institute of Technology
CEDAR	Coupling, Energetics, and Dynamics of Atmospheric Regions
CDC	Computer Data Corporation
CIMA	Telescope General Purpose Interface
CRSR	Center for Radiophysics and Space Research
CSIRO	Commonwealth Scientific and Industrial Research Organization
CSUAC	Cornell Sydney University Astronomy Center
CU	Cornell University
DARPA	Defense Advanced Research Projects Agency
DDR&E	Director, Defense Research & Engineering
DEC	Digital Equipment Corp.
DoD	Department of Defense
DODIRF	Department of Defense Ionospheric Research Facility
DPS	Division of Planetary Science of the American Astronomical Society
EALFA	Extra-Galactic ALFA Surveys
EISCAT	European Incoherent Scatter Association
EPSCOR	Established Program to Stimulate Competitive Research
ERD	Electronics Research Division
ERP	Effective radiated power
ESA	European Space Agency
ESSA	Environmental Science Services Administration
ESSCO	Division of CPI Composite Structures that makes Ground Based Radomes
FFT	Fast Fourier Transform
FRB	Fast Radio Burst
GALFACTS	Galactic ALFA Continuum Transit Survey
GALFA-RRL	GALFA-Radio Recombination Lines
GALSPEC	Galactic Spectrometer
GASS	GALEX Arecibo SDSS Survey
GLIPAR	Guide Line Identification Program for Anti-Missile Research
HALCA	Highly Advanced Laboratory for Communication and Astronomy (Japanese VLBI Space Mission)
HF	Heating Facility
HI	Atomic Hydrogen
HRMS	High Resolution Microwave Survey
HSA	High Sensitivity Array
HST	Hubble Space Telescope

ICA	Interagency Committee for Arecibo
ICBM	Intercontinental Ballistic Missiles
IDA	Institute for Defense Analyses
IDI	Imaging Doppler Interferometer
IEEE	Institute of Electrical and Electronics Engineers
ISTS	Institute of Space & Terrestrial Sciences
IR	Infrared
IRAC	Interdepartmental Radio Advisory Committee
IRBM	Intermediate Range Ballistic Missile
IRE	Institute of Radio Engineers
ISR	Incoherent Scatter Radar
ITS	Institute of Telecommunication Sciences
LIDAR	Light Detection and Ranging
LIGO	Laser Interferometer Gravitational Observatory
LTV	Ling-Temco-Vough Corp.
MBA	Main Belt Asteroid
MATS	Military Air Transport Service
MIT	Massachusetts Institute of Technology
NAIC	National Astronomy and Ionosphere Center
NASA	National Aeronautics and Space Administration
NASEM	National Academies for Science, Engineering and Medicine
NATK	North American Tower/Tower King
NBS	National Bureau of Standards
NEAs	Near Earth Asteroids
NEOs	Near Earth Objects
NEROC	North East Radio Observatory Consortium
NOAO	National Optical Astronomy Observatory
NRAO	National Radio Astronomy Observatory
NRL	Naval Research Laboratory
NSB	National Science Board
NSE	National Steel Erectors
NSF	National Science Foundation
NSO	National Solar Observatory
OAR	Office of Aerospace Research
ONR	Office of Naval Research
OVRO	Owens Valley Radio Observatory
PALFA	Pulsar Alfa Surveys
PRIFA	Puerto Rican Infrastructure Financing Authority
PRTC	Puerto Rican Telephone Co.
RADC	Rome Air Development Center
Rayspan	Raytheon Spectrum Analyzer
REU	Research Experience for Undergraduates
RFP	Request for Proposals
RMS	Radio, Millimeter and Sub-Millimeter

rms	root mean square
SAS	Space and Atmospheric Sciences
SDS	Scientific Data Systems
SDSS	Sloan Digital Sky Survey
SERENDIP	Search for Extraterrestrial Radio Emissions from Nearby Developed Intelligent Populations
SETI	Search for Extraterrestrial Intelligence
SOHO	Solar and Heliospheric Observatory
TOGS	Turn On GALFA Survey
TRG	Technical Research Group
Tsys	System Noise Temperature
UAD/RSI	Universal Antennas Division of Radiation Systems, Inc.
UCF	University of Central Florida
UMET	Universidad Metropolitana (UMet, a private university in Puerto Rico, now the Ana G. Mendez University.
UPR	University of Puerto Rico
USRA	Universities Space Research Association
VA	Vertex Atennentechnik GmbH
VC	Visiting Committee
VLA	Very Long Array
VLBA	Very Long Baseline Array
VLBI	Very Long Baseline Interferometry
VSQ	Visiting Scientist Quarters
WADC	Air Force's Wright Air Development Center
WJE	Wiss, Janney and Elstner Associates, Inc.
ZOA	Zone of Avoidance

Index

D. Campbell, *The Arecibo Observatory: A History of Innovation and Discovery*, Historical & Cultural Astronomy, https://doi.org/10.1007/978-3-031-75225-4

L

M

N